Footrot and Foot Abscess of Ruminants

Editors

J.R. Egerton, D.V.Sc.
Professor
Department of Veterinary Clinical Studies
University of Sydney
Camden, New South Wales, Australia

W. K. Yong, Ph.D.
Senior Scientific Officer
Regional Veterinary Laboratory
Department of Agriculture and Rural Affairs
Hamilton, Victoria, Australia

G. G. Riffkin, Ph.D.
Director
Regional Veterinary Laboratory
Department of Agriculture and Rural Affairs
Hamilton, Victoria, Australia

CRC Press, Inc.
Boca Raton, Florida

Library of Congress Cataloging-in-Publication Data

Footrot and foot abscess of ruminants / editors, J. R. Egerton, W. K.
Yong, G. G. Riffkin.
p. cm.
Includes bibliographies and index.
ISBN 0-8493-5861-2
1. Footrot in animals. 2. Foot—Abscess. 3. Hoofs—Diseases.
4. Ruminants—Diseases. I. Egerton, J. R. II. Yong, W. K.
III. Riffkin, G. G.
[DNLM: 1. Abscess–veterinary. 2. Bacterial Vaccines. 3. Foot
Diseases–veterinary. 4. Foot Rot. 5. Ruminants. SF 901 F689]
SF906.F66 1988
636.2′ 0897585–dc19
DNLM/DLC

88-7371
CIP

Direct all inquiries to CRC Press, Inc., 2000 Corporate Blvd., N.W., Boca Raton, Florida, 33431.

International Standard Book Number 0-8493-5861-2

Library of Congress Card Number 88-7371
Printed in the United States

GENERAL INTRODUCTION

Footrot and foot abscess are important diseases of the feet of sheep throughout the world.

Footrot is regarded as the most economically important contagious disease of sheep in Australia today, despite improved control measures. Also, it causes great suffering. The considerable expenditure on research and the dedication of scientists have been well rewarded, but each breakthrough has turned out to have its limitations. Eradication is impracticable under some circumstances, differentiation between benign and virulent outbreaks is sometimes difficult, no therapy is invariably effective, nor are the vaccines that are hitherto available. There are still problems to be solved. Nevertheless, the achievements are impressive and of great practical value. As this book shows, most of our knowledge on the subject has come from research in Australia and New Zealand, the countries most affected.

Ever since the discovery and general acceptance of the germ theory of disease some 100 years ago, there has been a tendency to oversimplify our concepts of the causation of contagious diseases: to consider the particular microbe implicated as *the* cause, disregarding other factors that also play a part. Reasearch shows that footrot and foot absecess result from a complex of causal factors: bacterial, environmental, and genetic.

Footrot and foot abscess differ from most other communicable diseases in being mixed bacterial infections. For example, in footrot the essential transmitting agent, *Bacteroides nodosus,* cannot produce disease by itself. To do so it requires the collaboration of other bacteria commonly present on the feet of sheep or in feces, in particular *Fusobacterium necrophorum.* Furthermore, these two nonsporing, oxygen-sensitive organisms depend for their survival on the protective effect of common aerobic bacteria. Although footrot is a mixed infection, the one organism that is essential lives only in lesions of the feet of ruminants, so eradication is possible. Another feature of footrot is that transmission occurs only under favorable environmental conditions, usually prolonged wet pasture and a mean ambient temperature above 10°C. Indeed, so much is the occurrence of the disease determined by these predisposing conditions that until 50 years ago they were themselves commonly believed to be the cause.

Progress has now reached the stage where it is appropriate to bring together in one volume papers describing the present state of knowledge about these complex and insidious diseases. This publication is a landmark which I believe indicates that at last we are getting on top of these diseases and future prospects are promising.

W. I. Beveridge, Sc.D., D.V. Sc.
Emeritus Professor
University House
Australia National University
Canberra, Australia

PREFACE

During the last two decades substantial advances have been made in research on footrot and foot abscess. The results of this research have been published in numerous journals and reports of conference proceedings and workshops. It seemed appropriate, therefore, to approach the people engaged in this research and to ask them to contribute to a composite and cohesive volume on the present state of knowledge of the major bacterial infections of the feet of ruminants.

Progress in this field of research would not have occurred without the major pioneering work of W. I. Beveridge and T. S. Gregory, who studied the etiology of footrot and foot abscess in Australian sheep in the 1930s. They focused attention on differences between the disease and the agents responsible for them.

We thank all the authors for their participation in this project. We appreciate also, their willingness to accede to our editorial suggestions and requirements.

J. R. Egerton
W. K. Yong
G. G. Riffkin

THE EDITORS

J. R. Egerton, D.V.Sc. is Professor of Animal Health at the Faculty of Veterinary Science, The University of Sydney. He graduated B.V.Sc. from The University of Queensland in 1955 and was awarded the Doctor of Veterinary Science degree in 1985 by The University of Sydney. His career has included investigations of a number of infectious diseases of livestock and zoonoses. He has published 75 papers on these investigations and the majority are on topics related to footrot in sheep and cattle. His principal interests have been in establishing the pathogenesis of this infection, improving treatment methods, and devising prophylactic measures. His current research involves him with teams of scientists developing recombinant DNA footrot vaccines and in determining the heritability of resistance to footrot.

Weng Kwong Yong, Ph.D., is a Senior Scientific Officer of the Victorian State Department of Agriculture and Rural Affairs in Australia. He is based at the Regional Veterinary Laboratory in Hamilton.

Born September 1946 in Singapore, he obtained a B.Sc. degree with First Class Honours in microbiology from the University of Otago, New Zealand in 1972 and a Ph.D. degree in parasite immunology from the University of Queensland, Australia in 1976. He served as a National Research Advisory Council's post-doctoral fellow at the Wallaceville Animal Research Centre (Upper Hutt, New Zealand) from 1976—1978, as a teaching and research associate at the International Immunology Training and Research Centre (Amsterdam, The Netherlands) from 1979—1980, and as a National Health and Medical Research Council's research fellow at the Department of Microbiology (University of Queensland) in 1981. It was in 1982 that he assumed his present position.

Dr. Yong has more than 60 published articles, abstracts, and contributions to books to his credit in the areas of microbiology, parasitology, and immunology. He is a member of the Australian Society for Immunology.

George G. Riffkin, Ph.D., is the Director of the Regional Veterinary Laboratory in Hamilton, Victoria, which is one of the major centres of fine wool sheep production in Australia. Born in 1939 in England, Dr. Riffkin graduated B.V.Sc., M.R.C.V.S. from the University of Liverpool in 1962.

After 7 years in veterinary practice in England and new Zealand he joined the staff at Ruakura Animal Health Laboratory as a veterinary investigation officer with specific interest in parasitology and pathology. In 1972 he gained first class honours in an MSC qualifying year at the University of Queensland. He subsequently enrolled for and was awarded a Ph.D from the same University.

Dr. Riffkin has been at the Hamilton Laboratory for 10 years. He has more than 20 publications as primary or joint author in the field of sheep and cattle diseases. He is the Victorian Representative on the Australian National Footrot Co-ordination Committee and a member of the refereeing panel for the *International Journal for Parasitology.*

Dr. Riffkin's current research interests include the development of better laboratory tests for the diagnosis of footrot in sheep. He is also involved in breeding nematode-resistant sheep and the development of an effective vaccine for sheep. He lives on a farm in Western Victoria running 2500 fine wool Merinos.

CONTRIBUTORS

Bruce Leo Clark, MACVSc.
Senior Principal Research Scientist
Division of Animal Health
CSIRO
Parkville, Victoria, Australia

P. D. Claxton, Ph.D.
Technical Director
Arthur Webster Pty, Ltd.
Sydney, Australia

J. R. Egerton, D.V.Sc.
Professor
Department of Animal Health
University of Sydney
Camden, New South Wales, Australia

Thomas C. Elleman, Ph.D.
Principal Research Scientist
Division of Biotechnology
CSIRO
Parkville, Victoria, Australia

David L. Emery, Ph.D.
Principal Research Scientist
Division of Animal Health
CSIRO
Parkville, Victoria, Australia

Peter A. Hoyne, B.App.Sc.
Experimental Scientist
Division of Biotechnology
CSIRO
Parkville, Victoria, Australia

John S. Mattick, Ph.D.
Principal Research Scientist
Division of Molecular Biology
CSIRO
Sydney, New South Wales, Australia

Neil M. McKern, Ph.D.
Senior Research Scientist
Division of Biotechnology
CSIRO
Parkville, Victoria, Australia

Eric Keith Moses, M.Sc.
Scientific Officer
Regional Veterinary Laboratory
Department of Agriculture and Rural Affairs
Hamilton, Victoria, Australia

G. G. Riffkin, Ph.D.
Director
Regional Veterinary Laboratory
Department of Agriculture and Rural Affairs
Hamilton, Victoria, Australia

Julian I. Rood, Ph.D.
Senior Lecturer
Department of Microbiology
Monash University
Clayton, Victoria, Australia

T. M. Skerman, Ph.D.
Scientist
Wallaceville Animal Research Centre
Ministry of Agriculture and Fisheries
Upper Hutt, New Zealand

David James Stewart, Ph.D.
Principal Research Scientist
Division of Animal Health
CSIRO
Parkville, Victoria, Australia

J. A. Vaughan, B. App. Sci
Senior Technical Officer
Division of Animal Health
CSIRO
Parkville, Victoria, Australia

David M. West, Ph.D., FACVSc.
Senior Lecturer
Department of Veterinary Clinical Sciences
Massey University
Palmerston North, New Zealand

W. K. Yong, Ph.D.
Senior Scientific Officer
Regional Veterinary Laboratory
Department of Agriculture and Rural Affairs
Hamilton, Victoria, Australia

TABLE OF CONTENTS

Section I: The Diseases

Section I

THE DISEASES

J. R. Egerton, W. K. Yong, and G. G. Riffkin

Ruminants are grazed in tropical, subtropical, and temperate environments. Their productivity is dependent on heritable and environmental factors. Among the environmental factors which limit production is disease. Ruminants, like other animals which provide *homosapiens* with food, fiber, draft, transport, recreation, and pleasure, are susceptible to a wide variety of diseases, infectious and noninfectious. In many regions, the major infectious diseases, those responsible for high morbidity and mortality, have been controlled by preventive techniques. Extensive and sophisticated eradication programs, usually preceded by prevalence-reduction strategies, have been developed to eliminate bacterial diseases like brucellosis, and tuberculosis, and contagious bovine pleuropneumonia from national herds.

Less emphasis has been placed, nationally, on chronic endemic diseases which cause substantial economic loss but which are not usually directly responsible for public-health problems. These diseases, some due to helminth parasites, some due to microorganisms, have been more recognized and perhaps more intensively investigated in countries like Australia and New Zealand which are heavily dependent on livestock products for national export income.

Lameness is a production-limiting syndrome in grazing ruminants in Australia and New Zealand and in other temperate regions of the world. The elucidation of the major causes of lameness in sheep and cattle, their treatment, control, and prevention have all been major research objectives in the Australasian region for the last 50 years. The chapters which follow in this book describe the results and conclusions of some of that research.

The principal diseases which cause lameness in sheep and cattle in Australia and New Zealand are due to bacterial infections of the tissues which comprise the digits of the host animals. This may be in contrast to the situation in a number of European countries where aseptic laminitis is considered to be the major disease of the feet of cattle. Dairy cattle in countries like Sweden, Norway, Denmark, and Holland are of much greater economic importance than sheep, and the prevalence and importance of laminitis is related to the special housing and diets which are characteristic of dairying in those Northern European countries.

Infections of the digits in both cattle and sheep certainly occur wherever these animals are kept and they are recognized by both farmers and veterinary scientists. In the past, assessment of the relative importance of these infections in the different species and in different countries has been complicated by a failure to distinguish between the different clinicopathological entities which occur and the use of popular names — usually derived from farmers' terms — to describe the diseases under discussion.

The nomenclature of diseases of the digits has been especially confused in the case of cattle, whereas in sheep the major diseases, footrot and foot abscess, were recognized as separate entities by Gregory* in Australia nearly 50 years ago. This separation of two major causes of lameness was facilitated by the identification of *Bacteroides nodosus* as an essential agent of footrot in sheep.** Some attempt has been made in recent years to achieve uniformity in the description of disorders of the digits of cattle and sheep. A series of symposia attended by veterinary clinicians and research scientists led, in 1984, to the publication of a useful

* Gregory, T. S., *Aust. Vet. J.*, 15, 160, 1939.

** Beveridge, W. I. B., *Bull. Counc. Sci. Ind. Res. (Melbourne)*, No. 140, 1941.

color atlas illustrating the common diseases. This was accompanied by terminology acceptable to those attending the symposia.

This section of the book is concerned principally with two diseases. The first results from invasion of some epidermal tissues of the digits of ruminants by *B. nodosus* and other bacteria (Chapters 1 and 2). In sheep, the description applied to this disease by the International Symposia was *dermatitis interdigitalis contagiosa,* whereas in cattle the same entity was described as *dermatitis interdigitalis.* Evidence for the contagiousness of this disease in cattle will be described in Chapter 2, so the use of different descriptions for the same disease in two species does not seem to be justified. For the sake of simplicity, the common English term for disease associated with *B. nodosus,* **footrot,** will be used irrespective of the host being considered.

The second disease results from deeper invasion of digital tissues by *Fusobacterium necrophorum* and other bacteria (Chapters 3 and 4). The bacterial invasion usually results in an extensive inflammatory response, pain, swelling, and necrosis of affected tissues. In sheep, this latter disease has been distinguished from footrot for at least half a century, whereas in cattle the distinction has been less clear. The International Symposia concluded that *Phlegmona interdigitalis* was an acceptable description of the disease which is commonly called **foot abscess** in sheep. In this book foot abscess will be used to refer to this disease in both cattle and sheep.

Chapter 1

FOOTROT OF SHEEP

D. J. Stewart

TABLE OF CONTENTS

I. DEFINITION ETIOLOGY

A. Definition

Despite recent advances in prevention and treatment, footrot is still one of the most economically important endemic diseases affecting the sheep industry in temperate climate regions enjoying good rainfall. Footrot is a contagious disease of ruminants and is transmitted by *Bacteroides nodosus*. Sheep and goats are the most susceptible, although cattle and deer are also affected under natural conditions with less virulent *B. nodosus*. Outbreaks usually have a seasonal occurrence when there are warm, moist conditions and lush pasture growth.

The disease results from a chronic bacterial infection which is confined to the epidermal tissues of the interdigital skin and horn of the hoof. The history of footrot in Europe, the U.S. and Australia has been reviewed by Mohler and Washburn[1] and Beveridge.[2,3] It was endemic in Britain during the 18th century[4] and its presence was reported in France, Germany, Italy, Australia, and the U.S. during the early 19th century. The contagious nature of the disease was first recognized in France by Gohier early in the 19th century.[5] Footrot was probably introduced into Australia with some of the first importations of sheep, as it was reported in 1802 that many die of the disease when not well cared for.[6]

It is thought that 50 to 60 million sheep are at risk in the footrot-endemic areas of Australia. The disease is of considerable economic importance, costing the sheep industry many million dollars per annum in terms of decreased productivity and labor-intensive treatment. However, no accurate assessment has been made in any country of the national economic loss resulting from this disease, due to a lack of data on prevalence, production losses in different breeds, indirect losses, and treatment costs for all endemic regions. Footrot results in loss of body weight, wool growth, decreased lambing percentages, and reduced value of stock for sale. Treatment costs include expenditure on labor, equipment, vaccines, drugs, and chemicals. In the U.S., footrot in range flocks creates problems with moving sheep from high range to lower pastures, as well as indirectly resulting in deaths from predation and decreased conception rates. Apart from the social stigma in having sheep with footrot, and animal welfare considerations, indirect losses include those due to management changes, lost market opportunities, isolation of sheep, a need for surveillance, costs associated with maintaining secure fences, and disruption of normal farm operations.

Economic evaluation studies by the University of New England, in conjunction with the Department of Agriculture of New South Wales, have compared costs due to footrot in the Northern Tablelands of New South Wales where a regional eradication program is being undertaken (the New England Protected Area) with those in the adjacent Central Tablelands, which does not have such a program.[7] The results have suggested an average production loss valued at 1979 prices of $A0.58 and $A1.29 per sheep, respectively, and control costs represented $A1.26 and $A1.36 per sheep, respectively, in an infected Merino flock. The average annual cost per infected farm in the two regions, valued at 1979 prices, for precautionary measures was $A891 and $A942, respectively, and for restrictions on the movement and marketing of stock, $A6925 and $A3830, respectively. Indirect costs per infected farm from changes in flock and farm management amounted to $A3759 and $A6218, respectively.[7] In 1984, in South Australia, which has a state-wide eradication program, it was estimated that the average cost to eradicate footrot from an infected flock was $A7 per sheep, based on equal use of replacement of the affected mob and of treating sheep under quarantine. Thus, for a property of 1400 sheep the average cost would be $A10,000.[8] Currently, the costs of the disease are probably much higher. In a trial presently being conducted by the Department of Agriculture in New South Wales, wool-production losses alone, in Merino sheep affected with footrot compared with unaffected sheep, were estimated at $A2.50 per head.[9] Based on 1985 figures, the cost of treating the disease in another flock, including labor, drugs, and equipment, has been estimated at about $A5 per head.[10] In the U.S., the cost of footrot in a range operation was calculated to be $U.S.4.41 per ewe annually over a 3-year period between 1980 to 1982. Treatment costs represented only 3% of the total expenses of the disease.[11]

B. Etiology

The casual bacterium *B. nosodus* is a Gram-negative obligate, anaerobic bacillus, with knob-like ends.[2] The disease is a mixed bacterial dermatitis and is considered to result from the synergy of several bacterial species that are individually incapable of causing the disease. Egerton et al.[12] and Roberts and Egerton[13] consider that *Fusobacterium necrophorum,* although part of the normal flora, is just as important as *B. nodosus* in the pathogenic process (see pathogenesis and pathology). However, *B. nodosus* is the specific causative pathogen since it is the only organism of the footrot flora capable of reproducing the disease when applied as a pure culture, lives only in diseased hoof, and survives for no longer than 7 to 14 d in feces, soil, or pasture.[2] More recent experiments have shown that viable organisms can only persist in muddy soil for a maximum of 4 d.[14] A healthy sheep normally acquires *B. nodosus* from an animal showing clinical signs of footrot.

II. CLINICAL SIGNS

The clinical signs observed in sheep affected with footrot depend on a number of factors.

A. Virulence of *B. nodosus*

Basically there are two clinically recognizable forms of footrot in sheep, virulent and benign, the severity of the disease depending on the invasiveness of *B. nodosus*. Both produce inflammation of the interdigital skin (Plate 1 A to C),* but virulent footrot is further characterized by extensive separation, which commences at the heel (Plate 1D), extends to the sole and toe, and frequently extends down the abaxial wall (Plate 1E), with the formation of a distinctive, foul-smelling, necrotic exudate (Plate 1F). In benign footrot (a form of scald), occasional separation of the horn of the heel bulbs and posterior sole region occurs, but there is little accumulation of necrotic material under the separated horn[15,16] (Plate 1G).

* Plate 1 follows page 54.

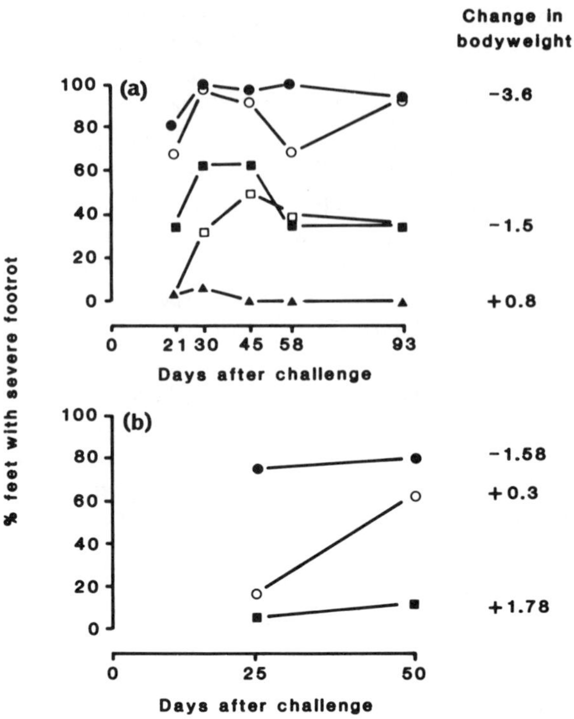

FIGURE 1. Percentage of feet with severe (>3 c) lesions of footrot in groups of sheep challenged with different strains of *B. nodosus*. All feet in (a) were scarified at challenge; those in (b) were not scarified. ● 334; ○ 336 (virulent); ■ 333, □ 335 (intermediate); ▲ 337 (benign).

A further characteristic of benign footrot is that it causes only mild lameness and rapidly regresses after topical treatment or after the advent of dry environmental conditions,[16] whereas virulent footrot causes severe chronic lameness and tends to persist after the onset of dry conditions unless intensive treatment is undertaken.

It has now become apparent that there are strains of *B. nodosus* intermediate in virulence between those causing benign and virulent footrot.[17-21] Furthermore, virulent strains under field conditions cause a significantly greater loss in body weight and decrease in the rate of wool growth than do benign strains. The intermediate strains lie between those two extremes.[19] Further research in our laboratory has confirmed and extended these findings.[22,23]

In the first pen experiment, groups of eight unvaccinated Merino sheep were challenged with various isolates of *B. nodosus* by the application of pure cultures to the scarified interdigital skin. The benign isolate (strain 337) produced lesions confined to the interdigital skin, with the exception of transient underrunning lesions in two feet (6%), whereas the two virulent strains, 334 and 336, both produced underrunning lesions in 94% of feet (Figure 1a). Two isolates of intermediate virulence, strains 333 and 335, produced persistent underrunning lesions in 35% of feet. Sheep challenged with strains 334 and 336 showed the largest mean weight loss (3.6 kg), those with 333 and 335 an intermediate amount (1.5 kg), and with 337, an average gain of 0.8 kg.[22,23]

In the second pen experiment, groups of 24 Merino sheep, representing two different sire lines, were challenged with three of the strains used in the first experiment (Figure 1b). The interdigital skin of their feet was not scarified prior to challenge. Strain 334 was the most virulent, 95.8% of sheep (75% of feet) having severe footrot 24 d after challenge. At the second inspection (50 d after challenge) 100% of sheep and 80.2% of feet were severely affected. This group showed a mean loss of 1.58 kg in bodyweight. Strain 336 was less virulent, the number of animals having severe footrot increasing from 33.3% (16.7% of feet) to 83.3% (62.5% of feet) between the first and second inspections. This group had a mean bodyweight increase of 0.30 kg. Strain 333 was the least virulent, the number of sheep affected with severe footrot rising from 8.3% to 34.8%, representing 5.2% and 12.5% of feet, respectively. This group gained the most bodyweight (mean 1.78 kg). As judged from the production of severe lesions of footrot, at the first inspection, strain 334 was significantly more virulent than strains 333 and 336. At the second inspection, strains 334 and 336 were not significantly different, but were significantly more virulent than 333. However, the difference at the second inspection between strains 334 and 336 did become significant when mean foot scores were compared. On the basis of severe lesions of footrot, strains 334 and 336 were considered to be virulent and 333 of intermediate virulence. The mean changes in bodyweight for the groups challenged with strains 334 and 336 were significantly different and both were significantly different from that of the group challenged with 333.[22]

Not only is there variation in virulence between strains but also within strains, so that the pathogenicity of "low virulence" (333) and "high virulence" intermediate (336) strains can be substantially increased by scarification of the interdigital skin at challenge. For practical purposes "high virulence" intermediate strains should be classified as virulent and "low virulence" strains as the true intermediate strains.

Definitions of the nomenclature used for classifying the three clinical entities of footrot in a susceptible breed such as the Merino and under optimum environmental conditions[19,21,24] are outlined below.

1. Virulent Footrot

This is a persistent or chronic condition with severe extensive necrotic underrunning of the laminae of the abaxial wall of the hoof (Plate 1E) in a high percentage of sheep, causing severe production losses. There is rapid spread of the disease and rapid development of lesions during favorable conditions. Usually both digits and more than one foot are affected and the lesion usually has a characteristic odor. Although spontaneous recovery (self cure) occurs in a proportion of cases, in others, lesions do not regress and *B. nodosus* can persist in some sheep for a year or more if not treated. Beveridge[2] observed that a sheep remained chronically affected over a period of at least $3^1/_2$ years.

Lameness is a prominent feature of virulent footrot (Figure 2A,B) and can be seen even in the initial stages of infection at 10 to 14 d, when an acute interdigital dermatitis is present and increases as underrunning of the horn of the hoof proceeds. At this stage, the feet feel warmer than usual. After several weeks, the hooves become long and misshapen, possibly due to the increased blood flow to the digits. The animal may carry the leg, kneel when grazing (Figure 2B), and when driven, drop to the rear of the flock. If more than one foot is affected with severe underrunning lesions, movement is painful and the animal prefers sternal recumbency unless forced to move. At this stage, feed intake may be markedly reduced and can virtually cease.[2,24] Consequently, body condition may be rapidly lost and wool growth declines,[17,19,25] probably due to suppressed follicular synthesis,[25] with a resulting break in the wool. In penned sheep, virulent footrot depressed the rate of wool growth at least 10% and affected the gain in bodyweight to a lesser extent.[25] Under field conditions, there can be a considerable loss in bodyweight (5 to 10 kg) in sheep affected with virulent footrot, although this is not evident in all sheep[9,17,19,22,23] (Figure 2C). Over a 12-month period in an experimental flock, virulent footrot resulted in a mean decrease in greasy fleece weight of 0.5 kg (~10%) and a higher proportion of fleeces from affected sheep with lower tensile strength.[9] When recumbency is prolonged, chronically affected sheep develop pressure necrosis and ulceration of the sternum, and deaths occur from a combination of emaciation and systemic bacterial infections.

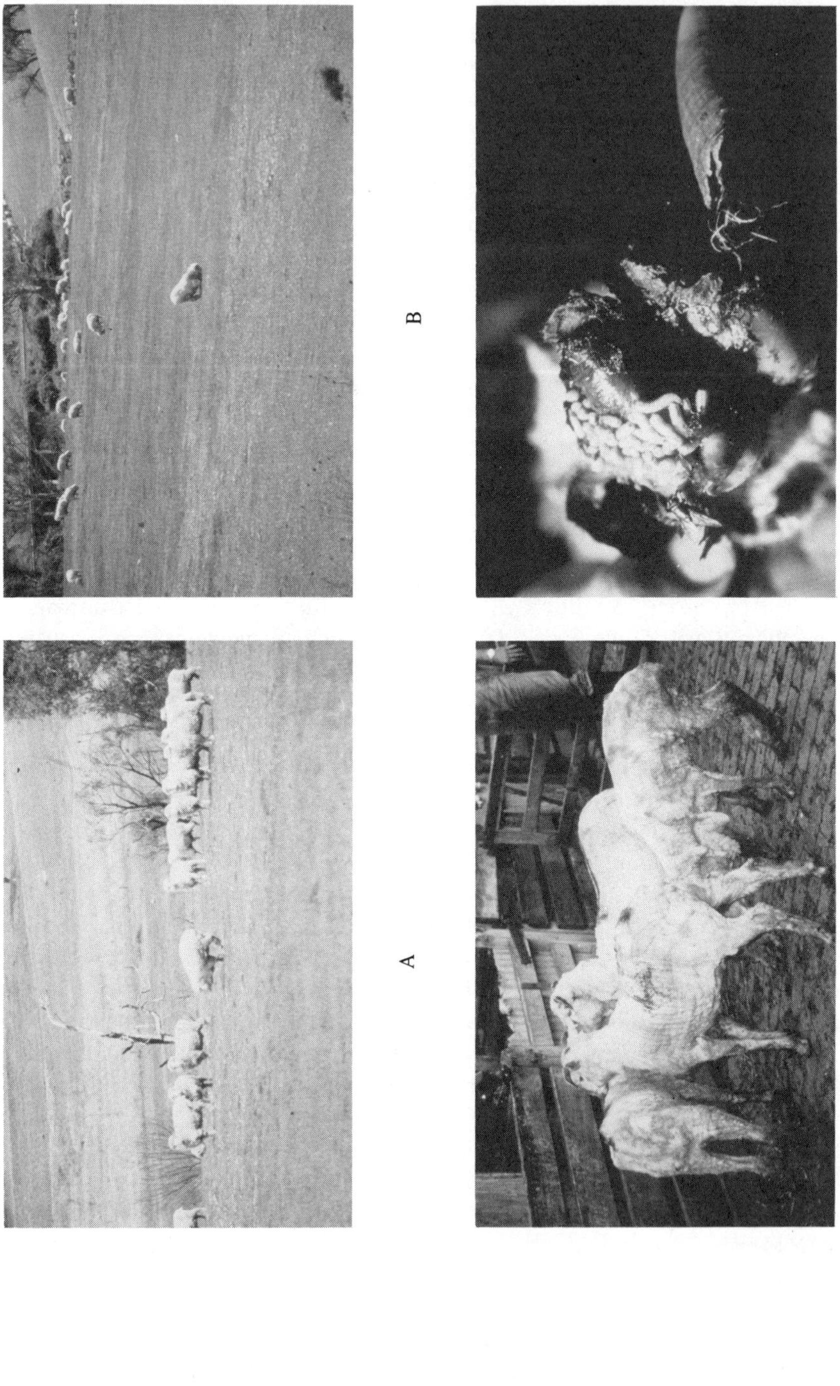

FIGURE 2. Virulent footrot. (A, B) Lameness is a prominent feature of virulent footrot. (C) Merino sheep, affected with severe underrunning lesions of virulent footrot, usually lose a considerable amount of bodyweight and have decreased wool production (sheep at left and right), although some with similar severe lesions can maintain productivity (sheep in center). (D) Fly strike (myiasis). Lesion of footrot infested with maggots (larvae) of the primary sheep blowfly *Lucilia cuprina*.

In late spring and summer in Australia, the most severe underrun lesions, particularly those containing foul-smelling necrotic exudate (Plate 1F) are most attractive to fly strike (myiasis) by *Lucilia cuprina*.[26-28] The struck foot is characterized by the presence of larvae (maggots) at all stages of development from small (<3 mm) to large (>8 mm) (Figure 2D). These strikes can persist for at least 22 weeks in the covert state. Hoof strikes always remain within the hoof and although body wool is stained by exudate from the hoof, which attracts considerable oviposition, most of this is unsuccessful in initiating body strike.[28]

2. Intermediate Footrot

This is a condition caused by strains of intermediate virulence that may not be readily distinguished from virulent footrot in individual sheep. It has recently been recommended by a Working Party on Footrot,[24] that the term ''subvirulent'' should be used to describe strains causing intermediate footrot. Although ''intermediate'' is also not a suitable description for the clinical condition caused by subvirulent strains, the terminology is now in such common usage that it was decided to recommend no change at present.[24] On a flock basis, it is a milder disease with only a small percentage of sheep having severe lesions extending to the abaxial edge of the sole of the hoof and rarely underrunning the abaxial wall. Lameness is commensurate with the severity of the lesion and self-cure tends to occur as the season dries. However, a few severely affected sheep will remain chronically affected. There may be some adverse effect on productivity depending on the severity of the lesions.

3. Benign Footrot

This is a less persistent condition in which interdigital dermatitis is the only lesion observed (Plate 1A, B) or there is very rarely underrunning to the edge of the sole, with a minimum amount of necrotic exudate (Plate 1G) in a small proportion of feet. Ridging or grooving of the soft horn of the posterior abaxial wall is a common feature and a large number of the flock can be affected. Lameness is usually only mild and not readily apparent, except when sheep first rise after resting, and rapidly disappears as the season dries off and following footbathing. The effects on production are minimal.

B. Breed of Sheep

Virulent strains are generally less pathogenic in British-breed sheep than in Merinos. Beveridge[2] noted that crossbreds, principally Dorset and Suffolk blood, were appreciably less susceptible than Merinos to footrot, and Egerton et al.[29] reported that the incidence, severity, duration, and extent of infection was higher among Merinos than Border Leicester × Merino crossbreds on different properties. However, the possibility was not ruled out that isolates differing in virulence could have been involved. In pen experiments using a virulent challenge strain, Skerman et al.[30] have confirmed these observations in that Romney Marsh sheep developed less severe lesions and became affected in fewer feet than Merinos. Under field-challenge conditions Romney, Border Leicester, and Dorset Horn sheep were found to be more naturally resistant to virulent footrot than Merinos. The British breeds had a much greater proportion of lesions confined to the interdigital skin and these were of short duration, whereas in the Merinos (Peppin and Saxon strains) a much larger number of feet developed severe underrunning lesions.[31] However, damage to the interdigital skin by scarification or bandaging increased the pathogenicity of the virulent-challenge strain for British-breed sheep and produced lesions of similar incidence and severity in all breeds. Other investigations have confirmed the importance of epidermal integrity as a primary-defense barrier in natural resistance to footrot. The marked differences in incidence of infection in resistant and susceptible lines of sheep, following natural exposure to the disease, was no longer evident after the skin-horn junction was scarified and artificially challenged with exudate from infected-lesion material.[32] It is conceivable that under more natural conditions, injury to the interdigital skin from *Strongyloides papillosus* larvae,[2,33] stones, sticks, thistles, or stubble would produce a similar effect in pastured sheep.

Natural resistance to footrot is a genetically heritable trait in that progeny of selected sire lines within the Targhee[32] and Romney[34] breeds differ in their resistance to footrot. The Targhee breed is a stabilized cross between a long-woolled British breed and the Rambouillet (French Merino).[35] Differences in susceptibility of unvaccinated Merino sheep to footrot have been noted by Stewart et al.[17,19] and Egerton et al.[36] In sheep naturally infected with a virulent strain, this was reflected in fewer necrotic lesions involving the abaxial wall and these sheep were able to maintain bodyweight, in contrast to other more severely affected animals. Variation in susceptibility to footrot was more apparent in sheep challenged with intermediate strains and some sheep were much more resistant than others to the development of severe underrunning lesions.[17,19] Egerton et al.[36] have reported that there is a spectrum of resistance to footrot among sheep in flocks, although the degree of virulence of the field strains of *B. nodosus* was not recorded. Some sheep did not become infected, whereas others remained chronically infected. Intermediate levels of resistance were reflected firstly in the time taken to become affected and secondly in the time taken to heal after becoming affected. In another Merino flock over a 2 year period, Marshall[9] has confirmed that a small number of sheep do not get footrot and in the remainder there are marked differences in the number of feet affected and the severity of lesions.

C. Environmental Factors

The full expression of the virulence of *B. nodosus* is dependent on environmental factors. This was observed in an experiment in which separate groups of sheep were challenged with strains differing in virulence and there were two replicate field plots for each strain. More severe lesions and greater transmission rates were observed in one of the replicate plots for each of two intermediate strains and one virulent strain. These plots could be uniformly irrigated and contained an even sward of lush, clover-dominant pasture growth. The other plots contained grass-dominant pasture and either dry or waterlogged areas.[19]

Thus, there is a complex interaction between the variation in natural resistance to footrot which occurs across and within breeds and the spectrum of virulence for *B. nodosus*. The full expression of virulence is further complicated by environmental factors such as topography, moisture, temperature, and type of pasture.

III. DIFFERENTIAL DIAGNOSIS

A. *B. nodosus* Infections

The clinical diagnosis of virulent footrot is straightforward when there are a number of sheep with extensive underrunning lesions since their nature and odor are so characteristic. However, diagnosis should be based on the clinical examination of the flock and not on individual sheep.

The differential diagnosis of footrot requires that information on the history of the flock be obtained, including their origin, contact with other ruminants, and the environment in which they have been kept.[24] Sheep should be examined in the paddock, preferably while grazing, to ascertain prevalence of lameness. At least 10 individual lame sheep, preferably in poor condition, should be selected, placed in a sitting position, and all feet examined for interdigital, skin-horn, soft-horn, and hard-horn lesions. If suspect lesions are found, paring of the horn may assist in diagnosis.[24]

Examination of affected feet reveals some or all of the following changes. The interdigital skin and skin-horn junction may be inflamed showing erythema, erosion, exudative necrosis, and hyperkeratosis (Plate 1C). There may be underrunning of the soft horn of the heel and sole extending under the hard horn of the toe and abaxial wall (Plate 1D to F). The underlying tissues may be erythematous and covered by a layer of white-grey necrotic exudate (Plate 1F). In more chronic cases, there may be pockets of this necrotic material and several layers of separated horn.

In benign footrot, the lesions are characterized by inflammation virtually confined to the interdigital skin (Plate 1A, B). The varying degress of inflammation range from hyperkeratosis of the posterior interdigital skin and skin-horn junction to more severe lesions of both these sites with erythema and exudative, erosive hyperkeratosis. These changes are usually much milder than in virulent footrot. The horn of the axial wall of the heels can possess a corrugated appearance. Slight separation of the soft horn of the axial wall at the heel is common but only rarely does this separation extend completely across the sole (Plate 1G).

With intermediate footrot, the lesions more closely resemble those of benign footrot except that there are a small percentage of sheep (up to 10%) with underrunning of the sole of the hoof to a varying extent.

Diagnosis is more difficult in the early stages of virulent, intermediate, and benign footrot when only interdigital lesions are present. In these situations, consecutive visits several weeks apart are required to examine the progression of lesions in previously identified sheep after a period in the paddock or after placing them in pens with wet mats. However, this is a rather inefficient and tedious method of diagnosis.

The presence of a *B. nodosus* infection in a flock can usually be determined clinically without the need to detect the organism microscopically. However, if confirmation is required for regulatory purposes, *B. nodosus* can be detected in smears from active lesions when stained by the Gram method since the organism is recognizable microscopically by its characteristic morphology.[2] Alternatively, a specific fluorescent-antibody staining method[37] can be used, but usually the Gram-staining method is sufficient. These simple staining techniques do not distinguish between the different clinical entities of footrot, so that samples of lesion material should be collected with sterile cotton swabs and inserted in transport medium. Alternatively, the swabs should be inoculated onto agar plates and placed in anaerobic containers[38] and then forwarded without delay for laboratory culture.

Substantial progress in laboratory diagnosis has been achieved in the past few years in that *B. nodosus* causing benign footrot can be confidently distinguished from the more virulent forms of the disease on the basis of differences in proteolytic activity,[39-47] and colony morphology.[22,23,39,41,48-51] The extracellular proteases of *B. nodosus* are used as a diagnostic tool to differentiate virulent and benign strains. The elastase test[40,41] is a reasonably reliable method for differentiation and is routinely used by regional veterinary laboratories in New South Wales, South Australia, and Victoria. The test can be used to differentiate at least some strains of intermediate virulence which can have reduced or delayed elastolytic activity.[22,23] However, others possess no activity. The incorporation of 0.15% $CaCl_2$ in the elastin-agar medium enables at least some of those strains to produce some clearing of the elastin particles.[22,23] The major disadvantage of this test is that it is slow, requiring 20 to 28 d to obtain a result from time of sampling. The degrading-proteinase test will also discriminate between virulent and benign strains[39] and a result can be obtained within 14 d from time of sampling when a modified test is used, utilizing heating instead of prolonged incubation.[17,44,46,47] This test does not discriminate among virulent strains and those of intermediate virulence.[17,18,20-23] The zymogram test provides qualitative information on the protease-isoenzyme composition of a particular strain.[43,45] Virulent strains and strains of intermediate virulence produce similar protease isoenzymes which can be differentiated from most of those produced by benign strains utilizing the zymogram technique.[18,20,22,23,45,51,52] This test, which also takes about 14 d for a result, is becoming widely accepted as an obligatory diagnostic test for distinguishing benign strains from the more virulent ones. Not all intermediate strains have thermostable proteases with a virulent isoenzyme pattern. One strain possessed a thermolabile protease and a zymogram pattern which differs slightly from both benign and virulent strains.[22,23] This strain also has some elastolytic activity when $CaCl_2$ is incorporated in the elastin-agar test.[22,23]

Colony morphology is useful for discriminating the various virulence categories: benign and intermediate strains have lower motility and smaller colonies than virulent strains.[21,51] Although colonies of some intermediate and virulent strains superficially resemble those of benign strains, particularly on hoof medium with 4% agar, differences become apparent on hoof medium with 2% agar and on blood Eugonagar. Thus, virulent strains can be separated from most of the intermediate, which in turn can be differentiated from the benign.[22,23]

Together, the results of colony morphology and proteolytic tests such as elastin agar, degrading proteinase, and zymogram tests allow better discrimination of virulent, intermediate, and benign strains of *B. nodosus*.[22,23] More rapid and definitive diagnostic tests based on monoclonal antibodies are currently being developed which will undoubtedly speed the decision-making process.

There are several other diseases causing lameness in sheep which should also be considered in the differential diagnosis of footrot.

B. Ovine Interdigital Dermatitis (OID)

OID is a mild disease characterized by inflammation of the posterior interdigital skin of sheep living on wet pastures. The interdigital skin is erythematous and swollen and often covered with a moist film of grey necrotic material. In some cases, areas of skin are eroded and ulcerated.

Usually there is no lameness and the disease is transient, disappearing when the pasture dries out.[53] However, a more severe necrotic interdigital dermatitis exposing the s.c. tissues and causing acute lameness in Perendale ram hoggets was reported by West.[54]

Fusobacterium necrophorum and *Corynebacterium pyogenes* are present in the tissues and there is intense epidermal invasion by *F. necrophorum,* the principal pathogen, as a result of water maceration and injury from abrasions and excessive cold, such as frosts.[53] OID resembles benign footrot. However, benign footrot tends to be more persistent[55] and the main distinguishing characteristic of the latter disease is that *B. nodosus* can be demonstrated in smears.[53]

C. Foot Abscess

In Australia and New Zealand, like footrot, outbreaks of foot abscess occur in wet seasons on lush, boggy pasture. It is seen most commonly in the hind feet of overweight ewes in late pregnancy,[56] and in rams.[57] This disease, called heel abscess (infective bulbar necrosis) by Roberts et al.[56] to differentiate it from toe abscess, affects the soft tissues, including the digital cushion in the bulbar region of the heel.[56] However, West[57] considers that the term, foot abscess, is more appropriate, as the disease can be defined as an infection of the distal interphalangeal joint and is characterized by more suppuration than necrosis in the later stages. The disease is due to a mixed bacterial infection caused principally by *F. necrophorum,* although other organisms such as *C. pyogenes* assist in the development of the lesion.[56] Unlike *B. nodosus,* which occurs only in diseased hooves, *F. necrophorum* and *C. pyogenes* occur in the feces of normal sheep and are always present in the environment. It is most likely that foot abscess develops as an extension of infection from OID penetrating the soft tissues,[56,58] which then enters the joint capsule between the second and third phalanx.[57]

The disease is sporadic in occurrence, and unlike footrot, it is unusual for a large number of the flock to be affected (usually $<10\%$), but outbreaks involving up to 40% of the flock have been reported.[59] Foot abscess causes marked lameness which may persist for several months. Usually only one foot and one claw are involved, the affected foot is carried, and sternal recumbency is common. On closer examination of early cases, a sinus may be observed in the inflamed and edematous interdigital space from which bloody necrotic material, and later pus, can be expressed. In more chronic cases, there is marked painful swelling and inflammation above the coronet. Purulent material is discharged from sinus tracts along the abaxial coronary border. In a proportion of cases, exaggerated movement of the digit indicates that the axial collateral ligaments and the interdigital ligament have been impaired. The majority of affected feet usually heal within about 8 weeks except for some permanent fibrous thickening, slight deformity, and slight lameness.[56,57]

D. Lamellar Suppuration (Toe Abscess)

This disease is distinct from foot abscess[56] and occurs in all classes of adult sheep and usually affects the front feet. Persistent wetness and mechanical trauma cause breakage of overgrown horn and subsequent injury to the sensitive laminae, allowing entry of pyogenic organisms into the soft tissue of the toe. Also, shelly-toe cavities may become infected, resulting in toe abscess. There is accumulation of pus and separation of horn at the toe, usually in one claw, which is painful if squeezed. Affected sheep are very lame and the leg is often carried. The infection may spread upwards causing heat and swelling and discharge of pus from a sinus at the anterior coronet region.[60]

E. Shelly-Toe (Dry Separated Wall)

Shelly toe is a fault of little importance, commonly found in the feet of sheep and may be accompanied by lameness. It results from separation of horn of the abaxial wall from the sole of the hoof near the toe, causing the formation of a semilunar cavity which becomes packed with mud, grass, and feces.[15]

F. Strawberry Footrot

This disease has only rarely been recorded in Australia and is due to infection of leg wounds in sheep by the actinomycete *Dermatophilus congolensis*. The proliferative dermatitis associated with strawberry footrot is characterized by multiple raised scabs between the coronet and the knee or hock. Removal of the scabs reveals shallow ulcers or granulomatous swellings with pus formation. The disease can be diagnosed on clinical appearance and the microscopic detection of filaments and coccoid forms of *D. congolensis* in stained smears.[60] Spontaneous recovery from this disease usually occurs in 5 to 6 weeks without treatment.[61]

G. Post-Dipping Lameness

Laminitis resulting from infection with *Erysipelas insidiosa* is now rarely seen since sheep dips contain a bacteriostatic agent. The organism multiplies in contaminated dipping fluids and it is believed that *E. insidiosa* gains entry from small wounds and abrasions on the feet, causing severe lameness several days after exposure, which results from the development of cellulitis, with loss of hair, coronitis, and laminitis but without joint involvement. The foot is hot and slightly swollen and very painful. The sheep usually recover spontaneously although there is fever, malaise, and inappetance. Some 90% of the flock may be affected but usually the prevalence is about 25%.[62,63]

H. Toxic Laminitis (Founder)

This disease is of nutritional or metabolic origin and is due to overfeeding of lambs on grain or in feed lots following the sudden transfer to high-concentrate diets. It is characterized by inappetence, depression, ruminal stasis, and acute lameness which occurs in one or more feet resulting in reluctance to move and recumbency. The lameness results from aseptic inflammation and edema of the sensitive laminae of the hooves, manifested by heat and pain around the coronets. There is usually no difficulty in diagnosis following a history of grain engorgement, lactic acidosis symptoms, and an examination of feet to exclude infective processes.[62,64]

I. Contagious Ecthyma (Scabby Mouth)

This disease is caused by a poxvirus infection which commonly produces skin lesions on the lips and muzzle of lambs and weaned sheep and which may extend into the buccal cavity. Lesions may also be seen on the udders and teats of ewes suckling lambs with lip lesions. Foot lesions are less common, but when present they are found in the interdigital cleft, coronet region, and posterior surface of the pastern, where they may extend up the legs towards the knee and hock. The primary lesions begin as small papules, vesicles, and pustules which rupture and become covered with coalescing, adherent thick scabs. The disease is self-limiting and resolves spontaneously. Occasionally, secondary bacterial infection of primary lesions in the feet may cause a purulent dermatitis resulting in acute lameness and blowfly strike, but usually there is only mild lameness and the feet lesions heal spontaneously. Although outbreaks can occur at any time, they are most common in summer when abrasions and trauma caused by mature, dry, prickly pasture or dry roughage assists entry of the virus into the skin. The disease also occurs in feedlot weaner lambs.[60,61,64,65]

J. Ulcerative Dermatosis (Lip and Leg Ulceration)

This contagious virus disease of sheep is characterized by the development of granulating ulcers covered with scabs, rather than proliferating lesions, on the skin of the upper lip, legs, feet, and external genitalia. The leg lesions may cause lameness and occur in the interdigital space and above the coronet to the knee and hock joint. Genital lesions involve the external preputial opening, the vulval lips of ewes, and less commonly the glans penis. Ulcerative dermatosis occurs in sheep of all ages and there is no systemic illness. Entry of the virus is facilitated by abrasions and trauma to the skin or by veneral transmission at mating.[60,61,64] The clinical description of ulcerative dermatosis resembles that for contagious ecthyma. Although ulcerative dermatosis has been mentioned as a separate entity, a specific virus has not, to date, been nominated as causing this disease.

K. Bluetongue (Catarrhal Fever)

Transmission of the arthropod-borne orbivirus causing bluetongue is dependent on the presence of suitable insects vectors such as *Culicoides* species. The disease is usually only clinically apparent in sheep. Cattle and goats can also become infected but they usually do not show any clinical evidence of the disease. In sheep, the disease is characterized by high fever (40 to 42°C), respiratory distress, nasal discharge, salivation, redness of the muzzle, nasal and buccal cavities, swelling of the lips, face, and ears, and erosions in the buccal cavity. The tongue may be swollen and cyanotic. Vomiting is common and secondary aspiration pneumonia and death may occur. Foot lesions manifested by intense lameness and recumbency only occur in some animals and are characterized by coronitis and laminitis. The appearance of a red to brown band encircling the coronet is an important diagnostic sign of bluetongue and becomes evident as the mouth lesions begin to heal. During convalescence, hoof walls may separate as in toxic laminitis. The acute systemic illness and mortality associated with bluetongue does not occur with contagious ecthyma, although mouth lesions may be similar. Bluetongue has a marked seasonal incidence, usually in late summer and autumn when the vector population is highest.[60-64,66,67] In northern Australia, the presence of bluetongue virus was discovered in 1977 but there has been no evidence of clinical disease in any livestock species in the field to date.[67,68]

L. Foot and Mouth Disease

Foot and mouth disease (FMD) is a highly contagious vesicular disease of cloven-hoofed animals caused by a picornavirus and is characterized by fever and vesicular eruption in the mouth and on the feet, high morbidity, and usually low mortality rate. The method of spread is by inhalation and digestion through direct contact and indirectly by fomites, swill feeding, contaminated milk, and by wind. The bulk of published evidence indicates that aerosol is the natural route.

In sheep, there is marked variability in the extent of lesions and lesions are usually not as severe as in cattle. However, serious outbreaks with heavy production losses have been reported in Turkey in 1962 to 1963 and 1964 to 1965. Thus, FMD is sometimes difficult to recognize and in addition, rapid healing occurs once vesicles have ruptured. The disease occurs in all the world's animal production areas with the exception of Australia, North America, U.K., Japan, New Zealand, and South Korea.

There is a short incubation period (3 to 4 d) and a short febrile period (40 to 41°C) when there is inappetence. The sheep are reluctant to move during the period of vesicle-eruption development. Lesions are relatively uncommon in the mouth and when they do occur are most likely on the dental pad and gums. These are thin walled and rupture quickly, leaving a raw surface, whereas on the tongue there are blanched areas of necrosis which slough and heal rapidly. Lesions can occur on one or more feet but rarely all feet are involved and lameness can be quite severe. Vesicles may appear in the interdigital area and around the coronary band, especially at the bulbs of the heel. The skin appears blanched and vesicles easily rupture with digital pressure leaving a raw weeping surface which is prone to secondary bacterial infection. As a result, in areas where footrot is endemic, the distinction between the two diseases can prove difficult. Vesicles may also occur on the teats leading to mastitis and also on the vulva and prepuce. FMD may cause pregnant ewes to abort and sudden death in young lambs up to 3 months of age due to myocarditis and skeletal muscle necrosis.[67,69-71]

IV. EPIDEMIOLOGY

A. Predisposing Environmental Factors

Outbreaks of footrot are usually confined to regions that have a sufficiently high annual average rainfall, usually above 500 mm, to produce a lush sward of pasture during the warmer, more humid months of the year. Thus, good pasture, dampness, warm weather, and susceptible hosts, together with a source of infection, favor outbreaks of the disease.

Under Australian conditions, spread of footrot has a seasonal occurrence with outbreaks occurring in the late autumn (April/May), particularly when good rains have fallen, but more commonly in late spring (October/November). Outbreaks can also occur in the summer on irrigated pastures. Transmission does not occur after pasture dries and, although during droughts the prevalence of the disease is substantially reduced, the disease does not disappear.[72]

In favorable seasons, 70% or more of an unvaccinated flock will develop footrot in an outbreak if no prophylactic precautions are taken.[28,36,73] In very wet years, in endemic footrot areas, the disease may extend to districts where the infection is not commonly seen.[74]

Footrot spreads under restricted environmental conditions requiring favorable conditions of moisture and temperature. Historically, it was long believed that these predisposing factors actually produced footrot.[75] Moisture is the most important factor, with at least 3 months of rainfall averaging 50 mm per month in the cooler part of the year being required for the spread of the disease. In the summer, falls of at least 125 mm/month are required. Prolonged rainfall promotes a lush growth of clover-rich pasture and keeps the soil wet for long periods.[76] Furthermore, Beveridge[2] noted that footrot is much more prevalent on improved pasture than on natural pasture and is most prevalent on country supporting a dense sward of pasture containing clover, than on country of lighter carrying capacity. From the late 1940s in Australia, pasture improvement and top dressing with superphosphate fertilizer created an ideal microenvironment for the spread of footrot, firstly by increasing the stock-carrying capacity and secondly by allowing better retention of surface moisture. Transmission of footrot ceases as the environment dries out, with the consequent dehydration of the interdigital skin, even though there may be sufficient moisture for continued plant growth.[76]

Transmission of footrot only occurs when the mean ambient temperature rises above 10°C. In winter, if the temperature is consistently below this level, disease outbreaks are not observed to any great extent because the low temperature inhibits transmission.[76] This finding has provided the potential for predicting outbreaks because the period when the temperature consistently rises above the minimal level is reasonably constant from year to year. In endemic districts in Australia, the period for transmission can be as short as 3 to 4 weeks but is usually about 6 to 12 weeks if sheep are exposed continuously to wet pasture during warm weather.[29,36,76] For example, in southern New South Wales, the appearance of new cases coincides with the seasonal rise in mean daily temperature and with spring rainfall in late August, with the highest prevalence of about 70% occurring in late October.[36] Further south in Victoria, there is usually a slow increase in the incidence of footrot in early spring but in October the number of affected sheep increases rapidly, reaching a maximum in either November or December.[29] Thus, the prevalence of footrot peaks according to fairly well-defined seasonal patterns each year, usually late spring.

B. Host Susceptibility

Breed resistance is discussed more fully under the section on clinical signs. Sheep of all ages are susceptible, with British breeds and their crosses being more naturally resistant to footrot than the Merino.[29-31] Susceptibility to footrot is a heritable characteristic and this natural resistance has been increased by selective breeding in the Targhee[32] and Romney Marsh.[34]

Lambs are considered to be less susceptible to footrot, usually developing mild lesions which are easily cured,[2] although infections caused by virulent[77] or benign[78] strains have been observed in lambs only a few weeks old. Furthermore, young sheep 4 to 12 months of age appear to be somewhat more resistant than adults and lesions are usually less severe.[2] Skerman[34] reported that ewes greater than 4 years old had higher footrot ratings and prevalence than younger-age cohorts. Although requiring further confirmation, there does appear to be an age-related increment in susceptibility to footrot.[34] Beveridge[2] and Littlejohn[77] considered that rams were more susceptible than ewes and this apparent sex effect has recently been verified, since footrot is substantially more pronounced among ram lambs than ewe lambs.[34]

Sheep vary in their susceptibility to footrot at different times of the year. In winter, there is a marked fall in the skin temperature of the digits due to a constriction of the arteries supplying the region. Thus, feet are less susceptible because the low skin-surface temperature inhibits the invasion of *B. nodosus,* as well as reducing the multiplication of the organism in lesions, thereby reducing the amount of infective material shed onto the pasture. In summer, the interdigital skin is dehydrated and consequently is a very sound barrier to invasion by the footrot bacterial flora.[76]

C. Infection Source

After footrot occurs, the end result is either chronic disease, inapparent infection, or complete recovery. The last category obviously no longer constitutes a source of reinfection.[79] Natural healing is a normal feature of footrot with the proportion doing so depending on the breed involved.[29-31] The chronically affected carrier animals constitute the major reservoir of infection, but can be eliminated from flocks because they are relatively easily recognized. These overt lesions may vary from relatively mild, superficial moist lesions persisting in the interdigital skin, through to those that exhibit marked underrunning with necrosis and fly strike. The inapparent or symptomless carrier that does not show lameness also constitutes a hazard in epidemiological terms. One type harbors a covert infection in a small focus under healthy horn and this can only be identified by careful paring. In some but not all cases, the affected foot can be identified by its slightly misshapen appearance. These animals are a potential source of reinfection if the sequestered infection breaks out on the surface. Another type is the one in which *B. nodosus* exists in the hyperkeratotic *stratum corneum* in clinically normal, apparently healed interdigital skin of recovered sheep as demonstrated histologically by Morgan et al.[55] These sheep had been affected with benign footrot and may explain the difficulty in eradicating this disease from properties. It remains to be determined if virulent *B. nodosus* can also be demonstrated in apparently normal feet. An alternative explanation for the greater difficulty in eradicating benign footrot is that the transient, underrunning lesions observed experimentally with one strain could induce a small, persistent focus of infection in the horn of the hoof and act as a reservoir for the carryover into the next favorable transmission period.[22,23] On at least three separate farms, benign strains have been isolated from the underrun-toe region.[22,23]

In endemic areas, sheep are the most likely carriers, but since goats can be infected with virulent strains[22,80] and benign strains[22,47,80] of *B. nodosus,* they are also a potential reservoir. However, sheep in contact with infected goats for an extended period under favorable conditions failed to develop lesions.[80] Several benign strains of *B. nodosus* have been isolated from farmed red deer and these caused benign footrot in experimentally challenged sheep.[81] *B. nodosus* are also commonly associated with foot lesions in cattle but the majority of cases caused benign footrot in sheep.[40,41,82-85] These strains possessed proteases characteristic of benign organisms in the elastase,[40,41] the degrading-proteinase,[86] and zymogram[45] tests. There is one report of severe underrunning lesions being induced in a sheep with material from a bovine lesion. However, pure cultures of *B. nodosus* from that sheep induced only benign lesions when further passaged in sheep.[87] In a field experiment, Stewart et al.[17] produced underrunning lesions in the feet of some sheep with a bovine isolate whereas the remainder developed only interdigital skin lesions. This isolate possessed virulent protease activity[17,22,23] and was considered to be of intermediate virulence for sheep.[17,19] Natural transmission of virulent *B. nodosus* from sheep to cattle has not been successful,[85] although Beveridge[2] succeeded in producing lesions in the scarified interdigital skin of a calf. Thus, it is highly probable that cattle are a reservoir for benign footrot of sheep,[86] although their role as a reservoir for intermediate and virulent footrot requires further confirmation. However, in relation to virulent footrot, most evidence suggests that it is unlikely that cattle constitute a major risk for sheep,[86] except by mechanical transmission.

D. Transmission

Infective material is transferred directly from exposed lesions onto soil, mud, and pasture, thereby contaminating the feet of other sheep. The social behavior of sheep walking in single file along foot pads probably contributes to spread of the disease.[2]

Being anaerobic and, consequently, oxygen sensitive, *B. nodosus* dies fairly rapidly away from its host, due to its oxygen sensitivity. It persists no longer than 4 to 5 d in lesion material.[2,14] This limited survival is probably assisted by common aerobic bacteria decreasing available oxygen. The basis for eradication of the disease is the total host dependency for multiplication and long-term survival of *B. nodosus*. Mechanical transmission, such as wheels of vehicles, people's boots, or the feet of animals other than sheep, could conceivably be the means of footrot spreading from one property to another, but this is considered to be improbable, as footrot is not highly contagious like some viral diseases.[59] However, instruments used in paring hooves could transfer infection from one sheep to another and therefore should be disinfected between sheep.[59]

On exceptional occasions, unexplained outbreaks of footrot have been reported in previously clean flocks in the absence of introductions of infected sheep, and some experimental data provided by the author lends some validity to these assertions. In one experiment, there was cross contamination between plots with a virulent strain but it was not determined how this occurred.[17,19] There was no cross contamination with three strains of lower virulence. Presumably, the most virulent strain produced more exudate from necrotic lesions and consequently more *B. nodosus* were shed onto the pasture. Hares and ibis moved freely between plots and blowflies and irrigation water could also have been responsible. Direct human transfer seems unlikely, since hands and boots were disinfected and clothing changed between plots.[19] On another occasion, observed by the author, footrot occurred in a flock of ewes in a paddock through which a group of infected sheep had been inadvertently transported by truck during a thunderstorm.[88] Presumably, infective material was washed from the truck, contaminating the pasture. However, the vast majority of outbreaks result from contact between clean and infected sheep.

Management factors such as lush clover-dominant pasture, irrigated pasture, poorly drained soils, or any factor which concentrated sheep in small areas such as high-stocking rates and access to creek flats or waterways increases the likelihood of transmission of the disease. Sheep in feedlots or housed indoors under wet, crowded conditions are also more at risk.

V. PATHOGENESIS AND PATHOLOGY

A. Pathogenesis

Before the pioneering work of Beveridge,[2,89] there was no precise knowledge of the etiology of footrot. Up until that time, *F. necrophorum* and spirochetes had been considered as causal agents but had not been shown to cause the disease. In these studies, Beveridge reported the isolation of a Gram-negative, nonsporing anaerobic rod, *Fusiformis (Bacteroides) nodosus* from lesions which reproduced footrot when cultures of the organism were applied to the scarified feet of sheep and which was comparable in all respects with the natural disease. Footrot of greater severity was more consistently produced when *Spirochaeta penortha* were included in the inoculum. In histological sections, *B. nodosus* were often difficult to demonstrate but where present, it usually occurred near the more superficial portion of the lesion with occasional organisms penetrating with the motile fusiform or *S. penortha* ahead of other bacteria. Bacilli resembling the motile fusiform were abundant, often occurring as a dense, felted mass and were observed penetrating further into the tissue along fissures. However, these rods and filaments were sometimes difficult to differentiate from *Fusobacterium necrophorum. S. penortha* were usually less frequently present than the motile fusiform. On the surface of the lesion, bacteria consisted mainly of cocci and corynebacterium. Beveridge concluded that although footrot was a complex mixed infection, *B. nodosus* are the primary causal organism with another anaerobic bacterium, *S. penortha,* possibly playing a specific role as an accessory causal agent. Beveridge regarded the motile fusiform as a constant secondary invader from the environment and it probably did not play a significant part in the pathogenesis of the disease. There was no evidence that *F. necrophorum* played a part in the pathogenesis of the disease and consequently he considered it to be a secondary invader. The continual survival of these anaerobic bacteria is probably aided by association with nonspecific aerobic bacteria, such as cocci and corynebacteria, which are also present in this mixed infection.

Thomas[90] confirmed the observations of Beveridge by reproducing footrot with pure cultures of *B. nodosus* applied to scarified feet or injected near the skin-horn junction of the heel without the intervention of *S. penortha* or the motile fusiform. An experimental infection at the base of the horn of the head with *B. nodosus,* which caused shedding, provided additional evidence for the primary role of *B. nodosus,* since cocci were the only secondary organisms observed. Histological sections showed abundant *B. nodosus* at the point of cleavage in the *stratum corneum* but *S. penortha* and fusiform bacteria were absent. He concluded that after injury to the skin-horn junction, *B. nodosus* alone was capable of causing typical severe footrot with progressive separation of the horn of the hoof. The presence of *S. penortha* and the motile fusiform, which grow in necrotic tissue, was dependent on primary invasion by *B. nodosus*.

In 1967, Parsonson et al.[53] clarified the etiology of ovine interdigital dermatitis (OID, a form of scald) and considered to be a predisposing cause of footrot. They showed that the necrotic lesions in the interdigital skin were associated with deep invasion of the epidermis by *F. necrophorum*. OID could be reproduced with *F. necrophorum* provided there was prior skin damage. A dense flora of aerobic diptheroid bacteria including *Corynebacterium pyogenes* on the surface of these lesions may contribute to the growth of *F. necrophorum* by decreasing oxygen tension, Eh, and destroying H_2O_2 by the production of catalase.[53]

In the light of the findings with OID, Egerton and his colleagues reinvestigated the invasive behavior and importance of *F. necrophorum* in the pathogenesis of footrot.[12,13] From their research they considered that the ability of *B. nodosus* to cause footrot without the assistance of *F. necrophorum* was doubtful. In experimentally produced footrot, a superficial invasion by *F. necrophorum* always preceded growth of *B. nodosus* in the lesion. This was supported by observations that cultures of *B. nodosus* alone did not induce infection in sheep obtained from a dry environment, when the interdigital skin was incised and sheep were immediately placed in a wet environment following challenge. Natural lesion material containing *F. necrophorum* successfully induced the disease under the same conditions. In the absence of injury to the interdigital skin, prior water maceration together with a factor supplied by fecal contamination was a necessary prerequisite for the induction of the disease with pure cultures of *B. nodosus*.[12,13] Maceration alone was not sufficient to render sheep susceptible to infection. A role for *Strongyloides papillosus* as the fecal factor was excluded in these experiments, but there was always a definite association between fecal contamination and the invasion of the interdigital *stratum corneum* by *F. necrophorum*. This was supported by the finding that induction of footrot in clean feet required the intradermal injection of *F. necrophorum* and *B. nodosus*. The inclusion of *C. pyogenes* facilitated the development of footrot, probably because it produces a factor that increases the infectivity and invasiveness of *F. necrophorum*.[13]

Histological evidence indicated that the establishment and proliferation of *B. nodosus* was always followed by deeper penetration with *F. necrophorum*. This was supported by the induction of an allergic, accelerated inflammatory response in *F. necrophorum*-vaccinated sheep occurring 4 to 6 d after challenge with *B. nodosus*. *B. nodosus* and *F. necrophorum* were the only organisms observed penetrating fresh epidermal tissue. However, there was virtually no inflammatory response when *B. nodosus* was the sole invader at the point of separation, whereas there was severe inflammation and destruction in response to invasion by *F. necrophorum,* which contributed to the severity of footrot.[12]

It was concluded from these studies that following colonization of macerated interdigital skin, *F. necrophorum* facilitate infection with *B. nodosus,* probably by inducing hyperkeratosis and parakeratosis, which results in numerous crevices where anaerobic bacteria can survive and multiply. In established lesions, *F. necrophorum,* because of its invasiveness and elaboration of exotoxin, is responsible for a large proportion of the inflammation and tissue damage. Although *B. nododus* has poor destructive and inflammatory capacity, it initiates the invasion and extension of the lesion into the hoof matrix. Furthermore, it has the capacity to persist in lesions for long periods, thereby sustaining the infection between brief but intense waves of growth of *F. necrophorum*. It also produces a heat-stable soluble factor which enhances the infectivity and invasiveness of *F. necrophorum,* converting tissue colonized by *B. nodosus* into niches for *F. necrophorum* where it can begin its characteristic waves of filamentous growth.[13] Thus, they considered that footrot is caused by a synergic association of the obligate parasite, *B. nodosus* which is the essential transmitting agent, in conjunction with *F. necrophorum,* a normal inhabitant of the environment derived from ovine feces. The initial establishment is facilitated by *C. pyogenes* and surface-located diptheroid bacteria. The disease only develops when the availability of *B. nodosus* coincides with conditions of warmth, moisture, and fecal contamination enabling *F. necrophorum* to colonize the interdigital skin.[12,13] Beveridge[59] considers that some injury to the interdigital skin by physical factors is also necessary for a portal of entry, since intact healthy skin is resistant to infection. The main factor is water maceration, which weakens the barrier properties of the *stratum corneum* allowing bacteria commonly present in the skin and feces to cause a local dermatitis.

More recent studies by Hine,[91] using electron microscopic and immunocytochemical techniques, have revealed that apart from *B. nodosus* and *F. necrophorum,* a filamentous rod-shaped bacteria is always associated with *F. necrophorum* in the tissues and appears to be able to penetrate keratinized tissues directly. The growth characteristics and structure of this organism are consistent with those of the motile fusiform described by Beveridge.[2] This "mother" organism can form a stable synergic relationship whereby it is covered with a coat of smaller rods and the growth of both organisms is enhanced.[91] The phenomenon resembles that described by Beveridge[2] in which he sometimes observed other rod-shaped bacteria clumped around *B. nodosus* in a radial arrangement. These latter baby-like organisms, associated with *B. nodosus,* have been cultured by Thorley[92] and are nonmotile, Gram-negative, short to long, pear-shaped rods.

Hine[91] has confirmed that *F. necrophorum* were always found deeper within the lesion of footrot where tissue destruction was taking place. Large numbers of the filamentous "mother" organism and *F. necrophorum* were always found in close association. The associated "baby" organisms occurred throughout the lesion material but also penetrated ahead of *F. necrophorum*. Thus, the two newly described synergically linked bacteria are closely associated with the developing lesion and may have a role in footrot.

B. Pathology

Early studies by Beveridge[2] located the advancing edge of the active lesion of footrot as occurring in the superficial layers of the epidermis of the hoof just beneath the keratinized layers but above the deeper, viable layers of the *stratum spinosum*. He considered that the inflammatory process caused progressive accumulation of exudate in the intercellular spaces producing cavities, progressive cellular atrophy, and detachment of the keratinized horn from the unkeratinized epithelium. Deane and Jensen[93] identified the layer of the epithelium in which degeneration occurred as the *stratum granulosum* and superficial layers of the *stratum spinosum*. In early lesions, there was hydropic degeneration, necrosis of epithelial cells, and formation of larger vacuoles with infiltration of these cavities with leukocytes. Advanced lesions were characterized by extensive liquefactive degeneration and necrosis of epithelial cells at the advancing edge of the lesion in the *stratum granulosum* and superficial cells of the *stratum spinosum*. Thus, their view, as opposed to that of Beveridge,[2] was that necrosis in this region was responsible for detachment of the hoof from the underlying epithelial structures in the foot.

In neither of these studies was a direct association established between the pathological changes and the presence of *B. nodosus* due to the early difficulty of demonstrating these organisms in sections,[2,93] although occasionally a few organisms were observed penetrating the epidermis together with the motile fusiform or *Spirochaeta penortha* ahead of other bacteria.[2] Thomas[90] was able to demonstrate this relationship. In early lesions, he described abundant *B. nodosus* invading the *stratum lucidum* beyond the point of cleavage of the epidermis. *B. nodosus* was not observed to penetrate the *stratum granulosum* and *stratum spinosum,* although cells in these regions showed cytoplasmic degeneration. In confirmation of the findings of Deane and Jensen,[93] advanced lesions showed marked pathological changes with cell lysis usually resulting in formation of vacuoles, cavities, and microabscesses.

Leukocytes, mainly neutrophils, were numerous in the dermis and penetrated throughout the *stratum spinosum* to the site of invasion of *B. nodosus*. Thomas[90] considered that neutrophils phagocytosed *B. nodosus,* since very few of these organisms were observed in areas of the *stratum lucidum* where neutrophils had penetrated.

The degeneration, detachment, and liquefactive digestion of partially keratinized cells of the *stratum lucidum* and *stratum spinosum,* as well as formation of fissures, sinuses, and necrotic foci in the fully keratinized horn, appeared to be directly related to the presence of *B. nodosus* and were unassociated with other bacteria.[90] Thomas[90,94,95] considered that the damage was consistent with digestion by keratolytic proteases elaborated by *B. nodosus.* The pathological changes in the *stratum spinosum* were probably due to secondary infection and pressure atrophy, the latter resulting from removal of the horn which provides support and protection to the living cells.[90]

Further conflicting results were later obtained by Egerton et al.[12] who concluded that the epidermis was disrupted by diffuse inflammation and cellular infiltration which caused separation of the horn and this was unassociated with the immediate presence of bacteria or keratolytic activity. Thus, they considered that there was no evidence of separation due to direct bacterial attack on the horn itself. However, more recent studies by Hine and his colleagues[96] and Stewart and Parsonson,[97] although to some extent still contradictory, have confirmed some of the observations of Thomas,[90] in that separation of the nonliving keratinized horn from the underlying living uncornified epidermal tissues of the foot is due to direct bacterial attack. The histological study of Stewart and Parsonson[97] has shown that only *B. nodosus* is consistently associated with the point of cleavage and the plane of separation is located in the interface zone between the *stratum lucidum* and *stratum granulosum.* Although *B. nodosus* does not appear to possess keratolytic activity *in vivo*[12] or *in vitro,*[44,98] this does not rule out the possibility that proteases or other factors[23] elaborated by *B. nodosus* may be involved in the epidermal-splitting process, since the lesion does not involve keratinized tissue. On the other hand, Hine and his colleagues,[96] based on immunocytochemical evidence, believe that metabolites produced by *F. necrophorum* and the unidentified, motile organism caused the intercellular-matrix breakdown, leading to sloughing of horn. A common feature of lesions of footrot was the intensity of the inflammatory response which usually extended from the underlying dermis along the rete pegs or dermal papillae into the epidermis.[97] In the epidermis, monocyte and neutrophil leucocytes were often in large numbers and occasionally phagocytosis of *B. nodosus* by neutrophils was noted.[97] Polymorphonuclear leucocytes probably also contribute to the disease by exacerbating tissue destruction caused by secondary bacterial invasion of the tissues.[96]

VI. TREATMENT, CONTROL, AND ERADICATION

A. Treatment

There are several treatment options since footrot is a relatively superficial bacterial infection located in the zone between the keratinized horn and the avascular, living epidermal tissues of the hoof, but still accessible to substances diffusing from the bloodstream. The options used either alone or in combination are discussed below. At present, footbathing combined with vaccination is the most widely used method of treating footrot. Although oral-zinc therapy is discussed, the initial claims for its effectiveness have not been substantiated.

1. Individual Topical Treatment

Surgical exposure of footrot lesions by paring the horn of the hoof, so as to facilitate contact with topically applied bactericidal agents, is the time-honored but most labor-intensive method of treating the disease.[59] For topical treatment to be effective, the surgical preparation of the foot is very important and the procedure was essentially described by Youatt 150 years ago.[2,4] Although very labor intensive, all feet must be carefully examined for evidence of lesions and any feet with either obvious or suspicious lesions meticulously pared. This requires skill, perseverance, and conscientiousness. The full extent of the infection must be completely exposed by carefully removing all underrun horn covering the lesion.[2,99,100]

Paring can be most effectively performed when sheep are properly restrained in a sheep-handling device which may vary from the simple manual tip-over type to the more complex rotating, hydraulic, or belt-delivery systems.[101] There are several commercial units, as well as owner-constructed types. A simple sheep handler incorporating an inclined loading ramp and manual tip-over can be built to design-specification drawings.[102] The gravity-case-roller type is also a relatively simple labor-saving device in which sheep are tipped on their backs into wooden boxes or cases and the boxes are moved by gravity over split rollers.[103,104] Sheep handlers facilitate easy inspection of all feet and avoid the back-breaking labor of turning up sheep and lifting them into cradles.[104]

Paring instruments required are hoof secateurs with free-moving, closely opposed, sharp, cutting blades and a sharp-pointed, strong knife.[2] Paring knives should be used after the work with the secateurs has been completed.[100] Hemorrhage is undesirable since it may mask the field and interfere with the action of the topical agent. On the other hand, it is better not to be too conservative, since insufficient paring may leave pockets of infection undetected.[2,99,100] Bleeding can be minimized by experience and by paring in a direction from the heel to the toe.[100]

After paring, topical bactericidal agents are applied to the exposed diseased tissue with a 2.5 cm paintbrush with short, stiff bristles,[59,105] a hand sprayer,[106] or as an aerosol spray. Excellent results were obtained in the past with 10% chloramphenicol in an alcoholic solution. Higher cure rates and fewer relapses were achieved after one topical application of this antibiotic than with 5 or 10% formalin.[99,105-109] Chloramphenicol is now contraindicated in food-producing animals because of findings in the U.S. confirming the link between the use of this antibiotic and aplastic anemia, leukemia, and deaths in humans. In Australia, earlier this year, the National Health and Medical Research Council recommended that all use of chloramphenicol in food-producing animals be discontinued.[110] Dichlorophen[111] has also been discontinued for topical treatment of footrot. The antibiotic, oxytetracycline hydrochloride[108,109,112,113] as an aerosol spray or quarternary ammonium antiseptic solutions such as 20% cetrimide (cetyltrimethyl ammonium bromide) in alcohol or as aerosol sprays[108,114] are satisfactory alternatives. However, the topical use of antibiotics such as oxytetracycline hydrochloride should be avoided in diseases like footrot where diverse species of bacteria are present. In human medicine, the topical use of antibiotics gives rise to a high incidence of antibiotic resistance much faster than parenteral and oral use of the same antibiotics.[110] Iodophore antiseptic solutions such as povidone-iodine paints or aerosol sprays are also available. Iodine is reasonably effective against *B. nodosus in vitro.*[113] The local antiseptic application of 10% zinc sulfate in water to affected feet has also been reported to give excellent results.[106] In the latter study, although only a small number of feet were treated, more were cured than with 10% formalin solution or 10% copper sulfate. Topical treatment is more effective under dry than wet conditions.[109,111,114] Under wet or muddy conditions, the treated sheep should be held in a dry concrete yard or in a shed on boards for at least an hour.[59,99,105,115]

2. *Flock Topical Treatment (Footbathing)*

Whereas direct application of topical agents is favored for small numbers of sheep, footbaths are the preferred, more practical method for mass treatment of flocks of sheep.[100] Footbaths containing antiseptic solutions have been used for many years for both preventing and treating footrot. For treatment, Murnane[116] recommended standing sheep in footbaths containing 5% copper sulfate (blue stone) or 2% formalin for at least 1 h. Beveridge[2] considered that 10% formalin was probably better than 30% copper sulfate and that 10% zinc sulfate had a moderately good curative effect, although probably inferior to the latter. He recommended that all sheep should be walked through a footbath containing 2% formalin or 10% copper sulfate weekly as a preventative measure. For treatment, sheep may be walked through a 10% solution of formalin or a 30% solution of copper sulfate every 2 or 3 d for up to 2 weeks. Fitzpatrick[117] confirmed the prophylactic value of walking sheep with unpared feet through a formalin footbath at weekly intervals, but not less frequently, for controlling the spread of footrot.

More recently, footbaths containing zinc sulfate have been shown to be effective, safe, practical, and economic. Demertzis et al.[118] showed that twice daily footbathing in 20% zinc sulfate with no paring of feet produced a marked improvement in treated sheep. Cross and Parker[119] demonstrated that daily footbathing in 10% zinc sulfate for 30 d, also without paring, greatly reduced the number of clinical cases affected with footrot and there was no evidence of discomfort when infected feet were in the antiseptic solution. Further interest in the use of zinc as a topical application was aroused when Malecki and McCausland[120,121] showed that *in vitro* the anionic surfactant sodium lauryl sulfate (sodium dodecyl sulphate) increased the penetration of zinc into ovine hoof horn approximately sixfold and this was further improved if ethanol was also included. Formaldehyde, the most commonly used topical agent, did not penetrate at all, possibly due to the formation of cross-linking methylene bridges within proteins, thereby making the horn impervious. This was supported by the observation that zinc uptake by ovine hoof horn was greatly reduced in the presence of formalin. They suggested that formalin acts as a surface disinfectant but it may also render the skin-horn junction less susceptible to attack by bacteria as a result of modifying the surface of the epidermis. Their approach in undertaking this work was to select a therapeutic agent that could penetrate the *stratum corneum* without the necessity for extensive paring.[120,121]

Promising results were obtained without foot paring in pen experiments and field trials with a combination of 20% zinc sulfate and 2% sodium lauryl sulfate (SLS).[122] Sheep were footbathed for 1 h on two occasions 5 d apart. In the pen experiment, a complete cure was obtained with the zinc sulfate/SLS treatment, whereas formalin footbathing for 10 min 5 d apart failed to cure or alleviate lameness. Sheep standing in the formalin footbath were also clearly distressed and repeatedly lifted their feet out of the formalin. The new formulation was also effective in field trials producing a high percentage of cures without any foot paring being done and promoted a rapid recovery from lameness within 3 to 7 d of the completion of treatment.[122,123] The formulation was not effective in treating affected sheep with misshapen and overgrown feet and these invariably had an active lesion in the toe region.[122] Paring of these chronic lesions increased the cure rate obtained with the new formulation.[124]

Skerman et al.[125] compared in Merino ewes and Corriedale ewe lambs three topical treatments with untreated control sheep. They were preliminary hoof paring with weekly footbathing for 5 min in 10% formalin or in 10% zinc sulfate + 0.2% anionic surfactant containing 40% SLS (Teepol®, Shell Chemicals), or twice-weekly footbathing for 5 min in the same formulation without foot paring. Within 21 d all treatments produced a substantial reduction in the number and severity of lesions in comparison with the controls and there were no significant differences between treatments. However, more feet lesions healed and there were fewer chronic infections among sheep treated by paring and zinc-surfactant footbathing. Mild infections involving the soft horn of the heel and sole were highly responsive to all procedures but against the most extensive lesions involving the abaxial wall, the healing response was significantly better with paring and footbathing in the zinc-surfactant solution than to paring and formalin footbathing. The former treatment also produced a better curative response against these chronic lesions than zinc footbathing without foot-paring, although the difference was not significant.

Although footbathing in the zinc-surfactant solution was also highly effective in preventing new infections of footrot, there was no residual effect following cessation of footbathing.[125] This contrasted with a report by Atkins[126] claiming that a commercial formulation (Footrite®, Hardman Chemicals Pty. Ltd.) containing 20% zinc sulfate and 2% SLS gave residual protection for several weeks. Consequently, this facilitated eradication of footrot in three flocks during a period favoring spread of the disease.[126] However, additional footbathing experiments, comparing zinc sulfate with Footrite®, need to be performed to substantiate the claim that the latter has a residual protective effect.

Skerman and his colleagues[127] performed a second trial with Corriedale ewes and compared footparing with weekly footbathing for 2 min in 10% formalin or in zinc-surfactant solution or footbathing in either preparation without paring. Final assessments were carried out 1 week after the fifth successive weekly footbathing. Again, all treatment significantly reduced the number of lesions. Paring plus zinc-surfactant footbathing had a significantly greater curative effect than paring and formalin footbathing, but differences among the other treatments were not significant. In this experiment,[127] foot paring plus formalin footbathing resulted in a greater proportion of extensive lesions being cured than in the first experiment.[125] A similar number of serious lesions healed in the group that was footbathed in formalin but not pared, suggesting that initial paring did not influence the final curative effect to any extent when five weekly footbathing treatments are used. For those footbathed in the zinc-surfactant mixture there was a higher but not significantly different curative response when paring was used.

A third experiment was conducted in Merino wethers to determine the effect of adding 0.2% Teepol® to the 10% zinc-sulfate solution.[127] There was no difference between the treatment groups after weekly footbathings for 3 weeks followed by twice weekly footbathing for a further 3 weeks, although there was a marked reduction in the number of active lesions in comparison to the controls. However, this footbathing solution contained about 20-fold less SLS than used by Malecki and his colleagues.[122,123]

In a fourth trial, standing sheep for 5 min in 10% zinc sulfate was shown to be highly effective in curing and preventing interdigital dermatitis and curing pared underrun lesions of footrot in Corriedale ewes when inspected 6 d after the third weekly footbathing. By comparison, copper 8-hydroxyquinolate had no detectable effect.[128]

Bulgin and co workers[129] have also found that footbathing in 10% zinc sulfate (containing 0.2% heavy-duty laundry detergent with non-ionic surfactants) for 1 h with an interval of 10 d was an effective treatment of chronic cases of footrot in Rambouillet ewes without the necessity for footparing. Foot paring at four weekly intervals and a daily walk through footbathing appeared to be much less effective than prolonged soaking of feet, although a direct comparison was not made.[129] These investigators had previously found that walking crossbreed sheep through 10% copper sulfate two to three times a week was relatively effective in controlling the spread of footrot, in that it reduced the incidence of interdigital dermatitis but in the absence of paring, footbathing had a minimal effect on underrunning lesions of the hoof.[11]

In a treatment trial conducted on wet pasture, Malecki and Coffey[123] found that foot paring prior to footbathing had a small but statistically significant detrimental effect on the number of sheep cured with Footrite®, suggesting that paring increased the susceptibility of sheep to reinfection during a period of spread of footrot. However, they cautioned that these results should not be interpreted as applicable to a nonspread period in summer when the feet are hard and dry and the infection is encapsulated.

The importance of paring in a nonspread period irrespective of the footbathing treatment was demonstrated by Plant and Claxton.[130] They found no difference with footbathing in

5% formalin, Footrite®, or 10% zinc sulfate over paring alone. Sheep were footbathed in 10% zinc sulfate for 60 min on two occasions or for 5 min on five occasions, in Footrite®, for 60 min on two occasions, or in formalin on five occasions, each of 5 min. The interval between all treatments was 5 d. Initially, formalin footbathing plus foot paring produced the highest curative response for severe chronic lesions involving the abaxial wall, but these later relapsed and 112 d after the start of the trial there was little advantage in using any of the treatments over paring alone except some beneficial effect on healing of interdigital skin lesions. There was little evidence for spread of footrot during the trial.

In vitro studies have shown that *B. nodosus* is most susceptible to copper sulfate (minimal inhibitory concentration (MIC) 4 μg/ml), next is zinc sulfate (MIC 8 μg/ml), followed by formalin (MIC 32 μg/ml).[113] However, copper-sulfate solutions as footbaths have several disadvantages, including staining of wool[105,107,116] and copper poisoning[100,105,107] and their use was discouraged in favor of formalin. Furthermore, the bactericidal properties of formalin do not deteriorate in the presence of organic-matter pollution, unlike copper sulfate.[107] Arsenic solutions such as arsenic trioxide have been used but, because they are extremely toxic and there is also the difficulty of safe disposal,[100] their use is not recommended. The selling of arsenical sheep dips has now been banned in Australia.

Up until recently, formalin for many years has been the favored foot antiseptic for flock prevention and treatment of footrot because it is reasonably effectively yet relatively inexpensive. However, it does have a number of obvious disadvantages. Formalin is toxic and prolonged and repeated contact with 10% formalin or exposure to stronger solutions causes excessive keratinization, ulceration, and necrosis of the interdigital skin, secondary pyogenic infection, foot abscess, lameness, and death.[131-133] Since water has a higher partial vapor pressure than a dilute solution of formaldehyde, evaporation can exacerbate this skin-damaging effect by increasing the concentration of formaldehyde over time.[134] Consequently, Ross recommended that solutions of formalin in the range 2 to 5% should be prepared, used, and discarded on the same day. Formalin has a pungent, irritating odor affecting the eyes and respiratory tract which precludes its use indoors. It may promote allergic reactions and contact dermatitis.[100,125,134] Formalin does not penetrate hoof keratin so that it acts only as a surface disinfectant, requiring foot paring to expose all infected areas.[120,121]

Zinc-sulfate formulations appear to be at least as effective as formalin and in economic terms the cost of technical-grade zinc sulfate compared favorably with formalin.[125] It is safe, relatively nontoxic, pleasant to use, and has a long storage life.[122,125] Footbathing in zinc sulfate can induce a pronounced curative effect against a wide range of lesions of footrot without preliminary paring. Footbathing on a regular weekly basis is effective in preventing new infections where conditions favored transmission of footrot. However, a combination of footbathing and paring is still required to obtain a worthwhile curative effect for extensive chronic cases with encapsulated lesions.

3. *Parenteral Antibiotic Therapy*

As early as 1951, Coppini[135] had achieved favorable chemotherapeutic results in treating footrot with a course of i.m.-administered penicillin, although Forsyth[136] found that a single small dose of procaine penicillin (300,000 units) was ineffective. Stewart[105] confirmed that *B. nodosus* was extremely sensitive *in vitro* to penicillin and chloramphenicol but less so to oxytetracycline hydrochloride and streptomycin. In 1956, Danks[137] cured 72 of 73 cases treated daily for 5 d with i.m. injections of 200 mg tetracycline hydrochloride, after paring of feet. Similar results were obtained by Sterk in 1960,[138] who reported curing footrot with a course of three i.m. injections of 150 mg of oxytetracycline hydrochloride and he considered that this was more effective than tetracycline hydrochloride. On the other hand, Marsh[61] considered that single tetracycline injections were quite effective but could not be depended upon to eradicate the infection. In Australia in the mid 1960s, the search for an effective, faster, and less labor-intensive treatment to replace traditional methods focused on devising a more practical approach for using parenterally injected antibiotics. Initial histological studies showed that an inflammatory exudate was present in the region of epidermis where *B. nodosus* penetrated so that a parenterally administered antibiotic should be able to diffuse from the blood capillaries to the affected site in the hoof as well as to infections too deeply situated for effective topical therapy.[139] The approach adopted was to use a single large dose of i.m.-injected mixture of bactericidal antibiotics, penicillin-streptomycin.[139] Parallel studies with another infection of the epidermis, cutaneous actinomycosis (lumpy wool) caused by *Dermatophilus congolensis,* showed that a single large dose of this antibiotic combination was highly effective in completely curing persistent cases of the disease.[140] In preliminary investigations, a dose of 70,000 units per kg of procaine penicillin and 70 mg of dihydrostreptomycin sulfate cured eight of ten experimental cases of footrot.[139] Chloramphenicol and oxytetracycline were less effective. Field trials with 10 ml of penicillin-streptomycin per adult sheep (2.5 million units penicillin and 2.5 g dihydrostreptomycin sulfate) suggested that parenteral antibiotic therapy combined with formalin footbathing and paring may be more beneficial. On properties where either of these were used, cure rates of 87 and 96% were achieved in contrast to only 70% where neither of the latter was used.[139] Subsequent experiments confirmed and extended these findings.[141] Antibiotics that were bactericidal for *B. nodosus in vitro* were generally more effective than bacteriostatic ones when given as a single dose. Of the single bactericidal antibiotics, only penicillin and erythromycin cured affected sheep and these were effective at dose rates giving bactericidal blood concentrations for about 18 h. The curative effect of penicillin was further improved when streptomycin was added. More recent *in vitro* antibiotic studies in Oregon with 18 strains of *B. nodosus* have confirmed that penicillin was the most effective of a range of antimicrobials tested, including the newer antibiotics, cephalosporin and clindamycin, which are widely used in human medicine for treating *Bacteriodes* species and other anaerobic infections.[113]

Diffusion of antibiotics to the skin surface was decreased in a wet environment, resulting in a marked decrease in the proportion cured. Under dry conditions, evaporation at the surface of the lesion resulted in a greater concentration of antibiotic at the skin surface than in the serum and consequently higher cure rates. In wet conditions, better results were obtained if sheep were housed on battens in a dry environment for 24 h after treatment.[141]

Maximum benefit is derived from antibiotic injections if affected feet are trimmed to a minor extent to improve drainage and footbathed in formalin.[141] However, Harris[142] found that a further improvement in cure rates occurred and subsequent assessment was easier and more accurate if the sheeps' feet were severely (thoroughly) pared rather than trimmed. He concluded that penicillin-streptomycin gave no better results than could be expected with more conventional methods whereby a cure rate of 70 to 80% is not uncommon. More recently, Malecki and Coffey[123] did not obtain any further increase in the number of feet cured with parenteral antibiotic therapy over that obtained with footbathing in Footrite®. However, procaine penicillin was used instead of penicillin-streptomycin and the trial was conducted during a period of transmission of footrot.[123] An interdigital s.c. injection of penicillin-streptomycin was no more effective than the i.m. route.[106]

In theory, systemic antibiotic therapy has a number of advantages over conventional topical treatment including greater certainty of freedom from relapses provided that the sheep are not treated during a period of transmission;[137] speed of treatment; treatment of both overt and undiagnosed covert lesions in the same animal; and a high cure rate following a single dose.[143] In practice, the need for conventional procedures such as paring and footbathing for maximum effect, together with the relatively small additional benefit obtained from parenteral antibiotic therapy has generally discouraged the more widespread adoption of the latter approach. Furthermore penicillin-streptomycin is a restricted drug and is relatively expensive because of the high dose rate required. Its use is more justifiable in valuable animals such as rams and stud sheep. There is also the potential hazard to human and animal health of the development of transferable drug resistance in intestinal flora with the widespread use of these antibiotics in sheep.

4. Oral-Zinc Therapy

Reports from Greece[118] and France[144,145] indicated that continuous oral medication of zinc in the feed or in salt licks was effective in curing and preventing footrot. These followed an observation by Demertzis and Mills[146] that healing of cases of bovine infectious pododermatitis in bull calves could be attributed to a daily intake of zinc given as a drench. The latter authors concluded that the condition was similar to bovine interdigital dermatitis which is caused by *B. nodosus.* However, investigators in the U.S.,[106,147] New Zealand,[148] and Australia[149] have not confirmed these reports. Cross[106] used single, high oral doses of either zinc sulfate or zinc metal whereas Cross and Parker[147] medicated the sheeps' feed with zinc sulfate for up to 6 months at a similar dose used by Demertzis et al.[118] Skerman et al.[148] chose two dose rates given daily to sheep for up to 7 weeks. The lower dose level, given as a zinc-sulfate drench, was chosen to rectify any dietary zinc deficiency whereas the higher level was administered orally as zinc oxide in gelatin capsules at the dose recommended for preventing liver damage due to facial eczema intoxication. Sheep were challenged on day 23 at which time plasma zinc levels had peaked in the higher-dose group but thereafter declined to values comparable to those of undosed animals. Egerton et al.[149] also used two dose rates given by drench as zinc sulfate, the lower dose, at a level recommended by Demertzis and Mills,[146] was administered to cattle for 4 weeks and to sheep for 2 weeks. The higher dose (four to five times greater than that used by Skerman et al.[148]), increased serum levels above those normally present in cattle and sheep. However, these serum zinc levels were well below the concentration required to achieve an inhibitory effect *in vitro* against *B. nodosus.*[113] It was concluded from these studies that oral zinc was of no practical benefit in reducing the prevalence or severity of *B. nodosus* infections in treated animals[106,147-149] and did not enhance the resistance of sheep to footrot.[147,148] In the light of their findings, Skerman et al.[148] also considered that the relationship between nutritional zinc deficiency and susceptibility to footrot was doubtful.

5. Vaccination

Provided an effective vaccine is used, immunization is the most efficient and cost-effective method of controlling footrot. The tremendous potential that vaccination offered for controlling this disease became evident when Egerton and his colleagues showed that such an approach was feasible for both preventing and curing footrot.[150,151] Although in Australia the first generation commercial vaccines which followed this initial development performed poorly and were withdrawn from sale in 1976,[152] improved vaccines containing representative strains from all the major serogroups and incorporated into potent oil-based adjuvants have been available since 1981. These commercial multistrain vaccines have been registered for use in New South Wales, Tasmania, and Victoria, having satisfied the minimum requirements for licensing. They include Footvax® (Coopers Animal Health Australia Ltd., formerly ICI Australia Pty. Ltd.) which was released in 1981, followed by Pro-vac® (Coopers Animal Health Australia Ltd., formerly Wellcome Australia Pty. Ltd.) in 1982, and Vaxall NoRot® (Arthur Webster Pty. Ltd., product marketed by Smith Kline Animal Health) in 1985.

The licensing standards require that commercial vaccines should have cure rates of not less than 65% and protection rates of 70%, 6 and 12 weeks, respectively, after vaccination. With cure rates there must be at least 20% affected initially, and for protection rates, 30% of infected sheep in the unvaccinated control group must be infected at the time protection is measured. The use of footrot vaccines has been banned in South Australia and Western Australia where statewide eradication programs have reduced the prevalence of the disease to a relatively low level. It is thought by veterinary authorities in these states that vaccines could mask detection of the disease.

While there is general agreement that commercial vaccines have performed satisfactorily, there is only limited published information available for a true assessment. Evaluation trials in the U.S. conducted in Idaho,[11] California,[153] and Colorado[154] with Pro-vac® and Footvax® have given encouraging results, particularly in relation to prophylaxis. In Idaho, using Pro-vac®, protection rates of about 70% were obtained with crossbred flocks on a small farm and a large commercial-range operation. Furthermore, in the range flock, vaccination without footbathing provided good control of footrot by substantially preventing the development of both interdigital dermatitis and underrunning lesions of the hoof. Footbathing alone in 10% copper sulfate two to three times a week was only effective in treating the former type of lesion and not the latter. The main drawback of the vaccine was the propensity for producing granulomas and abscesses, particularly in the cheek, although inoculation behind the ear did not cause problems with shearing 3 months after the second vaccination. Some 20% of vaccinates developed discharging abscesses which healed without treatment. These reactions were acceptable in view of the beneficial effects of vaccination.[11] In California, in a small flock, over 18 weeks the protection and cure rates with Footvax® were 83% and 42%, respectively. All vaccine-injection abscesses had resolved by the end of the trial, although persistent reactions at the first and second injection sites were present in 45% and 30% of sheep, respectively. These were only of cosmetic importance.[153] Isolates of *B. nodosus* recovered from infected feet in the flock in California (serogroups D and B)[153] and the range sheep in Idaho (A, D. F, H)[11] were represented in the vaccine. In Colorado, promising results were obtained with Footvax®, although in the first flock, sheep were only vaccinated once, whereas in the other, whole-flock vaccination was performed in conjunction with trimming, footbathing, and culling.[154] Following these efficacy studies, approval has been granted by the U.S. Department of Agriculture for importing footrot vaccine.[154]

On an Australian property in Southern Victoria, Lambell[155,156] obtained a protection rate of 84% with Pro-vac® in Merino × Border Leicester ewes and this persisted for at least 10 weeks after the second inoculation. He considered that vaccination was the method of choice for controlling footrot, particularly in ewes with lambs at foot. Ninety percent of sheep still had vaccination reactions 120 d after the second dose of vaccine. However, discharging abscesses were highest (26.5%) after the first inoculation. Swellings equal to or larger than 4 cm in diameter were present in 48% of sheep at the first site and 42% at the second site. These were still present in 23% of ewes 100 d after the second dose when they were shorn. The reactions did not get fly struck and did not interfere with shearing.

In the footrot-endemic area of the southwest slopes and Riverina of New South Wales, current footrot vaccines, in general, have achieved good results and there was acceptance by the sheep owners of the reaction at the vaccination site.[157]

Not all of these reports have been encouraging, particularly in relation to the therapeutic efficacy of the vaccine in Merinos and their crosses. Mulvaney et al.[158] in New Zealand examined the therapeutic properties of Footvax® on five properties and obtained disappointing cure rates in Merino and Merino-cross sheep ranging from −40 to + 44% after the second vaccination, which was well below the level of 65% required for registration in Australia. Injection-site reactions (2 to 3 cm in diameter) were recorded in 51% of sheep. The severity of the reaction caused some concern, with discharging abscesses persisting for up to 6 weeks with associated wool loss.[158] The highest cure rate of Mulvaney et al.[158] was very similar to the 42% obtained with Footvax® in California.[153] On the southwest slopes of New South Wales, a cure rate of 55% was obtained with Footvax® in Merino wethers. The trial was conducted under conditions unsuitable for transmission and most affected feet had under-running lesions with very few interdigital skin lesions being present. It was considered that the therapeutic efficacy was too low to recommend its use in treating established cases of footrot in eradication programs.[159] In Victoria, vaccination of severely affected Merino sheep with a similar product did not increase the number cured over footbathing alone,[123] although there would have been a better assessment of the effects of vaccination if the sheep had not also been footbathed 2 weeks after the second vaccination. The curative response of footrot vaccines may not be evident until 6 weeks after the second dose.[160] In New Zealand, in less susceptible breeds such as Corriedales and Romneys, Liardet et al.[161] obtained good cure rates (73 to 100%) with Footvax® with marked responses being evident after the first dose. Other investigators in New South Wales using Pro-vac® have obtained a better long-term curative response with vaccination and severe foot paring than footbathing and paring.[130] In the trial, vaccination without paring was not examined. However, in Idaho, a poor curative response was obtained with Pro-vac® in Rambouillet ewes in which a high proportion of feet were affected with underrunning lesions and the results were similar to that achieved with paring every 4 weeks and a daily walk through footbathing in 10% zinc sulfate. The combination of vaccination, foot paring, and footbathing was a more effective treatment but this was found to be labor-intensive and costly.[129]

Ideally, it should be possible to effectively control footrot using vaccines without the need for either footbathing or extensive foot paring other than for diagnostic purposes. At present, footrot vaccines are relatively expensive and their cost becomes prohibitive when used in conjunction with extensive supportive measures.[143] If the cost of vaccines is not matched by high efficacy, then the economic justification of using footrot vaccines becomes rather doubtful.[158] Experience with commercial vaccines has led to the opinion that vaccines are best used to prevent spread of footrot, since protection rates exceed cure rates.[153,155] In these circumstances, vaccination is considered to be a valuable adjunct to standard footrot-control measures and vaccines are sufficiently effective to make eradication practical through limited paring and culling programs.[11] However, effective vaccines should be capable of preventing and curing footrot equally as well.[151,162] Furthermore, preventive and curative effects have been observed in Merinos after a single dose of homologous monovalent vaccine containing an optimum amount of pili (50 μg) and a remarkable curative response was obtained with chronic lesions in sheep vaccinated once with 82 μg of pili.[162] Thus, provided that the vaccine contains sufficient pili and the challenge is homologous, the level of protection and the substantial cure of both early and advanced lesions achieved by a single dose imply that the primary dose can be effective when administered during the active phase of an outbreak or later when lesions are at the chronic stage. Thus, relatively inexpensive and efficacious vaccines will be of major benefit to the sheep industry. In this regard, the prospects are extremely promising for more improved third-generation vaccines based on the results achieved with experimental recombinant-DNA vaccines[162-165] so that there is now hope that vaccines and the strategy for their use can progress to a stage where they can be used with a minimum of labor-intensive supportive measures to facilitate control and eradication of footrot.

For currently available vaccines, there are a number of applications and recommendations for the control and eradication of footrot.

Control of an outbreak when conditions are favorable for continued transmission — Because it takes 4 to 6 weeks and often two vaccinations to achieve the required level of vaccinal immunity for a beneficial effect, prophylactic footbathing is recommended where footrot is rapidly spreading through a flock. Footbathing should not be necessary when sheep have been previously primed and receive a booster at the time the outbreak is detected. The booster dose should be given at least 4 to 6 weeks and not more than 24 weeks after the initial vaccination.

Control after transmission has ceased to achieve a curative effect — With current oil-adjuvant vaccines, two doses are generally required to achieve a curative effect since most of the therapeutic benefit occurs after the second dose.[130]

Continued control in previously vaccinated flocks — Vaccinal immunity appears to be of longer duration in British-breed sheep such as Romneys than Merinos. Thus, a high proportion of Romneys can be protected for up to 16 weeks following the second dose of vaccine.[30] On the other hand, immunity in Merinos is relatively shortlived[30,166] and when challenge conditions are severe or the period of transmission is extended, such as on irrigated pasture or in high rainfall areas, a booster dose may be required to provide more prolonged protection.

Annual vaccination is not warranted on economic grounds in flocks free of the disease but it can be performed before the anticipated period of transmission in flocks where eradication in the short term is not feasible.

Current footrot vaccines have a role in the prevention and control of footrot and they help ease footrot-related management problems during mating and lambing as well as assist in maintaining flock productivity. Lambell[155] considers that they are most cost effective when used on properties with British-breed sheep or their crosses, in sheep infected with virulent footrot, where the percentage of affected sheep is low at the time of vaccination, and the second vaccination is administered at the beginning of a period of anticipated spread. Footrot vaccines also have a role in reducing the prevalence of the disease to a level where eradication becomes feasible.[20,143] The effect of vaccination should be to minimize the number requiring additional treatment, help prevent relapses, reinfections, and new cases of disease. Thus, vaccination should also assist eradication programs by increasing the healing rate in any footrot cases which were undetected at the initial inspection and help promote resistance to footrot in the next transmission period if undetected carriers remain. Vaccination followed by culling of chronically affected sheep will improve the level of flock immunity with a reduced chance of breakdowns occurring and this should improve the possibility of eradicating the disease within a shorter period of time.

B. Control and Eradication Strategies

The principles for control and eradication of footrot, which were originally devised by Beveridge in 1941[2] and based on a sound knowledge of the epidemiology of the disease, are still applicable. Using these principles, successful eradication of footrot has been achieved from many flocks of sheep in various parts of the world, particularly in areas where there is a well-defined seasonal rainfall pattern. However, to date, no country or region has achieved total eradication. There are three methods for controlling and eradicating footrot.[2,60,75,143,167]

1. Complete Depopulation

With this approach, all sheep are sold direct to slaughter and the farm is left without any sheep for at least 2 weeks before restocking from a source free of footrot. This strategy may be legally enforced in footrot-control and quarantine areas where sheep owners fail to take appropriate action or are unable to eradicate the disease within a specified period. This approach is not practiced to any great extent outside these areas except in flocks purchased by dealers or where the prevalence of footrot is high in all mobs within a flock. Destocking saves time and labor deciding which animals are free of footrot, but is unacceptable if valuable animals are involved or flock replacements are expensive.

2. Selling Affected Animals

This method relies on identifying all affected sheep, with no treatment, and disposing of them, usually for slaughter. In the long term, treatment is probably the most difficult and expensive alternative. Selling means that they no longer constitute a threat to the clean sheep and their removal will ensure that valuable time and effort can be spent making certain that the clean mob remains so. In the Footrot Control Area of western Victoria, the sale of affected sheep direct to an abattoir is a major means of eradication and this approach was adopted in 46% out of 479 outbreaks.[168] A combination of treatment and slaughter accounted for a further 47%.[168] Selling infected mobs when some mobs remain clean often sees the swift eradication of the disease and results in less threat to other mobs, as well as those of neighbors.[169] In South Australia, 90% of owners sold their sheep immediately after footrot was diagnosed.[170] In the New England Protected Area a combination of treatment and disposal remains the main method of dealing with outbreaks, although more properties are now using treatment alone to eradicate footrot.[171]

Although drastic culling may be inappropriate for economic reasons or because of the genetic value of the stock, there appears to be substantial justification for the long-established practice of culling chronic cases from breeding flocks based on the heritability estimates of footrot susceptibility in the more naturally resistant Romney Marsh breed.[34] Thus, culling of chronically affected breeding sheep should increase the genetic resistance of the flock to footrot and may facilitate eradication. Selling of aged sheep is also desirable not only because they are the least economically valuable but they are also considered to be more prone to footrot.[34] Within the Merino breed, there also appears to be some variation in susceptibility to footrot, particularly to the severe underrunning lesions caused by intermediate strains. With more virulent strains, this suggested natural-resistance phenomenon was reflected in fewer necrotic lesions involving the abaxial wall and maintenance of bodyweight in contrast to other more chronically affected animals.[17,19] Other investigators have suggested that there is a spectrum of resistance to footrot among sheep in Merino flocks. A minority of sheep do not become infected; others which become affected early in outbreaks have relatively persistent lesions; others are late-onset cases and have transient infections, whereas those which have infections present before the outbreak remain chronically affected, being least likely to heal.[36] Observations in another Merino flock over a 2-year period have confirmed that a small number of sheep (~15%) do not get footrot, and in the remainder there are marked differences in the severity of lesions and the number of feet per sheep affected.[9] Although at present there are no heritability estimates for natural resistance to footrot in the Merino, culling of sheep with chronic footrot would appear beneficial not only for eradicating footrot but also assisting in reducing the liability of the flock to the disease in succeeding generations. Furthermore, long-term strategies for selection and breeding for increased resistance to footrot may ultimately reduce dependence on other control measures such as chemotherapy and vaccination although it cannot be expected to replace them.[34] For this approach to be successful it will be necessary to identify biochemical or immunological markers for genetic resistance to footrot. In this regard, the usefulness of class 1 ovine lymphocyte antigen (OLA)-type of the major histocompatibility complex is being examined.[172]

3. Treatment and Segregation

This method requires that all clinical cases are detected, treated, and segregated until cured before the onset of the next period favorable for transmission. However, there is the risk that undetected carriers will be retained to reinfect the remainder of the flock.

The method adopted for eradication depends on various factors including the nature of the enterprise, the size and age of the flock, the number of sheep affected, the severity of the disease, the value of the animals, availability of labor, and whether quarantine restrictions have been imposed by disease-control authorities. Each method has its limitations and often the best approach is a combination of methods, such as treatment of affected animals and disposal of nonresponding culls. Identification and elimination of all affected animals have proved very difficult in some environments, particularly if climatic influences are difficult to predict, and control, rather than eradication, is sought.

C. Phases for Flock Eradication

There are three phases for an eradication program if the third option of treating affected animals is adopted.[2,143,169] These measures are particularly suited to regions where footrot is reasonably predictable with a defined seasonal pattern of spread.

1. Control (Prevention and Reduction of Prevalence)

These measures must be applied before and during the period of transmission. The control phase attempts to minimize the number of sheep affected at the end of the transmission period. Unless the number of infected sheep can be reduced to a low level of around 5% or less, subsequent eradication is more difficult and liable to failure.[2] If an outbreak is confined to just one flock on a property, the affected mob should be isolated behind secure fences from other stock as soon as a diagnosis is made. In this phase, the three choices include either vaccination, footbathing, or a combination of vaccination and footbathing. Paring, a particularly labor-intensive procedure, is not recommended because it is of questionable value during this phase. There is still controversy about which option should be used with each method having its proponents.

Vaccines — Provided vaccines are strategically administered just before or at the beginning of the period of spread, they can be expected to achieve reasonable control. Vaccination is simple, more convenient, less labor intensive, and provides fewer management problems than footbathing. It also requires only two yardings.[143] Once transmission has commenced, then it is advisable to carry out footbathing for several weeks to allow vaccinal immunity to build up.[155] However, with ongoing improvement in footrot vaccines it will not be unrealistic to expect that vaccines will be effective prophylactically and therapeutically when administered at any time during an outbreak.[162]

Footbathing — Jopp and colleagues[102,173,174] have had extensive practical experience using zinc-sulfate footbathing, in preference to vaccination, in conjunction with improved sheep-handling facilities to eradicate footrot (40 to 60% prevalence) from large fine-woolled flocks on improved pasture in the high country of Central Otago, New Zealand. They only recommended vaccination where there is a shortage of labor so as to keep an outbreak to a more manageable level.[102] During an anticipated period of spread, sheep are footbathed once a week in 10% zinc sulfate.[102,173,174] Specially designed permanent footbaths are located at strategic places on the farm as these provide longer foot-antiseptic-contact time than portable baths.[174] Alternatively, sheep can be footbathed in facilities attached to the woolshed where they are held in 10% zinc sulfate for periods of 10 to 60 min (usually 30 min) at an interval of 1 week, with both time factors depending on the severity of challenge.[102] However, movement of infected sheep from the quarantine paddock to these facilities during the transmission period poses management problems with contamination of laneways, paddocks, and permanent yards. In the eradication program in the New England Protected Area of New South Wales, veterinary officers favor the use of temporary yards built of strong portable hurdles and gates as well as temporary footbaths, these being lent free of charge to stockowners.[104,175]

2. Eradication

This phase can be an arduous and costly process and owners may suffer discouraging setbacks but nevertheless the fact that it is possible to eradicate footrot has been thoroughly demonstrated over the years. During the transmission period, new infections are occurring too rapidly and sheep incubating the disease may be unrecognizable, making eradication impracticable at that time. Also, the feet are muddy and difficult to examine properly. Thus, elimination of infected sheep should be confined to the driest period of the year when transmission of the disease is not favored. This is usually summer in southern Australia when ground conditions are dry and the interdigital skin is dehydrated and not susceptible to infection with *B. nodosus*. At this time there will be fewer animals to treat, since a large proportion of the flock (up to 50% or more) will have healed naturally with only chronically affected animals remaining. An attempt is then made to cure these sheep so that only those that remain chronically affected and nonresponsive to treatment are sold for slaughter.

a. Supervision and Facilities

Eradication of footrot is greatly facilitated by professional supervision and adequate sheep-handling facilities.[102,174-176] Supervision can be provided by private veterinary practitioners but in Australian regional footrot-control and quarantine programs, this service is usually provided by trained government inspectors and veterinary officers. Supervision provides the flock owner with a better level of understanding of the epidemiology of the disease and is essential in the early stages of the program for inspecting feet, demonstrating and advising on technique, as well as overviewing paring and footbathing.

Planned-control programs specifically designed for a particular property and written by the supervisor are a valuable aid in eradication as they increase the confidence, incentive, and commitment of staff and farmers. However, they must be carefully planned with clearly defined goals which can be achieved within realistic time limits.[176]

Jopp et al.[173] consider that improved sheep-handling facilities incorporating a continuous-flow system designed for handling, inspection, and treatment of large numbers of sheep efficiently and effectively with a minimum of labor input are an essential prerequisite before embarking on a treatment program. It also facilitates surveillance, which can be carried out during routine management procedures. In this system, sheep are moved from the wash trough used for washing and softening feet, up a ramp into the sheep handler where they are upended, assessed, and pared under adequate lighting. Sheep are either placed in retaining pens if infected, or shifted into a holding pen prior to moving into the footbathing pen. The latter is constructed to hold from 40 to 200 sheep for up to an hour. After footbathing, sheep are moved onto wooden battens in the sheep shed or onto a concrete area for drying of feet.

It is essential that all feet of every sheep in the diseased flock following a complete muster be carefully examined by an experienced person. Some diagnostic paring will be necessary to detect pockets of infection, particularly in hooves which are long and overgrown or misshapen with irregular soles or bulging walls. Knives and secateurs should be disinfected between sheep to prevent possible transfer of infection.[58] Jopp et al.[173] use a galvanized pot containing disinfectant as well as an electric heating element for heat sterilization, which also assists in preventing the spread of caseous lymphadenitis on dirty instruments. Inspection and diagnostic paring is followed by segregation into two separate flocks consisting of clean (noninfected) and hospital (infected) mobs. The latter consists of obvious cases of virulent footrot, those with pockets of infection, those with interdigital dermatitis, as well as those with suspicious lesions.

b. The Segregated Clean Sheep

After inspection, the clean sheep can be footbathed and moved to a clean paddock which has been spelled for at least a week, although in a nontransmission period spelling is not mandatory unless paddocks contain wet boggy areas. Frequently, owners neglect the clean sheep and place too much emphasis on the diseased mob. It is essential to concentrate on keeping the clean flock free of infection and giving them higher priority than the hospital mob.[20,174,177] Thus, it is advisable to perform one, preferably two, further inspections with an interval of 3 to 4 weeks. Any infected sheep are culled or placed with the hospital mob since it is of obvious importance that diseased sheep are not present in the clean flock during the next transmission period.

c. The Segregated Diseased Sheep

All the sheep in the diseased mob should be appropriately identified by ear tagging or by scourable, stock-marking raddle, using an aerosol spray or branding fluid. Limited culling prior to treatment of the remaining hospital mob is advisable for badly affected, debilitated, or chronic cases. Following treatment, nonresponding animals should also be culled. If any of the infected mob is kept for treatment, then it is essential that the quarantine paddock is securely fenced. Treatment involves footbathing sheep in groups or antibiotic therapy of individual sheep.

Footbathing — There are several approaches. Jopp and colleagues recommend that after adequate paring to expose all pockets of necrotic tissue, sheep should stand in zinc sulfate for 10 to 30 min, three times with an interval of 1 week between bathings and are then reexamined 1 week after the last treatment. Those that have healed are moved to a convalescent flock, while the remainder are culled. The cured mob should be reinspected in a further 3 to 4 weeks but they should not be reintroduced to the clean flock until they have passed two successive clean examinations.[102,173] They have achieved adequate control and subsequent eradication using traditional methods without vaccination with an overall cost benefit to the farmer.[174]

Footrite®, a commercial formulation devised by Malecki and McCausland,[120,121] containing 20% zinc sulfate and 2% sodium lauryl sulfate, has also been successfully used to treat sheep infected with footrot.[122-124,126] Atkins[126] footbathed flocks of sheep on three properties twice for 1 h 5 d apart and then inspected them three times with a interval of 18 to 21 d after the first footbath followed by an autumn inspection. With this procedure, thorough paring is restricted to sheep with chronic footrot possessing severely misshapen and overgrown feet.[122] Those not cured at the first inspection were culled.[126]

Since formalin inhibits the penetration of zinc it is recommended that formalin should not be used within at least 6 weeks of Footrite® treatment.[124] Since Footrite® is relatively more expensive than formalin, a concrete-based footbath has been specifically designed to minimize wastage of antiseptic by preventing sheep from keeping their feet out of the solution and by decreasing losses from splashing.[124] A sloping-drainage exit ramp minimizes the antiseptic being carried out on legs and belly wool,[124] a feature also recommended by Jopp et al.[174] A bath sump allows the solution to be filtered free of fecal material, pumped, and stored in drums or tanks between use, thus minimizing loss through evaporation or dilution from rain.[124]

Formalin is still considered to be reasonably effective provided severe paring is carried out. Plant and Claxton[130] found that the curative responses to either 10% zinc sulfate or Footrite® were no better than formalin.

Antibiotic therapy — Egerton[143] considers that the most practical and economical way to eliminate all cases during the nontransmission phase is by parenteral antibiotic therapy provided that only minimal paring, removing overgrown horn, is used. Although drug costs are relatively high, this speeds up the procedure so that large numbers of sheep can be inspected and reinspected quickly. On the other hand, Jopp and colleagues reserve the use of parenteral or topical aerosol antibiotics for cases with severe footrot in two or more feet, as these will probably not respond to a single footparing and footbathing.[173]

3. Surveillance

The maintenance of clean flocks is dependent on the surveillance phase. An eradication strategy must incorporate regular foot inspections of the flock after removal of affected sheep, since carriers can be missed on previous inspections and response to treatment is not absolute. This must be continued until no further cases of footrot are found.[143] An annual inspection is advocated to ensure that possible carriers are kept to a minimum, as well as rigorous surveillance during routine management procedures.[174] Flocks in which there are no cases of footrot after one complete inspection can be regarded as provisionally free of the disease.[143] However, to be classed as free of footrot requires that the flock remain clean during a subsequent period favorable to transmission.[143]

It is vital to prevent reentry of the disease by having stock-proof boundary fences, by buying from known footrot-free properties, by detailed examination and treatment of introduced sheep, and their quarantine until they have been through a transmission period and subsequently classified as clean. In Australia, markets and saleyards have in the past, been notorious for spreading footrot to the properties of unsuspecting sheep buyers.

Although cattle have been implicated in the spread of benign footrot to sheep, they do not appear to constitute a reservoir for virulent footrot.[84-86] However, as a precaution cattle should be kept separate from sheep passed as clean. Clean sheep should be segregated from goats since the latter are also capable of being affected with benign[22,47,80] and virulent strains of *B. nodosus*.[22,80]

Although there can be problems with differential diagnosis, benign footrot causes only transient lameness and minimal loss of production in sheep in contrast to virulent footrot. Furthermore, benign footrot regresses rapidly following footbathing or the advent of dry conditions. Thus, at present it is considered not economically justifiable to attempt its eradication.

D. Regional Eradication Programs

Virulent footrot is a notifiable disease covered by legislation under the various Stock Diseases Acts in Western Australia since 1949,[178,179] South Australia since 1957,[170,180] the New England Protected Area (NEPA) of New South Wales since 1960,[171,175] and the Footrot Control Area (FCA) of western Victoria since 1968.[168,181] Regulatory control measures include notification of footrot in flocks and of introductions to the area, as well as prohibition of exposing infected sheep in any public place such as in saleyards or on roads. Infected properties are placed under quarantine and infected sheep must be segregated and treated. Sheep movements are restricted to transport by road or rail to designated markets for slaughter. Stock inspectors have the power to enter properties where footrot is suspected, and to require mustering, yarding, and assistance in examining sheep. Legal provisions also enable affected and in contact sheep to be ordered for slaughter. In the FCA of western Victoria, the Minister of Agriculture has the power to publish the names of flock owners under quarantine.

Successful eradication of virulent footrot and maintenance of flocks free of the disease on a regional basis require accurate diagnosis, low initial flock prevalence, and an adequate source of clean sheep for restocking. Thirty eight percent of breakdowns in the FCA of western Victoria were due to introduction of infected sheep from outside the control area.[168] Due to the constant introduction of sheep from regions where eradication programs are not in operation and problems with differential diagnosis of *B. nodosus* infection, it is considered unlikely that footrot can be totally eradicated from the FCA in western Victoria. Resources are therefore directed to control of the disease in the FCA by legislation and extension.[168] Despite this, by 1984, there had been a significant reduction in the prevalence of virulent footrot following the implementation of regional footrot control and eradication schemes in South Australia,[170] Western Australia,[179] the NEPA of New South Wales,[171] and the FCA of western Victoria.[181] In 1984, the upsurge in outbreaks in the FCA of western Victoria[181] and the NEPA of New South Wales[171] was attributed in part to restocking after the drought in 1982 to 1983 with carrier sheep from footrot-endemic areas. However, some of the outbreaks in the NEPA were apparently associated with footrot in goats and some were due to local spread.[171] In Western Australia in 1982, there were no known infected properties, but in 1984, 78 properties were quarantined. The source of infection for these outbreaks was not determined but it may have originated from large number of sheep introduced by dealers following the drought in eastern Australia.[179]

There appears to be sound economic justification for continuing these regional control programs. For the NEPA of New South Wales at a 5% rate of discount the net social benefit for the most likely scenario over the next 20 years is $A25.5 million with a benefit-cost ratio of 19:1. At a 10% rate of discount, the net social benefit is $A15.5 million with a benefit-cost ratio of 14:1.[7,182,183] An economic evaluation of the FCA of western Victoria has shown that there will be a net social benefit to the community of $A24 million with a benefit-cost ratio of 44:1 over the next 30 years at a 10% rate of discount and a supply shift occurring over 5 years. This supply shift implies pessimistically that footrot would return to high levels within 5 years if extra control measures were to cease. If a 5% rate of discount is used, the estimated benefits increase to $A40 million with a benefit-cost ratio of 35:1.[184] As a result of these studies, the FCA in western Victoria, where only 0.25% (13) of 5200 flocks, as at June 1986, were infected with footrot,[185] will gradually be expanded over the next 15 years to include the whole of the state.[186] By comparison, in 1971 there were 3.5% (180) properties under quarantine for footrot.[181] Outside the present FCA, although infected flocks are not quarantined, the prevalence of footrot has been reduced from about 50% to about 10% due to strictly enforced provisions under the Stock Disease Act prohibiting sale of infected sheep except directly for slaughter, and requiring flock owners to regularly inspect their sheep and undertake treatment of those infected.[168,181] In fact, a survey in 1986 conducted adjacent to the FCA in two regions with an annual average rainfall of more than or less than 650 mm revealed that the prevalence of footrot was 6% (± 5%) and 9% (± 5%), respectively.[185]

In New South Wales, enlargement of the Protected Area is not being considered at present since there are an estimated 5000 infected properties and in some districts the flock prevalence is more than 50%. Thus, the imposition of quarantine and eradication measures on all these properties would require a large increase in resources and labor. In endemic regions, footrot control is currently based on reducing the prevalence by an advisory program started in 1983, as well as regulatory activities centered on saleyards, stock routes, and problem properties.[171]

Despite the apparent success of control programs on a statewide or regional basis in reducing the flock prevalence of virulent footrot to 1% or less, the imposition of quarantine restriction can incur severe financial and managerial burdens on flock owners. As regional eradication programs progress, there is also an increasing dilemma with differential diagnosis of footrot which can impose additional hardships on sheep owners. Where doubt exists, flock owners can be placed in quarantine even though the inspector only suspects that the disease is present, but restrictions are lifted if benign footrot is present. In South Australia[8,170,176] and Tasmania,[187] regulatory control has resulted in an increase in the number of outbreaks in which only a small proportion of the flock develop underrunning lesions. In Tasmania, an eradication program introduced in 1939 was eventually abandoned in 1975 because of the difficulty in determining whether flocks were infected with virulent footrot.[187,188] This resulted in an increasing number of flocks being quarantined for virulent footrot previously diagnosed as benign. In hindsight these outbreaks were probably caused by the emergence of strains of intermediate virulence,[17,19] which in favorable seasons produce lesions indistinguishable from virulent footrot in a small proportion of sheep. In South Australia, the emergence of intermediate strains causing lower-profile clinical disease has resulted in confusion over diagnosis, difficulty in detection, and reluctance to notify, thereby hampering progress in eradication of the more virulent form of the disease.[8,176] However, development of improved laboratory diagnostic tests should speed the process of decision making and greatly assist the regulatory control and eradication of footrot.

REFERENCES

1. **Mohler, J. R. and Washburn, H. J.,** Foot-rot of sheep: its nature, cause and treatment, *U.S. Dept. Agric. Bur. Anim. Ind. Bull.,* No. 63, 7, 1904.
2. **Beveridge, W. I. B.,** Foot-rot in sheep: a transmissible disease due to infection with *Fusiformis nodosus* (n. sp.). Studies on its cause, epidemiology and control, *J. Counc. Sci. Ind. Res.,* Bull. No. 140, 14, 1941.
3. **Beveridge, W. I. B.,** The early history of foot-rot, in *Ovine Footrot,* Rep. Workshop University of Sydney, May 3, 1981, 200.
4. **Youatt, W.,** *Sheep: their Breeds, Management and Diseases,* Baldwin and Cradock, London, 1837.
5. **Gohier, J. B.,** *Memoires et Observations sur la Chirurgie et la Medecine Vétérinaires,* Tome premier, Lions, Lyon, 1813.
6. **Robertson, W. A. N.,** Milestones in the pastoral age of Australia, in *Australian and New Zealand Association for the Advancement of Science,* Report of the 21st Meeting, Walkom, A. B., Ed., Australian and New Zealand Association for Advancement of Science, Sydney, 1932, 295.
7. **Carmody, M. J.,** An economic evaluation of the footrot eradication program on the Northern Tablelands of N.S.W., *Agric. Econ. Bull. (Armidale, N.S.W.),* No. 25, 1981.
8. **Dobson, K. J.,** The impact of *Bacteroides nodosus* strains of intermediate virulence on footrot control, in *Footrot in Ruminants,* Proc. Workshop, Melbourne, 1985, Stewart, D. J., Peterson, J. E., McKern, N. M., and Emery, D. L., Eds., CSIRO Division Animal Health/Australian Wool Corp., Glebe, N.S.W., 1986, 23.
9. **Marshall, D. J.,** Personal communication, 1987.
10. **Egerton, J. R.,** Discussion of sessions 1 and 2, on the prevalence, control and eradication of footrot, in *Footrot in Ruminants,* Proc. Workshop, Melbourne, 1985, Stewart, D. J., Peterson, J. E., McKern, N. M., and Emery, D. L., Eds., CSIRO Division Animal Health/Australian Wool Corp., Glebe, N.S.W., 1986, 78.
11. **Bulgin, M. S., Lincoln, S. D., Lane, V. M., South, P. J., Dahmen, J. J., Gradin, J. L., and Smith, A. W.,** Evaluating an ovine foot-rot vaccine, *Vet. Med.,* 80, 105, 1985.
12. **Egerton, J. R., Roberts, D. S., and Parsonson, I. M.,** The aetiology and pathogenesis of ovine foot-rot. I. A histological study of the bacterial invasion, *J. Comp. Pathol.,* 79, 207, 1969.
13. **Roberts, D. S. and Egerton, J. R.,** The aetiology and pathogenesis of ovine foot-rot. II. The pathogenic association of *Fusiformis nodosus* and *F. necrophorus, J. Comp. Pathol.,* 79, 217, 1969.
14. **Laing, E. A. and Egerton, J. R.,** Some aspects of *B. nodosus* infection in cattle, in Ovine Footrot, Rep. Workshop University Sydney, May 3, 1981, 195.
15. **Thomas, J. H.,** The differential diagnosis of footrot in sheep, *Aust. Vet. J.,* 38, 159, 1962.
16. **Egerton, J. R. and Parsonson, I. M.,** Benign foot-rot — a specific interdigital dermatitis of sheep associated with infection by less proteolytic strains of *Fusiformis nodosus, Aust. Vet. J.,* 45, 345, 1969.
17. **Stewart, D. J., Clark, B. L., and Jarrett, R. G.,** Observations on strains of *Bacteroides nodosus* of intermediate virulence to sheep, in *Aust. Ad. Vet. Sci.,* Australian Veterinary Assoc., Sydney, 1982, 219.
18. **Stewart, D. J., Clark, B. L., and Kortt, A. A.,** Differentiation of *Bacteroides nodosus* infections in sheep and cattle, *22nd World Vet. Congr.,* World Veterinary Congress Organising Committee, Perth, 1983, 145.
19. **Stewart, D. J., Clark, B. L., and Jarrett, R. G.,** Differences between strains of *Bacteroides nodosus* in their effects on the severity of foot-rot, bodyweight and wool growth in Merino sheep, *Aust. Vet. J.,* 61, 348, 1984.
20. **Stewart, D. J.,** Review of footrot — diagnosis and the role of vaccination, in Refresher Course for Veterinarians, Proceedings, No. 67, University of Sydney, Post-Graduate Committee in Veterinary Science, Sydney, 1983, 487.
21. **Depiazzi, L. J. and Richards, R. B.,** Motility in relation to virulence of *Bacteroides nodosus, Vet. Microbiol.,* 10, 107, 1984/85.
22. **Stewart, D. J., Peterson, J. E., Vaughan, J. A., Clark, B. L., Emery, D. L., Caldwell, J. B., and Kortt, A. A.,** Clinical and laboratory diagnosis of benign, intermediate and virulent strains of *Bacteroides nodosus,* in *Footrot in Ruminants,* Proc. Workshop, Melbourne, 1985, Stewart, D. J., Peterson, J. E., McKern, N. M., and Emery, D. L., Eds., CSIRO Division Animal Health/Australian Wool Corp., Glebe, N.S.W., 1986, 81.
23. **Stewart, D. J., Peterson, J. E., Vaughan, J. A., Clark, B. L., Emergy, D. L., Caldwell, J. B., and Kortt, A. A.,** The pathogenicity and cultural characteristics of virulent, intermediate and benign strains of *Bacteroides nodosus* causing ovine foot-rot, *Aust. Vet. J.,* 63, 317, 1986.
24. **Anon.,** Report of Working Party on Footrot in Sheep and Cattle, Extract minutes Australian Animal Health Comm. Meet., (AHC 38) Item 5.1, October, 1986.
25. **Symons, L. E. A.,** Experimental foot-rot, wool growth and body mass, *Aust. Vet. J.,* 54, 362, 1978.
26. **Watts, J. E., Muller, M. J., and Dyce, A. L.,** The species of flies reared from struck sheep in South-Eastern Australia, *Aust. Vet. J.,* 52, 488, 1976.

27. **Stewart, D. J., Clark, B. L., Peterson, J. E., Griffiths, D. A., and Smith, E. F.,** Importance of pilus-associated antigen in *Bacteroides nodosus* vaccines, *Res. Vet. Sci.*, 32, 140, 1982.
28. **Wardhaugh, K. G. and Dallwitz, R.,** Covert flystrike, *Wool Technol. Sheep Breed.*, 32, 15, 1984.
29. **Egerton, J. R., Morgan, I. R., and Burrell, D. H.,** Foot-rot in vaccinated and unvaccinated sheep, *Vet. Rec.*, 91, 447, 1972.
30. **Skerman, T. M., Erasmuson, S. K., and Morrison, L. M.,** Duration of resistance to experimental footrot infection in Romney and Merino sheep vaccinated with *Bacteroides nodosus* oil adjuvant vaccine, *N.Z. Vet. J.*, 30, 27, 1982.
31. **Emery, D. L., Stewart, D. J., and Clark, B. L.,** The comparative susceptibility of five breeds of sheep to footrot, *Aust. Vet. J.*, 61, 85, 1984.
32. **Parker, C. F., Cross, R. F., and Hamilton, K. L.,** Genetic resistance to footrot in sheep, *Annu. Proc. Sheep Vet. Soc.*, 9, 16, 1985.
33. **Cross, R. F.,** Influence of environmental factors on transmission of ovine contagious footrot, *J. Am. Vet. Med. Assoc.*, 173, 1567, 1978.
34. **Skerman, T. M.,** Genetic variation and inheritance of susceptibility to footrot in sheep, in *Footrot in Ruminants,* Proc. Workshop, Melbourne, 1985, Stewart, D. J., Peterson, J. E., McKern, N. M., and Emery, D. L., Eds., CSIRO Division Animal Health/Australian Wool Corp. Glebe, N.S.W., 1986, 73.
35. **Michalk, D. L.,** Sheep production in the United States, *Wool Technol. Sheep Breed.*, 27 (No. 1), 21, 1979.
36. **Egerton, J. R., Ribeiro, L. A., Kieran, P. J., and Thorley, C. M.,** Onset and remission of ovine footrot, *Aust. Vet. J.*, 60, 334, 1983.
37. **Roberts, D. S. and Walker, P. D.,** Fluorescein-labelled antibody for the diagnosis of foot-rot, *Vet. Rec.*, 92, 70, 1973.
38. **Claxton, P.,** Isolation of *Bacteroides nodosus,* in Ovine Footrot, Rep. Workshop University of Sydney, May 3, 1981, 204.
39. **Depiazzi, L. J. and Richards, R. B.,** A degrading proteinase test to distinguish benign and virulent ovine isolates of *Bacteroides nodosus, Aust. Vet. J.*, 55, 25, 1979.
40. **Stewart, D. J.,** The role of elastase in the differentiation of *Bacteroides nodosus* infections in sheep and cattle, *Res. Vet. Sci.*, 27, 99, 1979.
41. **Skerman, T. M., Erasmuson, S. K., and Every, D.,** Differentiation of *Bacteroides nodosus* biotypes and colony variants in relation to their virulence and immunoprotective properties in sheep, *Infect. Immun.*, 32, 788, 1981.
42. **Every, D.,** Surface structures and proteases of *Bacteroides nodosus* in relation to virulence and immunoprotective properties in sheep, in Ovine Footrot, Rep. Workshop University of Sydney, May 3, 1981, 8.
43. **Every, D.,** Proteinase isoenzyme patterns of *Bacteroides nodosus:* distinction between ovine virulent isolates, ovine benign isolates and bovine isolates, *J. Gen. Microbiol.*, 128, 809, 1982.
44. **Kortt, A. A., O'Donnell, I. J., Stewart, D. J., and Clark, B. L.,** Activities and partial purification of extracellular proteases of *Bacteroides nodosus* from virulent and benign footrot, *Aust. J. Biol. Sci.*, 35, 481, 1982.
45. **Kortt, A. A., Burns, J. E., and Stewart, D. J.,** Detection of the extracellular proteases of *Bacteroides nodosus* in polyacrylamide gels: a rapid method of distinguishing virulent and benign ovine isolates, *Res. Vet. Sci.*, 35, 171, 1983.
46. **Depiazzi, L. J. and Rood, J. I.,** The thermostability of proteases from virulent and benign strains of *Bacteroides nodosus, Vet. Microbiol.*, 9, 227, 1984.
47. **Green, R. S.,** A method to differentiate between virulent and benign isolates of *Bacteroides nodosus* based on the thermal stability of their extracellular proteinases, *N.Z. Vet. J.*, 33, 11, 1985.
48. **Stewart, D. J.,** Studies on the Morphology and Antigenic Structure of *Fusiformis Nodosus,* Ph.D. thesis, University of Sydney, Sydney, 1975.
49. **Thorley, C. M.,** A simplified method for the isolation of *Bcteroidesa nodosus* from ovine foot-rot and studies on its colony morphology and serology, *J. Appl. Bacteriol.*, 40, 301, 1976.
50. **Egerton, J. R. and Laing, E. A.,** Characteristics of *Bacteroides nodosus* isolated from cattle, *Vet. Microbiol.*, 3, 269, 1978/1979.
51. **Depiazzi, L. J., Richards, R. B., McQuade, N. C., and Rood, J. I.,** Virulence, protease, pilus and motility characteristics of *Bacteroides nodosus,* in *Footrot in Ruminants,* Proc. Workshop, Melbourne, 1985, Stewart, D. J., Peterson, J. E., McKern, N. M., and Emery, D. L., Eds., CSIRO Division Animal Health/Australian Wool Corp., Glebe, N.S.W., 1986, 103.
52. **Gordon, L. M., Yong, W. K., and Woodward, C. A. M.,** Tempral relationships and characterisation of extracellular proteases from benign and virulent strains of *Bacteroides nodosus* as detected in zymogram gels, *Res. Vet. Sci.*, 39, 165, 1985.
53. **Parsonson, I. M., Egerton, J. R., and Roberts, D. S.,** Ovine interdigital dermatitis, *J. Comp. Pathol.*, 77, 309, 1967.

54. **West, D. M.,** Observations on an outbreak of foot abscess in sheep, *N. Z. Vet. J.*, 31, 71, 1983.
55. **Morgan, I. R., Piercy, D. W., and Egerton, J. R.,** The incidence of interdigital skin diseases in flocks of sheep free of virulent foot-rot, *Aust. Vet. J.*, 48, 23, 1972.
56. **Roberts, D. S., Graham, N. P. H., Egerton, J. R., and Parsonson, I. M.,** Infective bulbar necrosis (heel-abscess) of sheep, a mixed infection with *Fusiformis necrophorus* and *Corynebacterium pyogenes, J. Comp. Pathol.*, 78, 1, 1968.
57. **West, D.,** A study of naturally occurring cases of ovine foot abscess in New Zealand, *N.Z. Vet. J.*, 31, 152, 1983.
58. **Egerton, J. R., Roberts, D. S., Graham, N. P. H., and Parsonson, I. M.,** Ovine interdigital dermatitis and its relationship to foot-rot and foot abscess, *Aust. Vet. J.*, 42, 440, 1966.
59. **Beveridge, W. I. B.,** *Bacterial Diseases of Cattle, Sheep and Goats,* Vol. 4, Australian Government Publishing Service, Canberra, 1983, 108, 178.
60. **Egerton, J. R. and Graham, N. P. H.,** Diseases causing lameness in sheep, in *Veterinary Review,* No. 5, University of Sydney, Post-Graduate Foundation in Veterinary Science, Sydney, 1969.
61. **Marsh, H.,** *Newsom's Sheep Diseases,* 3rd ed., Williams & Wilkins, Baltimore, 1965, 121; 308.
62. **Blood, D. C., Radostits, O. M., and Henderson, J. A.,** *Veterinary Medicine: A Textbook of the Diseases of Cattle, Sheep, Pigs, Goats and Horses,* 6th ed., Baillière Tindall, London, 1983, 526; 661; 705; 733; 1197.
63. **Egerton, J. R.,** Lameness in sheep, in Refresher Course for Veterinarians, Proceedings No. 58: Refresher Course on Sheep, University of Sydney, Post-Graduate Committee in Veterinary Science, Sydney, 1981, 635.
64. **Jensen, R.,** *Diseases of Sheep,* Lea & Febiger, Philadelphia, 1974, 17; 135; 215; 315.
65. **Larsen, J.,** Contagious ecthyma: field aspects and possible confusion with vesicular disease, in *Veterinary Viral Disease: their Significance in South-East Asia and the Western Pacific,* Della-Porta, A. J., Ed., Academic Press, New York, 1985, 545.
66. **French, E. L.,** Bluetongue, in *Exotic Diseases of Animals: a Manual for Diagnosis, Service Publication (Animal Quarantine) Commonwealth Department of Health, No. 11,* 2nd ed., French, E. L. and Geering, W. A., Eds., Australian Government Publishing Service, Canberra, 1978, 54.
67. **Geering, W. A. and Gee, R. W.,** Exotic diseases of sheep, in Refresher Course for Veterinarians, Proceedings No. 58: Refresher Course on Sheep, University of Sydney, Post-Graduate Committee in Veterinary Science, Sydney, 1981, 717.
68. **Geering, W. A.,** Australia, in *Veterinary Viral Diseases: Their Significance in South-East Asia and the Western Pacific,* Della-Porta, A. J., Ed., Academic Press, New York, 1985, 161.
69. **Cottral, G. E., Smith, V. W., and Littlejohn, I. R.,** Vesicular diseases, in *Exotic Diseases of Animals: a Manual for Diagnosis, Service Publication (Animal Quarantine) Commonwealth Department of Health,* No. 11, 2nd ed., French, E. L. and Geering, W. A., Eds., Australian Government Publishing Service, Canberra, 1978, 210.
70. **Hunt, E. R.,** Pregnancy toxaemia in ewes, in Refresher Course for Veterinarians, Proceedings No. 58: Refresher Course on Sheep, University of Sydney, Post-Graduate Committee in Veterinary Science, Sydney, 1981, 211.
71. **Morgan, D. O.,** Foot-and-mouth disease: advances in prophylactic technology, in *Veterinary Viral Diseases: Their Significance in South-East Asia and the Western Pacific,* Della-Porta, A. J., Ed., Academic Press, New York, 1985, 255.
72. **Egerton, J. R.,** Footrot control in drought, *Aust. Vet. J.*, 60, 315, 1983.
73. **Egerton, J. R. and Morgan, I. R.,** Foot-rot in vaccinated and unvaccinated sheep. I. Incidence, severity and duration of infection, *Vet. Rec.*, 91, 447, 1972.
74. **Keogh, J.,** Ovine foot-rot problem in New South Wales, in Proceedings of the Refresher Course for Veterinarians (the J.D. Stewart course for 1971) on the Importance of Disease Control in the Livestock Economy, University of Sydney, Sydney, 1971, 140.
75. **Beveridge, W. I. B.,** Footrot of sheep: its epidemiology and control, *Bull. Off. Int. Epizoot.*, 59, 1537, 1963.
76. **Graham, N. P. H. and Egerton, J. R.,** Pathogenesis of ovine foot-rot: the role of some environmental factors, *Aust. Vet. J.*, 44, 235, 1968.
77. **Littlejohn, A. I.,** Foot-rot in sheep: some observations on epizootiology, economics, and control, *Vet. Annu.*, 8, 71, 1966/67.
78. **Stewart, D. J.,** Unpublished data, 1978.
79. **Egerton, J. R.,** Epidemiology and Control of foot-rot, in Proceedings of the Refresher Course for Veterinarians (the J.D. Stewart Course for 1971) on the Importance of Disease Control in the Livestock Economy, University of Sydney, Sydney, 1971, 130.

80. **Claxton, P. D. and O'Grady, K. C.,** Footrot in goats and characterisation of caprine isolates of *Bacteroides nodosus,* in *Footrot in Ruminants,* Proc. Workshop, Melbourne, 1985, Stewart, D. J., Peterson, J. E., McKern, N. M., and Emery, D. L., Eds., CSIRO Division Animal Health/Australian Wool Corp., Glebe, N.S.W., 1986, 119.
81. **Skerman, T. M.,** Isolation of *Bacteroides nodosus* from hoof lesions in a farmed red deer *(Cervus elaphus), N.Z. Vet J.,* 31, 102, 1983.
82. **Alexander, T. M.,** The differential diagnosis of footrot in sheep, *Aust. Vet. J.,* 38, 366, 1962.
83. **Toussaint Raven, E. and Cornelisse, J. L.,** The specific contagious inflammation of the interdigital skin in cattle, *Vet. Med. Rev.,* 2/3, 223, 1971.
84. **Wilkinson, F. C., Egerton, J. R., and Dickson, J.,** Transmission of *Fusiformis nodosus* infection from cattle to sheep, *Aust. Vet. J.,* 46, 382, 1970.
85. **Laing, E. A. and Egerton, J. R.,** The occurrence, prevalence and transmission of *Bacteroides nodosus* infection in cattle, *Res. Vet. Sci.,* 24, 300, 1978.
86. **Richards, R. B., Depiazzi, L. J., Edwards, J. R., and Wilkinson, F. C.,** Isolation and characterisation of *Bacteroides nodosus* from foot lesions of cattle in Western Australia, *Aust. Vet. J.,* 56, 517, 1980.
87. **Egerton, J. R. and Parsonson, I. M.,** Isolation of *Fusiformis nodosus* from cattle, *Aust. Vet. J.,* 42, 425, 1966.
88. **Stewart, D. J.,** Unpublished data, 1977.
89. **Beveridge, W. I. B.,** Foot-rot in sheep: a preliminary note on the probable causal agent, *J. Counc. Sci. Ind. Res.,* 11, No. 1, 1938.
90. **Thomas, J. H.,** The bacteriology and histopathology of footrot in sheep, *Aust. J. Agric. Res.,* 13, 725, 1962.
91. **Hine, P. M.,** Ovine footrot: histopathology of a synergic disease, in *Models of Anaerobic Infections,* Hill, M. J., Ed., Martinus Nijhoff, The Hague, 1984, 85.
92. **Thorley, C. M.,** Discussion, in Ovine Footrot, Rep. Workshop University of Sydney, May 3, 1981, 86.
93. **Deane, H. M. and Jensen, R.,** The pathology of contagious footrot in sheep, *Am. J. Vet. Res.,* 16, 203, 1955.
94. **Thomas, J. H.,** Proteolytic enzymes produced in liquid media by *Fusiformis nodosus, Aust. J. Agric. Res.,* 15, 417, 1964.
95. **Thomas, J. H.,** The pathogenesis of footrot in sheep with reference to proteases of *Fusiformis nodosus, Aust. J. Agric. Res.,* 15, 1001, 1964.
96. **Walker, P. D., Fernie, D. S., Hine, P. M., and Thorley, C. M.,** Anaerobic diseases of animals, in *Anaerobic Bacteria in Habitats other than Man,* Barnes, E. M. and Mead, G. C., Eds., Blackwell Scientific, Oxford, 1986, 303.
97. **Stewart, D. J. and Parsonson, I. M.,** Unpublished data, 1976.
98. **Broad, T. E. and Skerman, T. M.,** Partial purification and properties of extracellular proteolytic activity of *Bacteroides nodosus, N.Z. J. Agric. Res.,* 19, 317, 1976.
99. **Stewart, D. F.,** Treatment and control of foot-rot, in *Proceedings of Conference of Veterinary Officers on Foot-Rot and Other Diseases of the Feet of Sheep,* New South Wales Dept. of Agriculture, Young, N.S.W., 1953.
100. **Pryor, W. J.,** The treatment of contagious foot-rot in sheep, *Aust. Vet. J.,* 30, 385, 1954.
101. **Brunskill, A.,** Review of sheep handling devices and their application, in *Sheep and Wool Seminar: Livestock Officers (Sheep & Wool),* Tamworth, New South Wales Dept of Agriculture, 1980, 6.
102. **Mulvaney, C. J., Jackson, R., and Jopp, A. J.,** A revised concept of ovine footrot control, in *Footrot in Ruminants,* Proc. Workshop, Melbourne, 1985, Stewart, D. J., Peterson, J. E., McKern, N. M., and Emery, D. L., Eds., CSIRO Division Animal Health/Australian Wool Corp., Glebe, N.S.W., 1986, 63.
103. **Smeal, M. G.,** Controlling footrot — equipment used in the Armidale quarantine area, *Agric. Gaz. N.S.W.,* 80, 71, 1969.
104. **Clark, F. L.,** Ovine footrot — problem in Northern N.S.W., in Proceedings of the Refresher Course for Veterinarians (J.D. Stewart Course for 1971) on the Importance of Disease Control in the Livestock Economy, University of Sydney, Sydney, 1971, 145.
105. **Stewart, D. F.,** The treatment of contagious footrot in sheep by the topical application of chloromycetin, *Aust. Vet. J.,* 30, 209, 1954.
106. **Cross, R. F.,** Response of sheep to various topical, oral, and parenteral treatments for footrot, *J. Am. Vet. Med. Assoc.,* 173, 1569, 1978.
107. **Stewart, D. F.,** The treatment of contagious foot-rot in sheep — with particular reference to the value of chloromycetin, *Aust. Vet. J.,* 30, 380, 1954.
108. **Thomas, J. H.,** The eradication of contagious footrot of sheep, *Aust. Vet. J.,* 33, 263, 1957.
109. **Thomas, J. H.,** The treatment of contagious footrot of sheep, with special reference to the value of terramycin, *Aust. Vet. J.,* 34, 33, 1958.
110. **Anon.,** Guidelines on veterinary therapeutics, *AVA News,* No. 4, 8, 1987.

111. **Sinclair, A. N.,** Studies on contagious footrot of sheep, *Aust. Vet. J.*, 33, 202, 1957.
112. **Harriss, S. T.,** Terramycin and foot-rot in sheep, *Br. Vet. J.*, 111, 212, 1955.
113. **Gradin, J. L. and Schmitz, J. A.,** Susceptibility of *Bacteroides nodosus* to various antimicrobial agents, *J. Am. Vet. Med. Assoc.*, 183, 434, 1983.
114. **Forsyth, B. A.,** The treatment of contagious footrot in sheep by the topical application of a ''cetavlon'' tincture, *Aust. Vet. J.*, 33, 157, 1957.
115. **Penny, R. H. C.,** The topical application of 10 percent chloroamphenicol in spirit in the treatment of footrot in sheep, *Br. Vet. J.*, 111, 125, 1955.
116. **Murnane, D.,** Footrot in sheep, *J. Counc. Sci. Ind. Res.*, 6, 252, 1933.
117. **Fitzpatrick, D.,** The control of footrot in sheep in Victoria, *Aust. Vet. J.*, 37, 460, 1961.
118. **Demertzis, P. N., Spais, A. G., and Papasteriadis, A. A.,** Zinc therapy in the control of footrot in sheep, *Vet. Med. Rev.*, 1/78, 101, 1978.
119. **Cross, R. F. and Parker, C. F.,** Zinc sulfate foot bath for control of ovine footrot, *J. Am. Vet. Med. Assoc.*, 178, 706, 1981.
120. **Malecki, J. C. and McCausland, I. P.,** Improvement of topical chemotherapy of ovine footrot by increasing the penetration of therapeutic agents, in *Ovine Footrot,* Rep. Workshop University of Sydney, May 3, 1981, 165.
121. **Malecki, J. C. and McCausland, I. P.,** *In vitro* penetration and absorption of chemicals into the ovine hoof, *Res. Vet. Sci.*, 33, 192, 1982.
122. **Malecki, J., McCausland, I., and Lambell, R.,** A new topical treatment for footrot, in Refresher Course for Veterinarians, Proceedings No. 67: Sheep Production and Preventive Medicine, University of Sydney, Post-Graduate Committee in Veterinary Science, Sydney, 1983, 63.
123. **Malecki, J. C. and Coffey, L.,** Effectiveness of treatment programs based on footbathing with a zinc sulphate formulation for virulent *Bacteroides nodosus* infections in sheep, in *Footrot in Ruminants,* Proc. Workshop, Melbourne, 1985, Stewart, D. J., Peterson, J. E., McKern, N. M., and Emery, D. L., Eds., CSIRO Division Animal Health/Australian Wool Corp., Glebe, N.S.W., 1986, 51.
124. **Lambell, R. G., Atkins, J. W., and Brightling, A.,** The use of a zinc sulphate formulation for the eradication of footrot during a period unfavorable for the spread of the disease, in *Footrot in Ruminants,* Proc. Workshop, Melbourne, 1985, Stewart, D. J., Peterson, J. E., McKern, N. M., and Emery, D. L., Eds., CSIRO Division Animal Health/Australian Wool Corp., Glebe, N.S.W., 1986, 47.
125. **Skerman, T. M., Green, R. S., Hughes, J. M., and Herceg, M.,** Comparison of footbathing treatments for ovine footrot using formalin or zinc sulphate, *N.Z. Vet. J.*, 31, 91, 1983.
126. **Atkins, J. W.,** The use of a zinc sulphate formulation for the eradication of footrot during a period favouring spread of the disease, in *Footrot in Ruminants,* Proc. Workshop, Melbourne, 1985, Stewart, D. J., Peterson, J. E., McKern, N. M., and Emery, D. L., Eds., CSIRO Division Animal Health/Australian Wool Corp., Glebe, N.S.W., 1986, 43.
127. **Skerman, T. M., Moorhouse, S. R., and Green, R. S.,** Further investigations of zinc sulphate footbathing for the prevention and treatment of ovine footrot, *N.Z. Vet. J.*, 31, 100, 1983.
128. **Skerman, T. M., Green, R. S., Moorhouse, S. R., and Broadfoot, K. G.,** Field evaluation of ''Defeat'' a commercial preparation based on copper 8-hydroxyquinolate, for the control of ovine footrot, *N. Z. Vet. J.*, 32, 218, 1984.
129. **Bulgin, M. S., Lincoln, S. D., Lane, V. M., and Matlock, M.,** Comparison of treatment methods for the control of contagious ovine footrot, *J. Am. Vet. Med. Assoc,* 189, 194, 1986.
130. **Plant, J. W. and Claxton, P. D.,** Efficacy of paring, footbathing and vaccination in the treatment of footrot, in *Footrot in Ruminants,* Proc. Workshop, Melbourne, 1985, Stewart, D. J., Peterson, J. E., McKern, N. M., and Emery, D. L., Eds., CSIRO Division Animal Health/Australian Wool Corp., Glebe, N.S.W., 1986, 57.
131. **Pryor, W. J.,** Two foot syndromes in sheep following excessive foot bathing, *Aust. Vet. J.*, 35, 493, 1959.
132. **Littlejohn, A. I.,** The potential danger arising from the misuse of formalin in the treatment of foot-rot in sheep, *Vet. Rec.*, 90, 693, 1972.
133. **Hooper, R. S. and Jones, T. W.,** Corono-pedal abscessation following the excessive use of formalin as a treatment for foot-rot in sheep, *Vet. Rec.*, 90, 697, 1972.
134. **Ross, A. D.,** Formalin and footrot in sheep, *N.Z. Vet. J.*, 31, 170, 1983.
135. **Coppini, R.,** Ulteriori ricerche sulla chemioterapia della pedaina degli ovini, *Atti Soc. Ital. Sci. Vet.*, 5, 427, 1951.
136. **Forsyth, B. A.,** The experimental treatment of contagious footrot in sheep, *Aust. Vet. J.*, 29, 73, 1953.
137. **Danks, A. G.,** Contagious footrot in sheep, *North Am. Vet.*, 37, 738, 1956.
138. **Sterk, V.,** Prilog parenteralnoj aplikaciji sredstava za lecenje zarazne sepavosti ovaca, [a contribution to the parenteral application of the remedies for the treatment of the footrot in sheep], *Veterinaria (Sarajevo),* 9, 277, 1960.

139. **Egerton, J. R. and Parsonson, I. M.,** Parenteral antibiotic treatment of ovine footrot, *Aust. Vet. J.*, 42, 97, 1966.
140. **Roberts, D. S. and Graham, N. P. H.,** Control of ovine cutaneous actinomycosis, *Aust. Vet. J.*, 42, 74, 1966.
141. **Egerton, J. R., Parsonson, I. M., and Graham, N. P. H.,** Parenteral chemotherapy of ovine foot-rot, *Aust. Vet. J.*, 44, 275, 1968.
142. **Harris, D. J.,** Field observations on parenteral antibiotic treatment of ovine foot-rot, *Aust. Vet. J.*, 44, 284, 1968.
143. **Egerton, J. R.,** Control and eradication of ovine footrot, in *Footrot in Ruminants,* Proc. Workshop, Melbourne, 1985, Stewart, D. J., Peterson, J. E., McKern, N. M., and Emery, D. L., Eds., CSIRO Division Animal Health/Australian Wool Corp., Glebe, N.S.W., 1986, 35.
144. **Banting, A. de L., Delsaux, J. M., Dupeux, D., Jacques, M., Matrat, M., and Turpin, M.,** Un nouveau systeme de notation des lesions de pietin. Application lors d'une experimentation en milieu infecte, *Rev. Med. Vet.*, 129, 1657, 1978.
145. **Banting, A. de L.,** Footrot in sheep, *Vet. Rec.*, 105, 359, 1979.
146. **Demertzis, P. N. and Mills, C. F.,** Oral zinc therapy in the control of infectious pododermatitis in young bulls, *Vet. Rec.*, 93, 219, 1973.
147. **Cross, R. F. and Parker, C. F.,** Oral administration of zinc sulfate for control of ovine footrot, *J. Am. Vet. Med. Assoc.*, 178, 704, 1981.
148. **Skerman, T. M., Millar, K. R., Sheppard, A. D., Herceg, M., and Hughes, J. M.,** Failure of orally administered zinc to prevent experimentally induced footrot in sheep, *N. Z. Vet. J.*, 31, 54, 1983.
149. **Egerton, J. R., Laing, E. A., and Mulley, R. C.,** Failure of oral zinc therapy to alleviate *Bacteroides nodosus* infections in cattle and sheep, *Aust. Vet. J.*, 62, 85, 1985.
150. **Egerton, J. R. and Roberts, D. S.,** Vaccination against ovine foot-rot, *J. Comp. Pathol.*, 81, 179, 1971.
151. **Egerton, J. R. and Burrell, D. H.,** Prophylactic and therapeutic vaccination against ovine foot-rot, *Aust. Vet. J.*, 46, 517, 1970.
152. **Keogh, J.,** Footrot vaccine trials, *N. S. W. Vet. Proc.*, 12, 35, 1976.
153. **Glenn, J., Carpenter, T. E., and Hird, D. W.,** A field trial to assess the therapeutic and prophylactic effect of a footrot vaccine in sheep, *J. Am. Vet. Med. Assoc.*, 187, 1009, 1985.
154. **Ellis, R. P., Kimberling, C. V., Wynn, D. M., Swanson, V. B., and Butler, J. G.,** The efficacy of Footvax in the control of footrot, *SID Res. Dig.*, Winter, 7, 1986.
155. **Lambell, R. G.,** Field experience with footrot vaccines in Victoria, in *Footrot in Ruminants,* Proc. Workshop, Melbourne, 1985, Stewart, D. J., Peterson, J. E., McKern, N. M., and Emery, D. L., Eds., CSIRO Division Animal Health/Australian Wool Corporation, Glebe, N.S.W., 1986, 199.
156. **Lambell, R. G.,** A field trial with a commercial vaccine against footrot in sheep, *Aust. Vet. J.*, 63, 415, 1986.
157. **Walker, R. I.,** Experience with vaccines in the control of footrot in sheep, in *Aust. Adv. Vet. Sci.*, Australian Veterinary Assoc., Sydney, 1984, 114.
158. **Mulvaney, C. J., Jackson, R., and Jopp, A. J.,** Field trials with a killed, nine-strain, oil adjuvanted *Bacteroides nodosus* footrot vaccine in sheep, *N.Z. Vet. J.*, 32, 137, 1984.
159. **Kennedy, D. J., Marshall, D. J., Claxton, P. D., and Morton, A. G.,** Evaluation of the curative effect of foot-rot vaccine under dry conditions, *Aust. Vet. J.*, 62, 249, 1985.
160. **Reed, G. A.,** The role of footrot vaccines in Australia, in *Footrot in Ruminants,* Proc. Workshop, Melbourne, 1985, Stewart, D. J., Peterson, J. E., McKern, N. M., and Emery, D. L., Eds., CSIRO Division Animal Health/Australian Wool Corp., Glebe, N.S.W., 1986, 173.
161. **Liardet, D. M., Chetwin, D. H., Kingsley, D. F., and Hindmarsh, F. H.,** Results of field trials in New Zealand to confirm the protective and curative effects of a 10-strain ovine footrot vaccine, in *Footrot in Ruminants,* Proc. Workshop, Melbourne, 1985, Stewart, D. J., Peterson, J. E., McKern, N. M., and Emery, D. L., Eds., CSIRO Division Animal Health/Australian Wool Corp., Glebe, N.S.W., 1986, 181.
162. **Stewart, D. J. and Elleman, T. C.,** A *Bacteroides nodosus* pili vaccine produced by recombinant DNA for the prevention and treatment of foot-rot in sheep, *Aust. Vet. J.*, 64, 79, 1987.
163. **Elleman, T. C., Hoyne, P. A., Stewart, D. J., McKern, N. M., and Peterson, J. E.,** Expression of pili from *Bacteroides nodosus in Pseudomonas aeruginosa, J. Bacteriol.*, 168, 574, 1986.
164. **Mattick, J. S., Bills, M. M., Anderson, B. J., Dalrymple, B., Mott, M. R., and Egerton, J. R.,** Morphogenetic expression of *Bacteroides nodosus* fimbriae in *Pseudomonas aeruginosa, J. Bacteriol.*, 169, 33, 1987.
165. **Egerton, J. R., Cox, P. T., Anderson, B. J., Kristo, C., Norman, M., and Matticks, J. S.,** Protection of sheep against foot-rot with a recombinant DNA-based fimbrial vaccine, *Vet. Microbiol.*, 14, 393, 1987.
166. **Egerton, J. R. and Thorley, C. M.,** Effect of alum-precipitated or oil-adjuvant *Bacteroides nodosus* vaccines on the resistance of sheep to experimental footrot, *Res. Vet. Sci.*, 30, 28, 1981.
167. **Beveridge, W. I. B.,** Foot-rot of sheep: a modern approach to an old problem, *Vet. Rec.*, 68, 963, 1956.

168. **Tweddle, N. E.,** Footrot control programmes in Victoria, in *Ovine Footrot,* Rep. Workshop University of Sydney, May 3, 1981, 151.
169. **Roycroft, C. R. and Napthine, D. V.,** Practical footrot control, in Refresher Course for Veterinarians, Proceedings No. 67: Sheep Production and Preventive Medicine, University of Sydney, Post-Graduate Committee in Veterinary Science, Sydney, 1983, 317.
170. **Brownrigg, P. W.,** A historical review of footrot in South Australia, in *Footrot in Ruminants,* Proc. Workshop, Melbourne, 1985, Stewart, D. J., Peterson, J. E., McKern, N. M., and Emery, D. L., Eds., CSIRO Division Animal Health/Australian Wool Corp., Glebe, N.S.W., 1986, 1.
171. **Scott-Orr, H.,** The current footrot situation in New South Wales, in *Footrot in Ruminants,* Proc. Workshop, Melbourne, 1985, Stewart, D. J., Peterson, J. E., McKern, N. M., and Emery, D. L., Eds., CSIRO Division Animal Health/Australian Wool Corp., Glebe, N.S.W., 1986, 13.
172. **Outteridge, P. M., Windon, R. G., Dineen, J. K., Stewart, D. J., and Skerman, T. M.,** Breeding for parasite and footrot resistance in sheep using ovine lymphocyte antigen (OLA) markers, in *Merino Improvement Programs in Australia,* Proc. Natl. Symp. Leura, N.S.W., McGuirk, B. J., Ed., CSIRO Division Animal Health/Australian Wool Corp., Glebe, N.S.W., 1987, 377.
173. **Jopp, A. J., Jackson, R., and Mulvaney, C. J.,** An improved facility for the inspection, treatment and control of footrot, *N. Z. Vet. J.,* 32, 189, 1984.
174. **Jopp, A. J., Jackson, R., and Mulvaney, C. J.,** A perspective on ovine footrot control, *N. Z. Vet. J.,* 32, 211, 1984.
175. **Everett, R. E.,** Footrot eradication — Northern Tablelands of New South Wales, in Ovine Footrot, Rep. Workshop University of Sydney, May 3, 1981, 138.
176. **Vandegraaff, R. and Robinson, R. C.,** The use of planned and prescribed programs for the eradication of footrot from individual farms, in *Footrot in Ruminants,* Proc. Workshop, Melbourne, 1985, Stewart, D. J., Peterson, J. E., McKern, N. M., and Emery, D. L., Eds., CSIRO Division Animal Health/Australian Wool Corp., Glebe, N.S.W., 1986, 69.
177. **Parsonson, I. M.,** Personal communication, 1976.
178. **Richards, R. B. and Depiazzi, L. J.,** Research arising from the footrot eradication programme in Western Australia, in *Ovine Footrot,* Rep. Workshop University of Sydney, May 3, 1981, 80.
179. **Gwynn, R. V. R.,** The footrot situation in Western Australia, in *Footrot in Ruminants,* Proc. Workshop, Melbourne, 1985, Stewart, D. J., Peterson, J. E., McKern, N. M., and Emery, D. L., Eds., CSIRO Division Animal Health/Australian Wool Corp., Glebe, N. S. W., 1986, 19.
180. **Riley, M. J., Tolson, J. W., and Belfield, A. J. M.,** The control of ovine footrot in S.A., in Ovine Footrot, Rep. Workshop University of Sydney, May 3, 1981, 144.
181. **Napthine, D. V.,** Future directions for the control of footrot in Victoria, in *Footrot in Ruminants,* Proc. Workshop, Melbourne, 1985, Stewart, D. J., Peterson, J. E., McKern, N. M., and Emery, D. L., Eds., CSIRO Division Animal Health/Australian Wool Corp., Glebe, N. S. W., 1986, 21.
182. **Everett, R. E., Carmody, M. J., Hardaker, J. B., and Powell, R. A.,** Economic evaluation of the footrot eradication programme on the Northern Tablelands of New South Wales, in Ovine Footrot, Rep. Workshop University of Sydney, May 3, 1981, 186.
183. **Carmody, M. J., Hardaker, J. B., Powell, R. A., and Everett, R. E.,** An economic evaluation of the foot-rot eradication program in the New England region of New South Wales, *Rev. Market. Agric. Econ.,* 52, 9, 1984.
184. **Stott, K. J.,** Economic evaluation of the footrot control area in western Victoria, in *Footrot in Ruminants,* Proc. Workshop, Melbourne, 1985, Stewart, D. J., Peterson, J. E., McKern, N. M., and Emery, D. L., Eds., CSIRO Division Animal Health/Australian Wool Corp., Glebe, N. S. W., 1986, 27.
185. **Napthine, D. V.,** Personal communication, 1987.
186. **Anon.,** *Footrot of Sheep: Statewide Policy and Plan for Footrot Control,* Department of Agriculture and Rural Affairs, Melbourne, May, 1986.
187. **McConnell, J. D.,** Tasmanian control experience, in Ovine Footrot, Rep. Workshop University of Sydney, May 3, 1981, 164.
188. **Scanlan, W. A.,** Distribution, prevalence and control of ovine foot-rot in the various Australian States, in Ovine Footrot, Rep. Workshop University of Sydney, May 3, 1981, 119.

Chapter 2

FOOTROT OF CATTLE, GOATS, AND DEER

J. R. Egerton

TABLE OF CONTENTS

I. BACKGROUND

Footrot is the description usually applied to a disease which causes lameness in sheep. *Bacteriodes nodosus* is essential in the etiology of this infection and is its transmitting agent (see Chapter 1). Isolates of the same organism have more recently been implicated in the etiology and transmission of a similar disease in some other ruminants, especially cattle. The same epidermal structures are affected and the natural history of the diseases caused are similar in all affected animals. Footrot will therefore be used in describing these diseases in animals other than sheep.

II. DEFINITION AND ETIOLOGY

Footrot in cattle, goats, and deer is a mixed bacterial infection of the feet in which *B. nodosus* has an essential role. These infections are principally confined to epidermal components of the interdigital skin and they are characterized by inflammation, superficial ulceration, necrosis, fissuring, and hyperkeratosis.

Organisms with the morphology of *B. nodosus* were first reported in lesions from cattle feet by Alexander[1] and Skenman.[2] Gupta et al.[3] were unable to isolate the organism but transmitted the disease by transferring infected tissue to experimental animals. Subsequently, *B. nodosus* infection was transmitted deliberately from cattle to sheep and the pathogen was isolated from the disease which resulted in sheep.[4] In the same investigation, cultures of *B. nodosus* isolated from the affected sheep and assumed to have originated in cattle, were used to induce transient infections in calves and more persistent ones in sheep.

In Holland, Toussaint and Cornelisse[5] examined closely the clinical and bacteriological relationship of interdigital skin infections in cattle and sheep. Using lesion material, they were able to transfer infections from cattle to sheep and back to cattle. Organisms with the morphology of *B. nodosus* were readily demonstrable in diseased tissue of all affected animals. In later studies, Laing and Egerton[6] demonstrated that an isolate from cattle caused interdigital dermatitis in suitably predisposed cattle and sheep. In smears examined microscopically and in histology of affected skin,[7] *B. nodosus* occurred in association with other organisms considered to have a role in the pathogenesis of ovine footrot.[8] The contribution of *Fusobacterium necrophorum* and other prevalent organisms, such as the spirochaetal ones, in the etio-pathogenesis of footrot in nonovine ruminants, has not been extensively investigated. Clark et al.[8] studied the role of *F. necrophorum, B. nodosus,* and *B. melaninogenicus* in bovine foot infections. They isolated *B. nodosus* from superficial lesions but not from deeper abscesses in which *F. necrophorum* occurred. *B. melaninogenicus* was also present in these abscesses but did not seem to have a role in pathogenesis.

The circumstantial association of predispositon of feet to infection by wet environments and microbial contamination, however, suggests that, as in sheep, *B. nodosus,* by itself, is an inadequate pathogen in cattle,[5,9] goats, and deer.

III. CLINICAL SIGNS

A. Cattle

Lameness may be the first indication of footrot, but all affected cattle are not necessarily lame. Affected feet may be swollen, especially in the skin of the interdigital area (Plate 2).* In the acute stages of the disease, there may be extensive serous exudate from affected skin. There is no purulent or necrotic discharge — no leucocytes are apparent in material examined microscopically. Milk production may be adversely affected in severely affected dairy herds.[4,5]

* Plates 2, 3, and 4 follow page 54.

The reports from Holland[5] and Australia[6,8,9] suggest that there are a number of serious sequelae to chronic *B. nodosus* infection of the interdigital skin. One of these is separation of the horn overlying the heel region. This separation leads in turn to infection of the corium, which in turn results in the production of imperfect horn. The loss of integrity of horn in the heel and sole may predispose to the formation of ulcers of the sole (Plate 3). Proliferation of the interdigital skin is observed in some chronic infections. These may be granulomatous in their early stages. Eventually healing and epithelialization of these overgrowths result in the occurrence of so called interdigital fibroma (Plate 4). Another chronic manifestation of the interdigital dermatitis which characterizes *B. nodosus* infection in cattle is a papillomatous or verrucose proliferation in the heel region of the affected claws. Furthermore, chronic infection of the corium underlying the hard horn of the claw may result in exaggerated overgrowth of the toe region with resultant distortion of the whole hoof and abnormal redistribution of the weight-bearing function of affected hooves.[5]

B. Goats and Deer

There are few detailed reports of the clinical signs and lesions observed in goats. Lameness occurs in affected animals and where bucks are affected, reproductive performance is inhibited. As with sheep, goats affected in both fore limbs may bear weight and move about on their carpal-metacarpal joints.

Infection seems generally to be restricted to the interdigital skin, irrespective of the virulence of the isolate involved.[10,11] The inflammation in this tissue may be quite intense and be accompanied by copious serous exudate. The affected area may be ulcerated on the surface and grey. Underrunning and separation of the horn of the heel, sole wall, and toe of the hoof does occur but not nearly so frequently as in sheep.

Interdigital lesions were the usual manifestation of *B. nodosus* infection described by Skerman in deer,[12] although the organism was isolated from an underrunning infection of the horn in one stag.

IV. DIAGNOSIS

A. Microscopic Examination

Footrot in cattle, goats, and deer is essentially an interdigital dermatitis caused by a mixed bacterial infection which is restricted to epidermal tissues. While fissures may be present in the dorsal aspect of the interdigital cleft, there is no deep invasion or penetration of the interdigital tissues. There is no purulent or extensive necrotic discharge from sinuses in the interdigital cleft or proximal to the coronary band.

Diagnosis of footrot may be confirmed by demonstrating organisms with the characteristic morphology of *B. nosodus* in a Gram-stained smear (Figure 1). It will not be prevalent. In acute cases, there will be many fine, filamentous spirochaetal organisms. Confirmation of the diagnosis requires isolation and characterization of *B. nodosus* by the techniques described in Chapter 5.

A number of other infections cause tissue damage and, consequently, lameness in ruminants. In cattle, foot abscess (interdigital phlegmon, necrobacillosis) is common in both dairy and beef cattle. Foot abscess (see Chapter 4) is characterized by necrotizing invasion of subcutaneous and deeper tissues of the hoof leading to acute pain and lameness and usually to the occurrence of sinuses which discharge necrotic material. Filaments of *F. necrophorum* are readily demonstrated in this material. Foot abscess occurs in goats also but does not seem to have been reported in deer.

Cattle, goats, and other ruminants develop acute laminitis and lameness following engorgement on high-energy rations. Acute laminitis is not usually associated with necrosis of affected parts but there may be sufficient vascular damage in some cases to result in

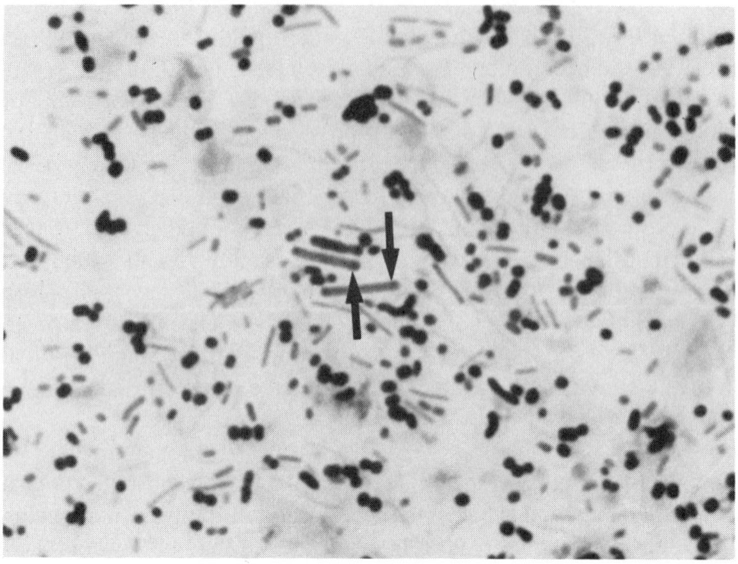

FIGURE 1. Gram-stained smear from interdigital skin lesion showing *B. nodosus*, arrowed. Many other species of bacteria occur in affected tissues. (Magnification × 3000.)

partial or complete sloughing of hooves. Secondary infection of the sensitive laminae may then occur but *B. nodosus* is absent.

B. Characteristics of *B. nodosus* Isolates from Cattle, Goats, and Deer

The cultural, biochemical, pathogenic, and antigenic features of ovine isolates of *B. nodosus* have been described in Chapter 1. The special characteristics of isolates from other ruminants will be described here.

1. Cultural Characteristics

A number of authors have described differences in the colony morphology of isolates of *B. nodosus* from hosts other than sheep. Thus Thorley[13] and Skerman et al.[14] referred to fimbriate-beaded, mucoid, and circular colonies. Virtually all isolates from cattle so far examined have either circumscribed circular or mucoid colonies.[14] Those from goats may be either fimbriate or fimbriate-mucoid.[10] The only isolates of deer so far characterized were of the nonfimbriate, circular type.[12]

2. Biochemical Characteristics

B. nodosus from cattle, goats, and deer may be differentiated by a number of tests which identify the proteolytic and elastolytic activity of isolates. Richards et al.[15] have described qualitative differences in the thermostability of proteases elaborated by virulent sheep and bovine isolates, respectively. Other investigators have described differences in the array of isoenzymes produced by ovine compared with bovine isolates.[11,16,17] Stewart[18] described an elastase test which could be used to distinguish isolates of *B. nodosus*. Those from cattle were either mildly elastolytic or elastase negative.[11] About 50% of isolates from goats were elastase negative but two of seven tested, although intermediate in elastase activity, were virulent when tested in sheep.[10]

3. Pathogenicity for Other Ruminants

Because of the potential role of other ruminants as a reservoir of ovine footrot, the virulence of isolates from other ruminants has been closely examined. Although the first reported isolate from cattle was considered virulent for sheep,[4] all subsequent work suggests that bovine isolates at least are substantially avirulent in sheep.[4,5,6,9,11,14,15,16,19]

This is an important finding epidemiologically because cattle and sheep are often grazed together in countries like Australia and New Zealand. The opportunity for transfer of infection is therefore high. Because cattle infections are relatively harmless for sheep, they can, for practical purposes, be ignored in ovine footrot-eradication programs.

Goat isolates do not seem to be as generally innocuous for sheep.[10] Isolates from two of seven goat farms were virulent for sheep. Furthermore, failure of sheep footrot-eradication campaigns in New Zealand have been attributed to cross infection of sheep from feral goats.

On the other hand, Claxton and O'Grady[10] have described an occurrence where goats failed to transmit infection to sheep grazing together in the same environment.

4. *Antigenic Features*

Isolates of *B. nodosus* have been classified antigenically into eight major serogroups.[20,21] The basis of this classification is the agglutinin response of immunized animals to the fimbriae (or pili) on the surface of *B. nodosus* cells (see also Chapter 10). Bovine isolates of *B. nodusus* have been reported to be less fimbriate than virulent isolates although this is disputed by Depiazzi et al.[22] Sufficient fimbriae are usually present, however, to allow bovine (and caprine isolates) to be classified into one or other of the recognized serogroups.

V. EPIDEMIOLOGY

A. Geographical Distribution

Footrot (*B. nodosus* infection) has been reported in cattle in the U.S.,[3] Australia,[4,6,15,19,23] Western Europe,[24] Holland,[5] Sweden,[25] Bulgaria,[26] and Iraq.[27] There seems little doubt that the disease is as widely distributed in temperate regions of the world as the analogous infection of sheep. Similarly, while isolations from goats have been reported definitively only from Australia,[10,11] clinically, footrot has been diagnosed in goats in New Zealand,[28] the U.S., and Europe.

B. Prevalence

A number of studies suggest that a high proportion of animals in affected cattle herds may have *B. nodosus* infections of the interdigital skin and contiguous tissue.[6] although all infected animals are not necessarily lame or showing other clinical signs. Laing and Egerton[6] recorded infection prevalences of up to 60% in dairy cattle on pasture in the Sydney region and this prevalence was similar in different seasons. Abattoir surveys[15,23] have confirmed the wide distribution of bovine footrot in Australia although prevalences of infection were much lower than those recorded by Laing and Egerton.[6] Investigators in Sweden[25] and Iraq[27] have either observed or isolated *B. nodosus* from 29 to 57% of cattle feet with interdigital dermatitis, whereas in Holland the infection was described as widespread among Friesian dairy cattle.[5] Most prevalence studies have been made on dairy cattle because of the occurrence of clinical infections in these circumstances. Footrot occurs also in beef breeds on pasture.[23] Whether or not footrot occurs in association with foot abscess in intensively managed cattle in feed lots has not been reported.

C. Transmission

Although infections have been induced experimentally with cultures of *B. nodosus* of cattle origin, the natural spread of the infection from cattle to cattle is the subject of only one report.[6] In this study, cattle and sheep free of *B. nodosus* were pastured together with cattle affected with a characterized bovine isolate. Infection with the same isolate was

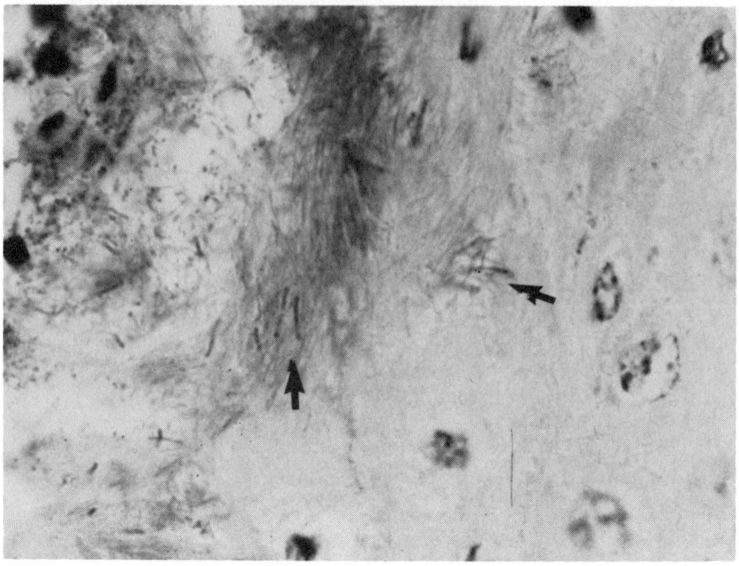

FIGURE 2. Giemsa-stained histological section of interdigital skin affected by footrot. *B. nodosus* cells are invading the germinal layer. Similar lesion to that in Figure 1. Complicated by embedded foreign body (a piece of gravel, arrowed).

subsequently diagnosed in cattle (and sheep) in contact with the affected animal. Cattle in a similar environment nearby but not in contact with affected animals, remained free of infection.

During these investigations it was established that *B. nodosus* could be isolated readily from the mud and soil of the area surrounding a dairy which had a continuing history of bovine footrot. The longest interval between contamination of the area and isolation of *B. nodosus* was 96 h.[29]

By analogy with the disease in sheep, it seems likely that transmission requires not only the presence of the causal agent, but some environmental predisposition of the interdigital skin. The feet of housed dairy cattle are normally exposed to wet, fecally contaminated environments, whereas those at pasture frequently enter and leave dairies through soft, and/or, muddy ground. The opportunities for both predisposition and contact with concentrations of *B. nodosus* are thereby enhanced.

VI. PATHOLOGY AND PATHOGENESIS

Footrot in cattle is essentially a dermatitis[4] of the interdigital skin. The epidermal layers are principally affected, although polymononuclear leucocytes accumulate in associated dermal papillae. Bacterial invasion of he *stratum corneum* and *stratum granulosum* is accompanied by degenerative processes in these epidermal layers. There is dissociation of epidermal cells following edema and extravasation. Loss of integrity of the *stratum corneum* is seen macroscopically as diffuse erosion and sometimes discrete ulceration. Where the interdigital skin is extensively affected, a longitudinal fissure sometimes occurs at the most proximal part of the cleft. In uncomplicated cases, bacterial invasion is restricted to the epidermis. *B. nodosus* has a characteristic morphology and is observed either singly in the keratinizing layers or in small bundles[4] (Figure 2). Other organisms, less readily identifiable by their morphology, also invade the affected epidermis. These may include *F. necrophorum* and *Treponema penortha.*

The environmental and microbial interactions which result in footrot in cattle have not been investigated closely. Laing[29] and Clark et al.[8] have studied the effect of combinations of *B. nodosus, F. necrophorum,* and *B. melaninogenicus* in the causation of interdigital dermatitis. Laing's conclusion was that *B. nodosus,* sometimes alone, and more regularly in association with the other two organisms, resulted in the development of interdigital dermatitis when applied intradermally. S.c. inoculation of *B. nodosus* was without effect, whereas injection of *F. necrophorum* and *B. melaninogenicus* s.c. resulted in lesions more characteristic of foot abscess. Clark et al. also concluded that *B. nodosus* was essential to the development of superficial interdigital dermatitis. Experimental infection of cattle feet is complicated by the uncertainty of ensuring freedom of infection in apparently normal subjects and in preventing cross contamination. Intradermal infections are also technically difficult. Also, isolates of *B. melaninogenicus* and *F. necrophorum* vary in virulence and those used were inadequately characterized in earlier studies.

The *stratum corneum* of the interdigital skin of cattle is particularly well developed and consists of a dense layer of highly stratified squamous cells. When healthy and dry, it is an effective barrier to growth and penetration by bacteria. The development of bacterial infection either in the *stratum corneum* or deeper layers of the epidermis presumably requires either extensive hydration of the *stratum corneum* or its breaching by mechanical forces. More complete understanding of the pathogenesis of both footrot and foot abscess in cattle will only be available when the relative importance of hydration, maceration, and injury (Plate 5).* in fostering bacterial growth and invasion, is established.

VII. TREATMENT, CONTROL, AND PREVENTION

A. Topical Treatment

Bacterial infection in footrot of cattle, goats, and deer is confined to the epidermal structures of the interdigital skin and therefore topically applied antibacterials should be effective in treatment. In those cases where infection has underrun horn, effective doses of topically applied treatments may not reach all affected tissue. Some preparative removal of this horn is required in these circumstances.

Solutions of formalin (5%) and zinc sulfate (10%) are effective but repeated treatments may be required, especially where environmental conditions favor reinfection.

B. Systemic Treatment

B. nodosus is susceptible *in vitro* to a wide spectrum of antibiotic agents, but not to the sulfanamide group.[30] Injections of antibiotic drugs are frequently used to treat footrot in cattle. This form of treatment is more likely to give good results where some components of the flora of footrot have invaded deeply into the dermal and subdermal tissues of the foot. It is unlikely that effective levels of antibiotics would diffuse to the avascular epidermis following parenteral inoculation of antibiotics at normal dose rates. In sheep, approximately ten times normal dose rates are required to achieve bactericidal concentrations and therapeutically effective levels of penicillin in affected epidermis.[30]

In 1973, Demertzis and Mills[31] published a report which suggested that zinc sulfate taken orally was effective in treating footrot of cattle. Subsequently, claims were made for both therapeutic and prophylactic benefits of ingested zinc in the management of footrot.[32] Other investigators were unable to substantiate these claims.[33]

C. Control and Prevention

Economically, footrot in cattle is most important in dairy cows. It is a cause of lameness

* Plate 5 follows page 54.

and may constitute a continuing reservoir of the bacteria which invade more extensively to cause foot abscess (see Chapter 4).

1. Footbaths

Management systems for dairy cattle not only expose cattle to the environments which predispose to footrot and facilitate transmission, but they also allow regular treatment of herds by incorporation of footbaths at points of entry or exit from milking parlors. These baths should be at least 15 cm deep and 5 m long. Formalin solution (3 to 5%), is an effective agent both for curing superficial footrot infection and in preventing the development of chronic cases and their sequelae. Zinc-sulfate solutions (10% w/v) are also effective. The latter have the advantage of being less painful to the animals and are devoid of the irritant vapors associated with formalin.

2. Vaccination

The bovine isolates of *B. nodosus* have been shown to possess the same range of fimbrial antigens identified among sheep isolates.[9] It is logical, therefore, that vaccines which have been shown to be effective in preventing and accelerating healing in ovine footrot have been tested for their capacity to control the disease in cattle.

Toussaint[37] and Wilkinson[38] both measured antibody response of cattle to monovalent, oil-adjuvanted *B. nodosus* vaccines but both were concerned at the degree of tissue response following s.c. infections. Antibody responses were not correlated with clinical effect but these investigations were done before the significance of antigenic diversity in *B. nodosus* was understood. Similarly, Laing[29] examined the response of clinically affected animals to monovalent alum-precipitated vaccines. There was a reduction in both prevalence and severity of infection in vaccinated, compared with untreated, animals in the same environments. Antibody levels recorded were lower than those considered necessary for protection in sheep.

In small groups of animals, Laing[29] showed that i.p. injections of oil-adjuvanted vaccine induced substantially higher agglutinin levels than similar doses of antigen given s.c. Oil-adjuvanted vaccines administered s.c. resulted in the development of unacceptable reactions at the infection site. There was no evidence of harmful affect, clinically or at autopsy, of the i.p. inoculations.

Various *B. nodosus* vaccines have been tested for their capacity to control bovine footrot but the results have been variable. Italian workers[34] suggest that *B. nodosus* vaccine not only prevented interdigital dermatitis but reduced the prevalence of sole ulcer also. In Poland,[35] vaccine was only moderately effective in controlling prevalence of disease and in reducing severity of infection. In Australia,[36] cattle in one herd were vaccinated with isolates of serogroup B and G from affected animals. Antibody response was relatively poor and there was no reduction of prevalence in vaccinated animals. Lesions were, however, less severe in those which were vaccinated. There was a suggestion in this investigation that animals vaccinated with *B. nodosus* had fewer cases of foot abscess. This finding, if confirmed, would support the hypothesis that chronic epidermal infections (i.e., bovine footrot) may be a precursor of foot abscess in some circumstances.

No conclusions can be made about the efficacy of *B. nodosus* vaccines in controlling infections with that organism in cattle. In most reports, no evidence has been presented on the antigenic identity of *B. nododus* in either the vaccines used or in the cattle being investigated. Given the serogroup specificity of immunity which follows immunization with *B. nodosus* (see Chapter 11) and the diversity of serogroups known to occur in the same flocks of sheep[21] and herds of cattle,[29] it is not surprising that variation in response to vaccination has been observed.

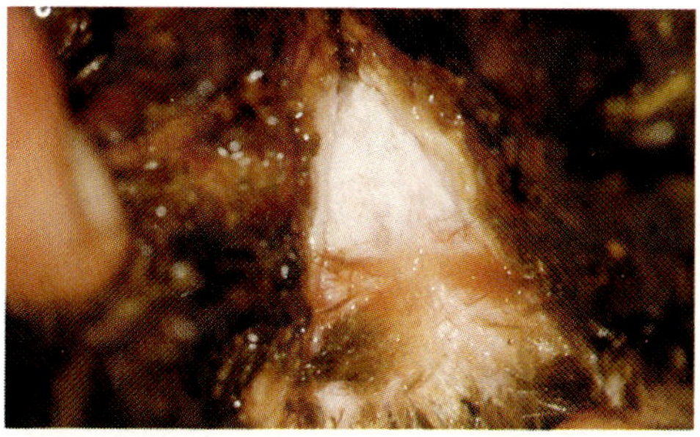

A

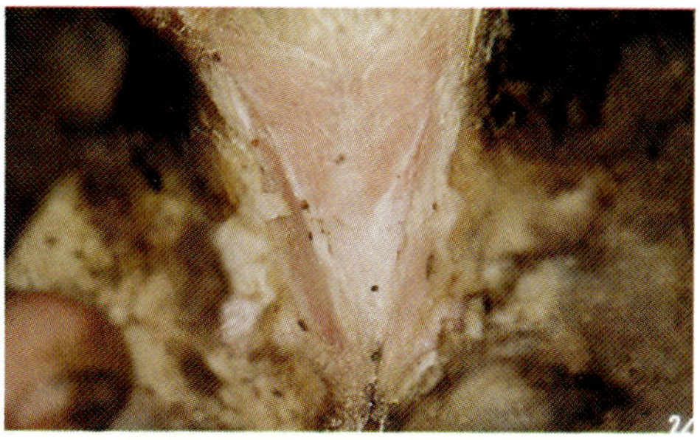

B

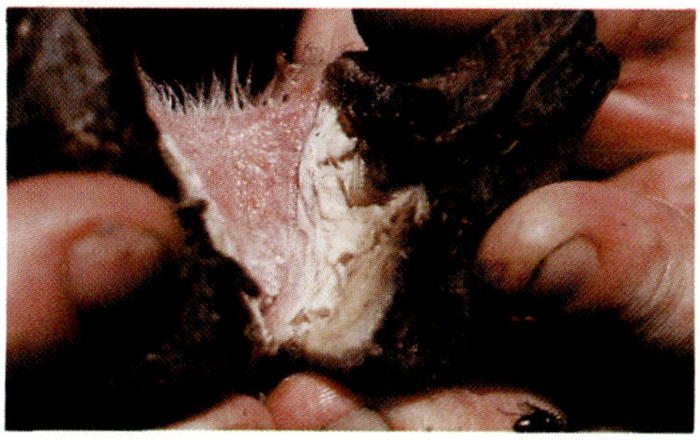

C

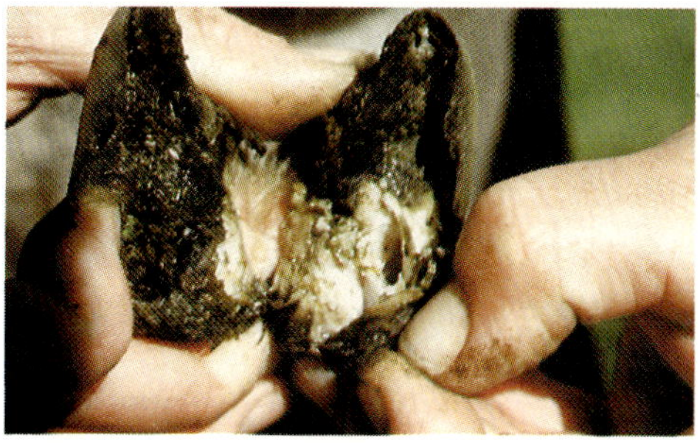

D

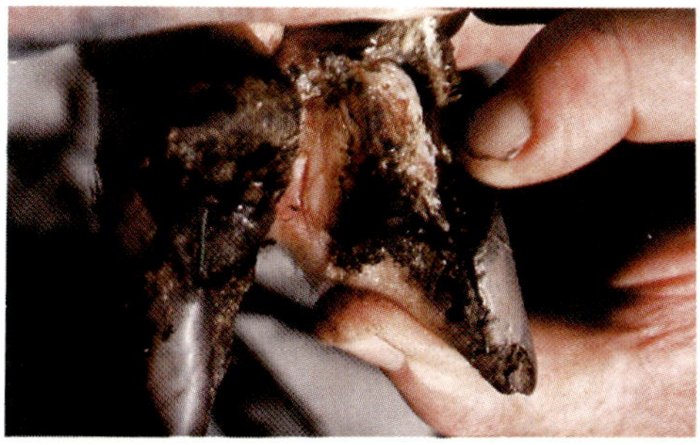

E

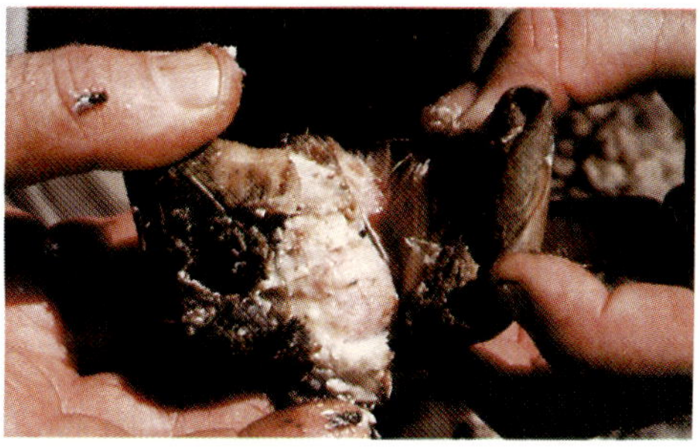

F

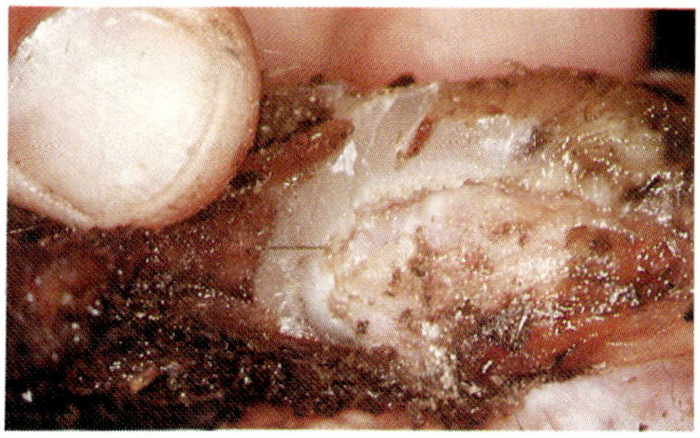

G

Plate 1. (A,B) Benign footrot. (A) Interdigital dermatitis of posterior interdigital skin and skin-horn junction, characterized mainly by hyperkeratosis with a small amount of exudate. (B) More pronounced dermatitis with erythema and erosion of the interdigital skin and soft horn of the posterior abaxial walls of the hoof. There is also slight separation of the horn of the heel. (C to F) Virulent footrot. (C) Severe dermatitis of interdigital skin and axial walls of the horn of the hoof, showing marked erythema, erosions, hyperkeratosis, and necrosis. (D) Separation of the soft horn of the hoof beginning at the heel and extending across the sole. (E) Underrunning of the hard horn of the hoof with the lesion extending down the abaxial wall. (F) In advanced virulent footrot, the underlying tissue beneath the separated horn is erythematous and covered by a distincitve foul-smelling, white-grey necrotic exudate. (G) In benign footrot, separation of the soft horn of the heel and sole is occasionally observed but there is little accumulation of necrotic exudate.

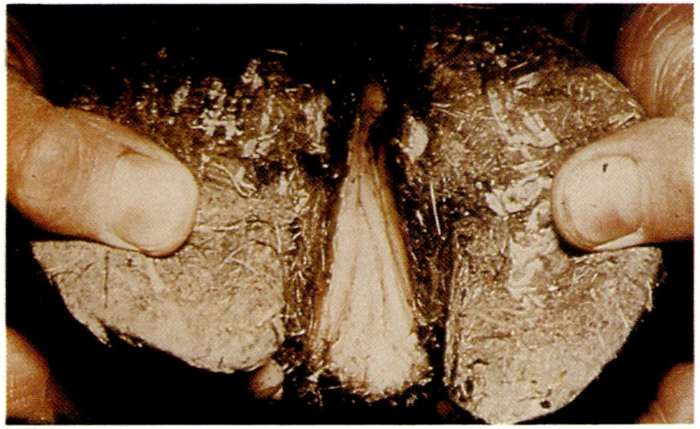

Plate 2. Interdigital skin of bovine hoof with footrot. This uncomplicated infection is characterized by swelling, serous exudate, and superficial erosion.

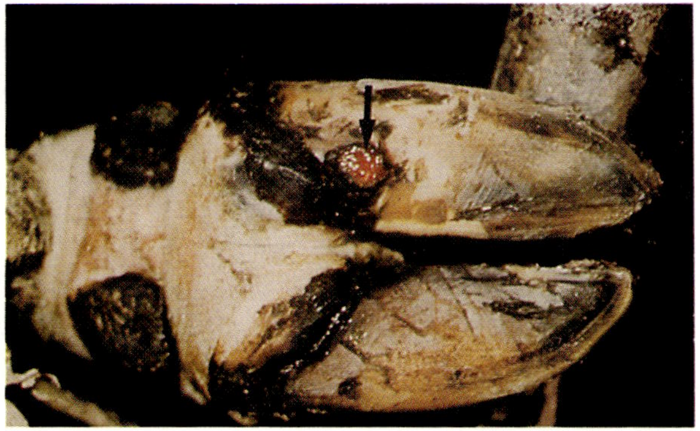

Plate 3. Chronic bovine footrot with necrosis of sole and underrunning of horn, pared to reveal associated ulcer formation, arrowed.

Plate 4. Bovine footrot, complicated by interdigital overgrowth, arrowed.

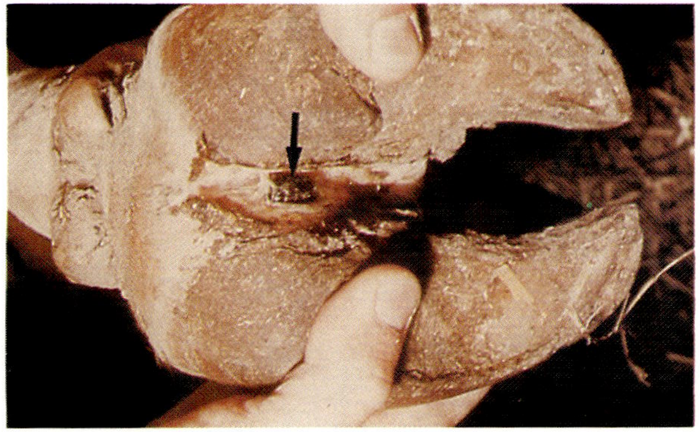

Plate 5. Similar lesion to that in Plate 2. Complicated by embedded foreign body (a piece of gravel, arrowed).

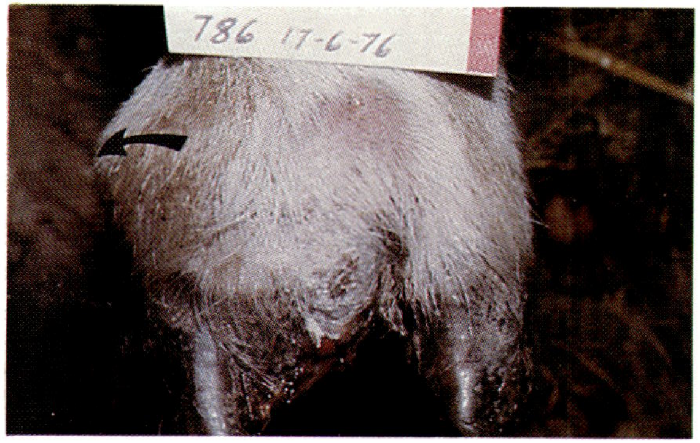

A

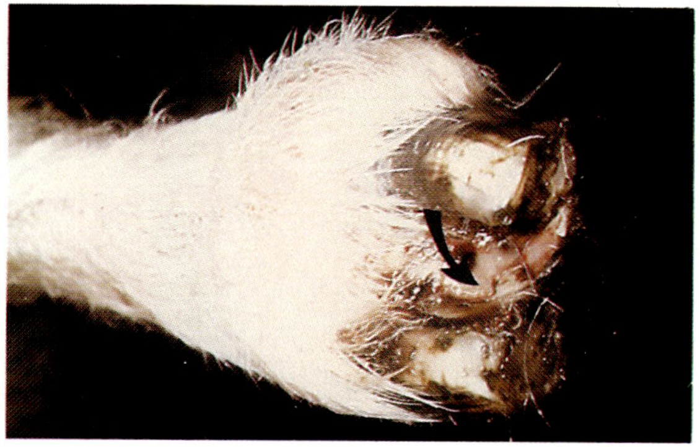

B

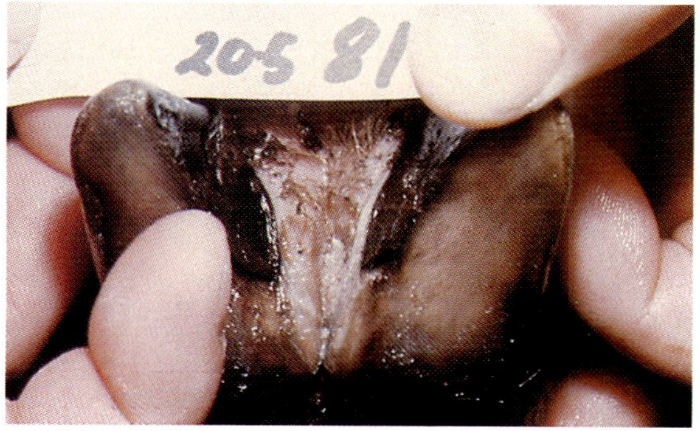

C

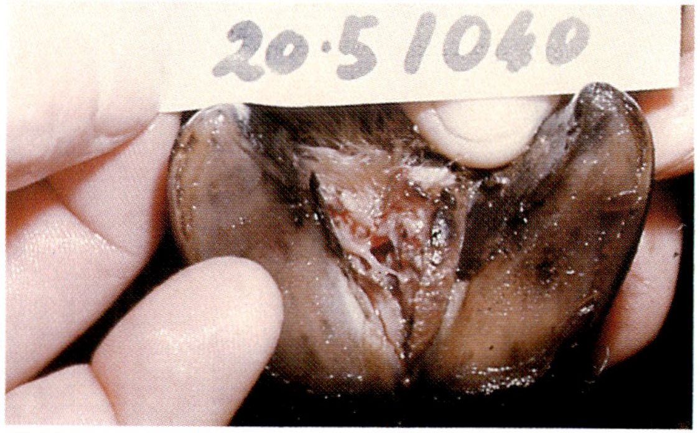

D

Plate 6. (A) Anterior view of sheep's foot affected with foot abscess. The affected digit is swollen and within 24 h, discharged necrotic material seen at the coronary region (arrow). (B) Foot abscess case with discharge (arrowed) from sinus in interdigital cleft. (C) Inflammation and erosion in an acute case of ovine interdigital dermatitis (OID). (D) Inflammation and necrosis in a case of OID which has progressed to deeper tissues of interdigital cleft.

VIII. CONCLUSIONS

A number of investigations in different countries have established the occurrence of *B. nodosus* infection (footrot) in cattle and, to a lesser extent, goats and deer. The pathology and epidemiology of the infection in these animals are similar to that described for sheep. Isolates of *B. nodosus* from cattle, while having the fimbrial antigens which characterize sheep isolates, have qualitatively and quantitatively different proteolytic enzymes. Caprine and cervine isolates have not been studied so extensively; those from goats are more likely to be virulent for sheep.

Clinically and economically, footrot is important in cattle and goats. It causes disease and production loss in both species. In the case of cattle, footrot may predispose to more damaging, deeper infections of the foot.

Footrot in cattle is preventable both by regular chemotherapy and perhaps by vaccination, although this approach needs more extensive and critical evaluation in both cattle and goats.

REFERENCES

1. **Alexander, T. M.,** The differential diagnosis of foot-rot in sheep, *Aust. Vet. J.*, 38, 366, 1962.
2. **Skenman, G.,** A case of possible *F. nodosus* infection in a cow, *Aust. Vet. J.*, 38, 306, 1962.
3. **Gupta, R. B., Fincher, M. F., and Bruner, D. W.,** A study of the etiology of footrot in cattle, *Cornell Vet.*, 54, 66, 1964.
4. **Egerton, J. R. and Parsonson, I. M.,** Isolation of *Fusiformis nodosus* from cattle, *Aust. Vet. J.*, 42, 425, 1966.
5. **Toussaint, R. E. and Cornelisse, J. L.,** The specific contagious inflammation of the interdigital skin of cattle, *Vet. Med. Rev.*, 2/3, 223, 1971.
6. **Laing, E. A. and Egerton, J. R.,** The occurrence, prevalence and transmission of *Bacteroides nodosus* infections in cattle, *Res. Vet. Sci.*, 24, 300, 1978.
7. **Egerton, J. R., Roberts, D. S., and Parsonson, I. M.,** The aetiology and pathogenesis of ovine footrot, *J. Comp. Pathol.*, 79, 207, 1969.
8. **Clark, B. L., Stewart, D. J., and Emery, B. L.,** A study of the role of *Bacteroides nodosus, Fusobacterium necrophorum* and *Bacteriodes melaninogenicus,* in The Aetiology of Bovine Foot Infections, Abstr. 22nd World Veterinary Cong., Perth, 1983, 145.
9. **Egerton, J. R. and Laing, E. A.,** Characteristics of *Bacteroides nodosus* isolated from cattle, *Vet. Microbiol.*, 3, 269, 1978/1979.
10. **Claxton, P. D. and O'Grady, K. C.,** Footrot in goats and characterisation of caprine isolates of *Bacteroides nodosus,* in *Footrot in Ruminants,* Proc. Workshop, Melbourne, 1985, Stewart, D. J., Peterson, J. E., McKern, N. M. and Emery, D. L. Eds., CSIRO Division Animal Health/Australian Wool Corp., Glebe, N.S.W., 1986, 119.
11. **Stewart, D. J., Peterson, J. E., Vaughan, J. A., Clark, B. L., Emery, D. L., Caldwell, J. B., and Kortt, A. A.,** Clinical and laboratory diagnosis of benign intermediate and virulent strains of *Bacteroides nodosus,* in *Footrot in Ruminants,* Proc. Workshop, Melbourne, 1985, Stewart, D. J., Peterson, J. E., McKern, N. M., and Emery, D. L., Eds., CSIRO Division Animal Health/Australian Wool Corp., Glebe, N.S.W., 1986, 81.
12. **Skerman, T. M.,** Isolation of *Bacteroides nodosus* from hoof lesions in a farmed red deer *(Cervus elephus), N. Z. Vet. J.*, 31, 102, 1983.
13. **Thorley, C. M.,** A simplified medium for the isolation of *Bacteroides nodosus* from ovine footrot and studies on its colony morphology and serology, *J. Appl. Bacteriol.*, 40, 301, 1976.
14. **Skerman, T. M., Erasmuson, S. K., and Emery, D.,** Differentiation of *Bacteroides nodosus* biotypes and colony variants in relation to their virulence and immunoprotective properties for sheep, *Infect. Immun.*, 32, 788, 1981.
15. **Richards, R. B., Depiazzi, L. J., Edwards, J. R., and Wilkinson, F. C.,** Isolation and characterisation of *Bacteroides nodosus* from foot lesions of cattle, *Aust. Vet. J.*, 56, 517, 1980.
16. **Every, D.,** Proteinase isoenzyme patterns of *Bacteroides nodosus:* distinction between ovine virulent isolates, ovine benign isolates and bovine isolates, *J. Gen. Microbiol.*, 128, 809, 1982.

17. **Gordon, L. M., Yong, W. K., and Woodward, C. A. M.,** Temporal relationships and characterisation of extracellular proteases from benign and virulent strains of *Bacteroides nodusus* as detected in zymogram gels, *Res. Vet. Sci.*, 39, 165, 1985.
18. **Stewart, D. J.,** The role of elastase in the differentiation of *Bacteriodes nodosus* infection in sheep and cattle, *Res. Vet. Sci.*, 27, 99, 1979.
19. **Wilkinson, F. C., Egerton, J. R., and Dickson, J.,** Transmission of *Fusiformis nodosus* infection from cattle to sheep, *Aust. Vet. J.*, 46, 382, 1970.
20. **Egerton, J. R.,** Surface and somatic antigens of *Fusiformis nodosus, J. Comp Pathol.*, 83, 151, 1973.
21. **Claxton, P. D., Egerton, J. R., and Ribeiro, L. A.,** Classification of *Bacteroides nodosus* by agglutination tests, *Aust. Vet. J.*, 60, 331, 1983.
22. **Depiazzi, L. J., Richards, R. B., McQuade, N. C., and Rood, J. I.,** Characterisation of virulence associated antigens for differentiating virulent and benign strains of *Bacteroides nodosus,* in *Footrot in Ruminants,* Proc. Workshop, CSIRO, Melbourne, 1985, D. J. Steward D. J., et al., Eds., CSIRO Division Animal Health/Australian Wool Corp, Glebe, N. S. W., 1986, 103.
23. **Morgan, I. R.,** A survey of cattle feet in Victoria for *Fusiformis nodosus, Aust. Vet. J.*, 45, 264, 1969.
24. **Thorley, C. M. and Day, S. E. J.,** Serotyping survey of 1296 strains of *Bacteroides nodosus* isolated from sheep and cattle in Great Britain and Western Europe, in *Footrot in Ruminants,* Proc. Workshop, Melbourne, 1985, Stewart, D. J., Peterson, J. E., McKern, H. M., and Emery, D. L., Eds., CSIRO Division Animal Health/Australian Wool Corp, Glebe, N. S. W., 1986, 135.
25. **Forshell, L. P. and Anderson, L.,** *Bacteroides nodosus* infection in bovine interdigital necrobacillosis, *Sven. Vet.*, 33, 551, 1981.
26. **Masalski, N.,** Outbreak of infectious interdigital dermatitis in cows associated with *Bacteroides nodosus* and *Treponema (Spirochaeta) penortha, Vet. Sbirka,* 81, 23, 1983.
27. **Hussein, H. A. and Hanna, A.-A.T.,** Isolation and identification of *Fusobacterium necrophorum* and *Bacteroides nodosus* from footrot in cattle in Bagdad, *J. Biol. Sci. Res.*, 15, 57, 1984.
28. **Manktelow, B. W.,** in *The Veterinary Handbook* Foundation for Continuing Education, New Zealand Veterinary Association, Palmerston North, New Zealand, 1984.
29. **Laing, E. A.,** The Epidemiology of Lower Limb Lameness in Cattle, M.V.Sc. thesis, University of Sydney, Sydney, 1979.
30. **Egerton, J. R., Parsonson, I. M., and Graham, N. P. H.,** Parenteral chemotherapy of ovine foot-rot, *Aust. Vet. J.*, 44, 275, 1968.
31. **Demertzis, P. N. and Mills, C. P.,** *Vet. Rec.*, 93, 219, 1973.
32. **Demertzis, P. N., Spais, A. G., and Papasteriadis, K. A.,** Zinc therapy in the control of footrot in sheep, *Vet. Med. Rev.*, 1, 101, 1978.
33. **Egerton, J. R., Laing, E. A., and Mulley, R. C.,** Failure of oral zinc therapy to alleviate *Bacteroides nodosus* infections in cattle and sheep, *Aust. Vet. J.*, 62, 85, 1985.
34. **Quintaralla, F., Zannetti, G., Licata, E., Tagliavini, M., and Pasini, A.,** Therapeutic use of a *Bacteroides nodosus* vaccine in various diseases of the bovine foot: preliminary note, *Atti Soc. Ital. Buiatria,* 17, 237, 1985.
35. **Cakala, S., Pilaszek, J., Lubiarz, J., Gogolewski, L., and Dembinski, Z.,** Prophylactic and therapeutic effects of Ristellan vaccine against footrot in cattle, in Proc. 13th World Cong. Diseases Cattle, Durban, South Africa, 1, 312, 1984.
36. **Clark, B. L., Stewart, D. J., Emery, D. L., Dufty, J. H., and Jarrett, R. G.,** Immunisation of cattle against interdigital dermatitis (footrot) with an autogenous *Bacteroides nodosus* vaccine, *Aust. Vet. J.*, 63, 61, 1986.
37. **Toussaint, R. E.,** Personal communication.
38. **Wilkinson, F. C.,** Personal communication.

Chapter 3

FOOT ABSCESS OF SHEEP

D. M. West

TABLE OF CONTENTS

I. DEFINITION

Foot abscess of sheep is an acute, often purulent infection, usually involving one digit of the foot.[1] In most cases, infection begins in the interdigital space[2,3] and extends into the deeper structures of the digit to involve the distal interphalangeal joint and associated ligaments and tendons.[4]

The major bacterial pathogens associated with ovine interdigital dermatitis are *Fusobacterium necrophorum* and *Corynebacterium pyogenes*[5] and it is these two organisms that are also considered to cause foot abscess, given suitable predisposing factors. This conclusion is based on the frequent isolation of both organisms from foot abscesses,[2,4] the known synergism between the two organisms,[6,7] and the distribution of morphologically similar organisms in stained sections of lesions of foot abscess. Furthermore, an identical condition can be reproduced by the inoculation of these two bacteria into the distal interphalangeal joint of sheep. However, relatively greater numbers of single species of either organism are required to reproduce foot abscess by this means than by using a mixture of the two organisms.[8] Thus, it is believed that *F. necrophorum* and *C. pyogenes* are the main causative organisms of foot abscess in sheep, but the possibility of other microorganisms playing an essential or supporting role cannot be dismissed.[9]

II. CLINICAL SIGNS

One of the most striking features of foot abscess is the acute lameness that accompanies the condition. In most cases, the affected foot is held off the ground when walking or running, and where more than one foot is affected, the sheep spends a considerable time lying down. Once infection becomes established in the distal interphalangeal joint, the foot abscess has developed. The condition usually follows a predictable pattern that inevitably results in some permanent damage and deformity to the affected digit.

In the early stages, there is marked edematous swelling and inflammation of the interdigital space, displacing the affected digit abaxially (Plate 6A)*. In places, the interdigital skin is necrotic and bloody necrotic material and/or pus can be expressed from these areas. In the latter stages, the discharge becomes more purulent in nature and contains less necrotic tissue. The discharge of pus is eventually blocked by the formation of granulation tissue so that it accumulates in the joint cavity. This pressure is transmitted to the skin above the abaxial coronet which becomes painful, inflamed, and swollen (Plate 6A). Sinus tracts form along the abaxial-coronary-border region at sites where the joint is relatively superficial. These sinus tracts usually become confluent, opening into the distal interphalangeal joint and discharging into the interdigital space (Plate 6B).

In approximately a third of the cases, the severe interphalangeal-joint infection causes rupture of the axial collateral ligaments and the interdigital ligament. When this occurs, the digit twists outwards, luxation of the distal phalanx and distal sesamoid bone occurs (Figure 1), and in the healing process, the distal phalanx does not return to its former position because of the pressure placed on it as the animal tries to walk. In cases where the ligaments remain intact, there is better alignment of the distal phalanx with the other phalanges and although some permanent thickening and deformity remains, the digit can function normally without becoming displaced abaxially.

Once the condition progresses into the healing stages, the pain and swelling regress and the sinus tracts stop discharging. In some cases, there is excessive granulation tissue formed in the interdigital space which may delay healing. However, usually the affected digit heals within about 8 weeks to a stage where there is no active infection in the digit and except

* Plate 6 follows page 54.

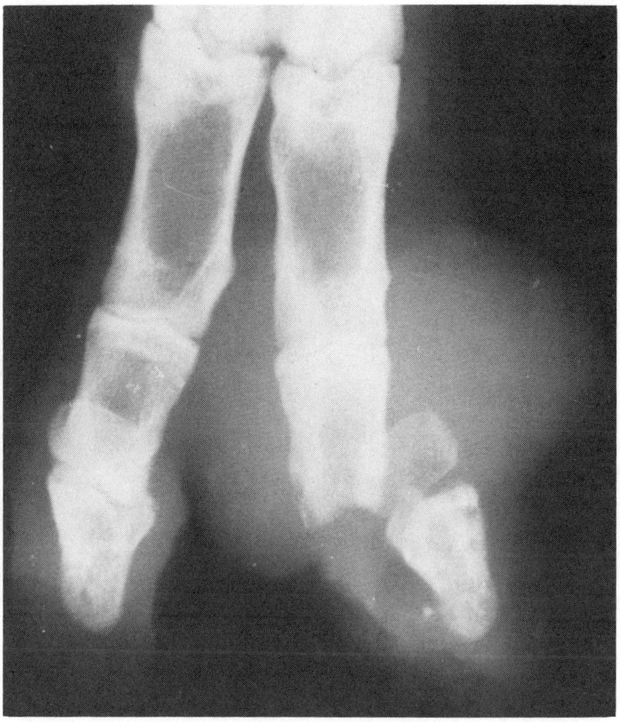

FIGURE 1. Radiograph of an acute foot abscess in which the distal phalanx and the distal sesamoid bone are luxated.

for some permanent fibrous thickening, the swelling has subsided. In most cases, slight digital displacement and lameness may be detected if the foot is examined carefully.

A. Radiology

Radiology of the affected foot is a particularly useful aid, both for confirming the diagnosis and for assessing the stage of the lesion and the possible outcome. In the early stages, radiographic changes are confined to soft-tissue swelling and widening of the joint space (Figure 2).[4] Within a few days, however, changes to the distal interphalangeal joint are obvious and include dissolution of bone, particularly the distal end of the middle phalanx, and proliferation of bone adjacent to the periosteum. Displacement of the distal phalanx may occur (Figure 1). In the healing phase, the soft-tissue swelling regresses and the outline of new bone that has proliferated becomes smooth and more clearly defined.

III. DIFFERENTIAL DIAGNOSIS

A. Toe Abscess (Lamellar Suppuration)

Confusion has arisen between differentiating the condition of toe abscess and that of foot abscess. Toe abscess, like foot abscess, is an acute suppurative condition commonly affecting only one digit. Unlike foot abscess, it arises sporadically in any flock and not as a complication of ovine interdigital dermatitis. The condition occurs mainly at the toe of the fore foot and the opportunist infection probably gains entry to the sensitive laminae region through minute fissures in the horn in the area of the white line, especially in misshapen or overgrown hooves. Occasional cases may also arise following injudicious paring of the feet.

In the majority of cases, the earliest sign of infection is an acute lameness which is so painful that the sheep refuses to bear any weight on the foot. Examination may fail to reveal

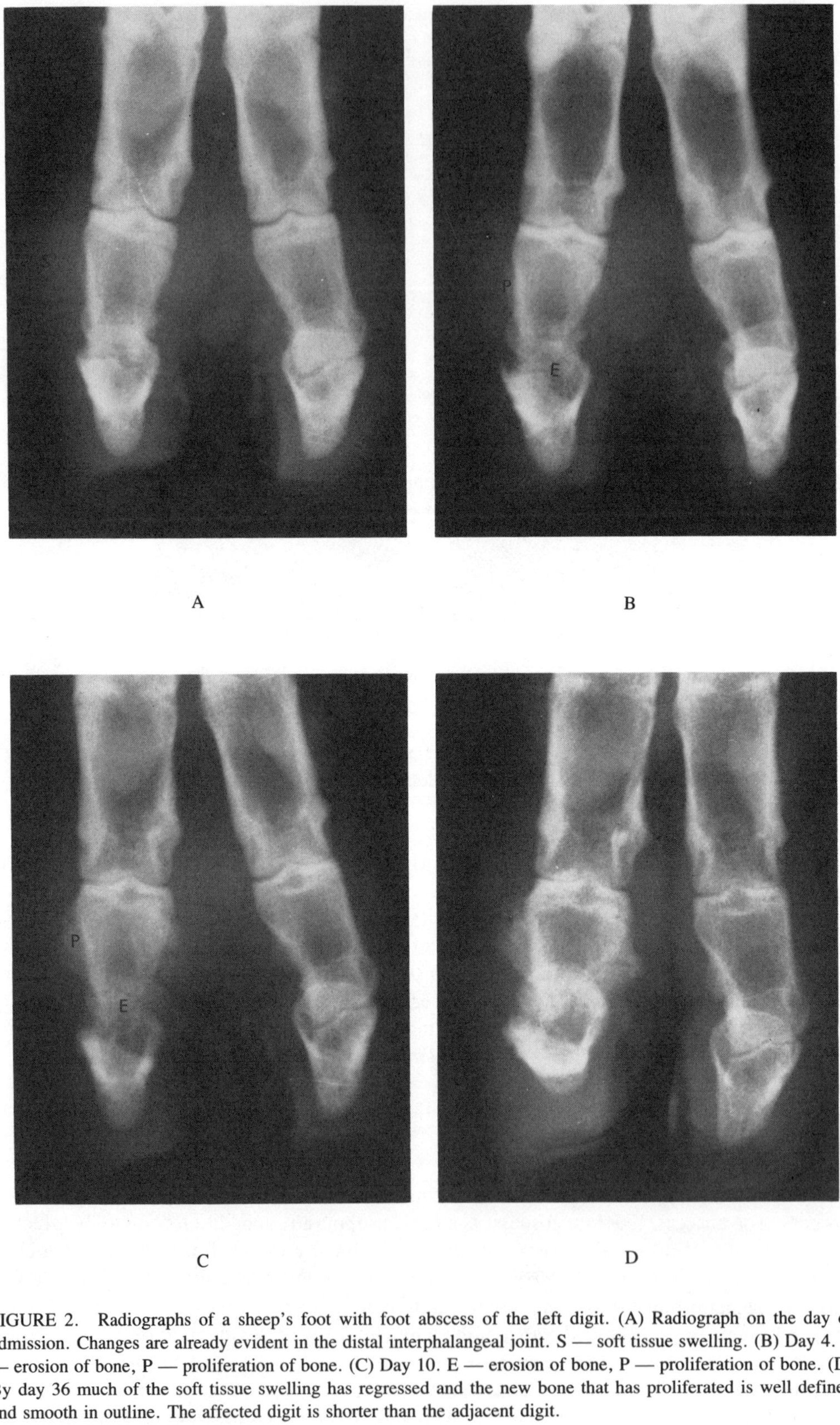

FIGURE 2. Radiographs of a sheep's foot with foot abscess of the left digit. (A) Radiograph on the day of admission. Changes are already evident in the distal interphalangeal joint. S — soft tissue swelling. (B) Day 4. E — erosion of bone, P — proliferation of bone. (C) Day 10. E — erosion of bone, P — proliferation of bone. (D) By day 36 much of the soft tissue swelling has regressed and the new bone that has proliferated is well defined and smooth in outline. The affected digit is shorter than the adjacent digit.

any obvious lesion but generally one claw is found to be hot and pressure applied to the wall causes the sheep to flinch. Careful paring away of the horn at the toe reveals the presence of imprisoned pus. If the original site of injury on the plantar surface becomes sealed by mud, feces, or exudate, infection can track upwards and produce a sinus above the coronet. *F. necrophorum* and *C. pyogenes* can often be cultured from the pus, along with other organisms found in the environment. In contrast with foot abscess, toe abscess responds rapidly to drainage of the lesion.

B. Footrot

Footrot is a contagious disease of the sheep's foot characterized by separation of a large portion of the hoof from the soft tissues due to a spreading infection immediately beneath the horn.[10] The infection commences in the interdigital skin, and horn separation is often seen first at the axial-coronary region but later extends down the axial wall, under the sole, and eventually the abaxial wall, so that finally the hoof is attached only at the coronet. In uncomplicated cases there is little suppuration. Both digits of the foot are usually affected, each sheep often having more than one, but rarely all four feet involved. Footrot is a flock disease and under suitable environmental conditions may spread rapidly until over 90% of the flock becomes infected. Footrot can be differentiated from foot abscess because of the separation of the horn, the minimum swelling and suppuration, the characteristic odor, the general bilateral involvement of the digits, and under suitable conditions, the rapid spread of the infection to involve other sheep in the flock. Confirmation of footrot is dependent on demonstrating the presence of *Bacteroides nodosus* in the lesion.

C. Ovine Interdigital Dermatitis

''Foot scald'' is a term that has been used to describe any interdigital dermatitis. Interdigital dermatitis caused primarily by *F. necrophorum* is now called ovine interdigital dermatitis[5] and interdigital dermatitis that is restricted to the interdigital integument, and caused by strains of *B. nodosus,* is now called benign or nonprogressive footrot.[11,12]

Ovine interdigital dermatitis is an acute necrotizing infection of the interdigital skin of sheep caused by *F. necrophorum* in association with *C. pyogenes.*[5] The disease may affect all ages of sheep and is most common in wet conditions and where high stocking rates are maintained. The condition can range in severity from a mild dermatitis to severe necrosis and erosion of the interdigital skin, thereby exposing the sensitive s.c. tissues to further infection.[3] (See Plate 6C and D.) Suppuration may also be present in severe cases. Under suitable environmental conditions, up to 90% of a flock may be affected, many sheep in all four feet.

Ovine interdigital dermatitis frequently precedes and predisposes to foot abscess[2,3,13] and both foot conditions may be present in a flock at the same time. Only during the transitional stage should clinical differentiation pose any difficulties; at other times, the confinement of ovine interdigital dermatitis to the interdigital space differentiates it from foot abscess.

D. Other Conditions

The following conditions should be mentioned under differential diagnosis because they cause lameness in sheep, but it is unlikely that they would be confused with foot abscess because they all have specific characteristic features.

1. Foot and Mouth Disease

This virus infection affects all cloven-hooved animals and characteristically produces vesicles on the coronet and in the interdigital space. Following rupture, the vesicles may be invaded by environmental organisms including *F. necrophorum.* Separation of the hooves starts first at the coronet, and the high morbidity and other systemic effects differentiate it from foot abscess.

2. Contagious Pustular Dermatitis

This virus disease is a proliferative dermatitis affecting mainly lambs. Scabs tend to build up and coalesce about the lips and coronet and sometimes in the interdigital space. Secondary infection of these lesions with *F. necrophorum* or *Dermatophilus congolensis* frequently occurs. The age group affected, the characteristic sites and proliferative nature of the lesion, and the high morbidity differentiate it from foot abscess.

3. Dermatophilus congolensis Infection

Infection of the coronet region with *D. congolensis* is seen occasionally during outbreaks of dermatophilosis. The infection appears in the form of multiple raised scabs between the coronet and the knee or hock. Characteristic filaments and coccoid forms of *D. congolensis* may be seen in stained smears made from the underside of scab material. *D. congolensis* is a frequent secondary invader of skin lesions such as contagious pustular dermatitis and may be further complicated by *F. necrophorum* infection.

4. Post-Dipping Lameness

Dipping sheep in dips contaminated with *Erysipelothrix insidiosa* organisms can lead to infection of skin wounds, with subsequent spread of infection to joints. There is no tendency for abscess formation and the high morbidity and history of recent dipping distinguishes it from foot abscess.

5. Laminitis

Laminitis is usually associated with grain feeding of sheep and the higher morbidity and lack of suppuration differentiates it from foot abscess.

IV. EPIDEMIOLOGY

A. Incidence and Economic Importance

The prevalence of foot abscess is low and seldom rises above 2%. Much of the importance of foot abscess undoubtedly arises because of the severe nature of the lesion and because ewes close to lambing and rams are more frequently affected. Production losses may therefore arise if sheep breeding programs are disrupted by rams developing foot abscess close to the period of mating or if ewes develop pregnancy toxemia as a consequence of the severe lameness associated with foot abscess.

Surveys conducted in New Zealand have shown that about three fourths of farms experience cases of foot abscess,[8,14] but the overall incidence is only 0.15%. Only 3% of respondents reported a prevalence greater than 1%. However on occasion, particularly following periods of heavy rain, severe outbreaks of foot abscess may occur which affect up to one third of a flock.[3] Similar outbreaks have been reported in Australia.[2,15]

B. Predisposing Factors

1. Environmental Factors

Predisposing factors exert a profound effect on the prevalence of foot abscess and of prime importance are wet and muddy underfoot conditions. It is only under such circumstances that outbreaks of foot abscess have been reported, usually in association with ovine interdigital dermatitis.[2,3,16]

Beveridge[10] showed that normal healthy skin was relatively resistant to infection and for *F. necrophorum* to invade the interdigital skin, it has been necessary to continuously expose sheeps' feet to water for 3 to 4 d[12] or to scarify the interdigital skin.[5] However, by itself, continuous moisture is not sufficient to reproduce foot abscess and keeping sheep in wet-underfoot conditions has not reliably reproduced the condition.[8]

2. Interdigital Skin Damage

It is apparent that some prior interdigital skin damage is necessary to facilitate the deeper bacterial invasion of the s.c. tissues and ultimately the distal interphalangeal joint. Experimentally, such tissue damage has been reproduced by applying liquid nitrogen to a small area of the interdigital skin for between 2 to 3 min.[8] Following this treatment, the sheep were housed in moist, dirty-underfoot conditions and foot abscess was consistently reproduced. While this may appear a rather extreme means of creating tissue damage, the lesions produced are similar to the severe interdigital necrosis and erosion that has been observed in naturally occurring cases of ovine interdigital dermatitis.[3] (Plate 6C and D.)

A number of causes of interdigital tissue damage have been suggested as predisposing to foot abscess. These include sheep walking on sharp stones, stubble, and thistles. It has even been suggested that friction between the digits may play an important role. A common finding during outbreaks of foot lameness and foot abscess is the accumulation of mud and straw into a plug which becomes hard when wedged between the claws. When removed, the underlying interdigital skin is inflamed, eroded, and necrotic and may predispose sheep to foot abscess.

Outbreaks of foot abscess have also been reported following the excessive use of formalin footbaths when treating footrot of sheep.[17-19] The suggested pathogenesis was that primary damage to the interdigital skin resulted from the misuse of formalin and that this permitted the bacterial invasion of the deeper tissues and the distal interphalangeal joint of the foot.

3. Animal Behavior

The attack rate of foot abscess is higher for rams than for ewes and a possible explanation for this has been advanced.[3] The behavior of rams, especially young rams, is different from that of ewes, and rams often congregate in groups and indulge in riding and sexual activity. As a result of these forays, the pasture is trampled and becomes muddy and heavily contaminated with feces. In addition, there is an increased risk of physical damage to the interdigital skin when mounting or being mounted by other sheep takes place.

V. PATHOGENESIS

It is generally accepted that most cases of foot abscess arise as a sequel to ovine interdigital dermatitis.[2,3] Both foot conditions often occur simultaneously in a flock and in several instances ovine interdigital dermatitis has been seen to precede foot abscess. In addition, the bacterial flora of the two conditions are similar.

Ovine interdigital dermatitis occurs commonly during wet underfoot conditions which facilitate the invasion of the interdigital skin by *F. necrophorum*.[5] *F. necrophorum* proliferates and produces exotoxins which destroy invading leucocytes and cause tissue necrosis. This allows other bacteria in addition to *F. necrophorum* to become established and multiply.[6] *C. pyogenes* produces a factor which stimulates the growth and invasiveness of *F. necrophorum*, thereby increasing tissue destruction.[7] Other bacteria may also contribute to tissue damage resulting from this mixed infection.[9]

A clearer understanding of the pathogenesis of foot abscess has come from studies on the anatomy of the digit and, in particular, the distal interphalangeal joint.[20] This joint is most vulnerable to infection or trauma on the axial or interdigital aspect where the joint capsule protrudes above the coronary border as the dorsal and volar pouches (Figures 3 and 4). At these two sites, the joint capsule is protected only by the interdigital skin layer and a small amount of s.c. tissue. Therefore, it is understandable that outbreaks of ovine interdigital dermatitis in which the interdigital skin becomes necrotic and eroded may result in cases of foot abscess. Indeed, any severe interdigital skin damage arising from such things as injury or excessive use of formalin footbaths increases the vulnerability of the foot and leads to the development of foot abscess.

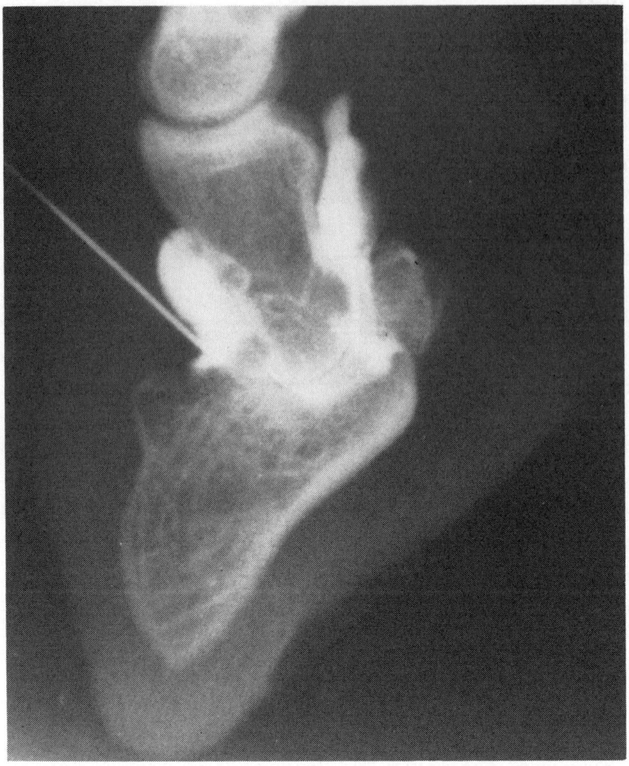

FIGURE 3. The digit of a Romney sheep radiographed immediately following the injection of contrast material into the dorsal pouch of the distal interphalangeal joint.

Occasionally, sporadic cases of foot abscess arise because bacteria gain entry to the distal interphalangeal joint from sources other than interdigital dermatitis. A common source is from infections in the lamellar region of the foot (toe abscess). In many instances, unintentional damage may be inflicted on the foot by the owner attempting to drain toe-abscess lesions and the trauma produced may be sufficient to allow infection to gain entry to the distal interphalangeal joint.[4]

VI. TREATMENT

Most recommendations for the treatment of foot abscess have consisted of drainage of abscesses, cleansing the affected area with disinfectant, and administering antibiotics. Because both the bacteria considered to be the main pathogens of foot abscess, namely *F. necrophorum* and *C. pyogenes,* show *in vitro* sensitivity to penicillin, either penicillin alone or in combination with streptomycin has been the main antibiotic used.

However, such treatment does not appear to markedly accelerate healing.[4,21] This is not surprising, given that the infection is centered on the distal interphalangeal joint. Purulent joint infections erode articular cartilage very quickly, resulting in permanent joint damage. In addition, the joint is contained within the hoof and is almost impossible to drain without creating further damage to joint structures such as ligaments. To significantly alter the outcome of infection, therapy would have to either prevent the infection becoming established in the joint, or prevent the infection involving the axial collateral ligaments with the possibility

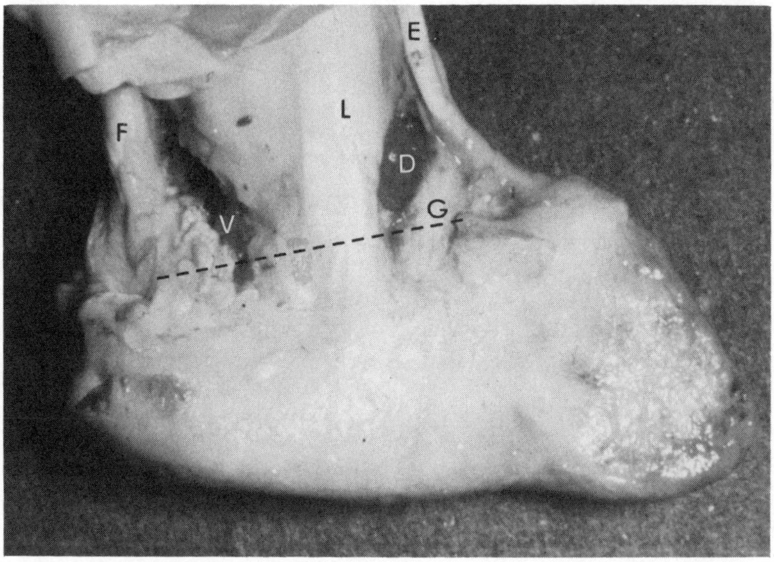

FIGURE 4. Dissection of the axial aspect of a sheep's digit following the injection of latex into the distal interphalangeal joint. The broken line indicates the approximate level of the axial coronary border. D — dorsal pouch of the distal interphalangeal joint, D — extensor tendon, F — flexor tendon, G — dorsal ligament, L — axial common collateral ligament, V — volar pouch of the distal interphalangeal joint.

of subsequent rupture. In other words, antibiotic therapy would have to be initiated before or very soon after the infection reached the joint.

Bandaging the foot to prevent splaying of the digits might be expected to reduce the chances of the axial collateral ligaments rupturing, by reducing pressure on the ligaments. Bandaging also reduces the risk of further bacterial contamination of the lesion from the environment.

Although sheep do not respond rapidly to therapy, the prognosis for functional recovery is good in most cases. The healing process is protracted but after a period of about 2 months most sheep can walk normally. The distal interphalangeal joint remains permanently damaged and upon careful examination digital deformity, and in some case lameness, may be detected.

VII. CONTROL

The sporadic occurrence of foot abscesses and the ubiquitous nature of the causative bacteria, make prevention of the disease difficult. Control has been based on minimizing the factors which predispose to outbreaks of the condition and in some instances the use of vaccination.

Outbreaks of foot abscess occur during wet conditions when property management is conducive to causing injury to the interdigital skin and allowing entry of the organisms. Such things as maintaining rams, and ewes in late pregnancy, on wet country or yarding sheep in wet muddy yards should be avoided. Where practical, temporary yards should be erected on a drier part of the farm. Care should be taken if sheep, especially late-pregnant ewes or rams, are grazed on crops during the wet winter months, as the muddy conditions may be ideal for the development of ovine interdigital dermatitis and foot abscess.

If, under such circumstances severe ovine interdigital dermatitis occurs in a flock, consideration should be given to instituting treatment before foot abscess develops. This may include careful footbathing of sheep in 10% zinc sulfate or 5% formalin solution and providing dry underfoot conditions. Providing dry conditions may be impossible in many situations but it may be appropriate to house the sheep for up to a week to allow the feet to dry.[3] In addition, selected use of antibiotics such as pencillin may limit the infection and prevent joint involvement.

Attempts to immunize animals against infection by *F. necrophorum* have been disappointing.[22] Although commercial *F. necrophorum* vaccines have been made and marketed there is little published record of their efficacy and farmers are unsure of, or have been dissatisfied with the protection given.[8]

Experimental use of these vaccines results in detectable levels of circulating serum antibodies but the protection afforded against infection is meager. An effective *F. necrophorum* vaccine would have wide application in the livestock industries and there remains considerable interest in developing a product such as this.[23]

ACKNOWLEDGMENT

I wish to acknowledge the permission of the *New Zealand Veterinary Journal* to reproduce some of the figures.

REFERENCES

1. **Gregory, T. S.,** Footrot in sheep, *Aust. Vet. J.,* 15, 160, 1939.
2. **Roberts, D. S., Graham, N. P. H., Egerton, J. R., and Parsonson, I. M.,** Infective bulbar necrosis (heel-abscess) of sheep, a mixed infection with *Fusiformis necrophorus* and *Corynebacterium pyogenes, J. Comp. Pathol.,* 78, 1, 1968.
3. **West, D. M.,** Observations on an outbreak of foot abscess in sheep, *N.Z. Vet. J.,* 31, 71, 1983.
4. **West, D. M.,** A study of naturally occurring cases of ovine foot abscess in New Zealand, *N.Z. Vet. J.,* 31, 152, 1983.
5. **Parsonson, I. M., Egerton, J. R., and Roberts, D. S.,** Ovine interdigital dermatitis, *J. Comp. Pathol.,* 77, 309, 1967.
6. **Roberts, D. S.,** The pathogenic synergy of *Fusiformis necrophorus* and *Corynebacterium pyogenes.* I. Influence of the leucocidal exotoxin of *F. necrophorus, Br. J. Exp. Pathol.,* 48, 665, 1967.
7. **Roberts, D. S.,** The pathogenic synergy of *Fusiformis necrophorus* and *Corynebacterium pyogenes.* II. The response of *F. necrophorus* to a filterable product of *C. pyogenes, Br. J. Exp. Pathol.,* 48, 674, 1967.
8. **West, D. M.,** A Study of Foot Abscess in Sheep, Ph.D. thesis, Massey University, New Zealand, 1981.
9. **Slee, K. J.,** A micro aerophilic coccus in pyogenic infections of ruminants, *Aust. Vet. J.,* 62, 57, 1985.
10. **Beveridge, W. I. B.,** Footrot in sheep: a transmissible disease due to infection with *Fusiformis nodosus* (n.sp.), *Bull Aust. Counc. Sci. Ind. Res. (Melbourne),* No. 140, 1941.
11. **Thomas, J. H.,** The differential diagnosis of footrot in sheep, *Aust. Vet. J.,* 38, 159, 1962.
12. **Egerton, J. R., Roberts, D. S., and Parsonson, I. M.,** The aetiology and pathogenesis of ovine footrot. 1. A histological study of the bacterial invasion, *J. Comp. Pathol.,* 79, 207, 1969.
13. **Egerton, J. R., Roberts, D. S., Graham, N. P. H., and Parsonson, I. M.,** Ovine interdigital dermatitis and its relationship to footrot and foot abscess, *Aust. Vet. J.,* 42, 440, 1966.
14. **Anon.,** Footrot in the N. Z. sheep industry — an analysis of survey results, New Zealand Meat and Wool Boards Economic Service Publ., Wellington, T79, 1983.
15. **Belschner, H. G.,** Footrot in sheep, *Aust. Vet. J.,* 15, 219, 1939.
16. **Murnane, D.,** Footrot in sheep, *J. Counc. Sci. Ind. Res.,* 6, 252, 1933.
17. **Pryor, W. J.,** Two foot syndromes in sheep following excessive foot bathing, *Aust. Vet. J.,* 35, 493, 1959.
18. **Littlejohn, A. I.,** The potential danger arising from the misuse of formalin in the treatment of footrot in sheep, *Vet. Rec.,* 90, 693, 1972.
19. **Hooper, R. S. and Jones, T. W.,** Corono-pedal abscessation following the excessive use of formalin as a treatment for footrot in sheep, *Vet. Rec.,* 90, 697, 1972.
20. **West, D. M.,** Anatomical considerations of the distal interphalangeal joint of sheep, *N.Z. Vet. J.,* 31, 58, 1983.

21. **Egerton, J. R. and Graham, N. P. H.,** Diseases causing lameness in sheep, in *Veterinary Review,* No. 5, University of Sydney, Post-Graduate Foundation in Veterinary Science, 1969.
22. **Simon, P. C. and Stovell, P. L.,** Diseases of animals associated with *Sphaerophorus necrophorus:* characteristics of the organism, *Vet. Bull. (London),* 39, 311, 1969.
23. **Emery, D. L., Vaughan, J. A., Clark, B. L., and Stewart, D. J.,** Virulence determinants of *Fusobacterium necrophorum* and their prophylactic potential in animals, in *Footrot in Ruminants,* Proc. Workshop, Melbounre, 1985, Stewart, D. J., Peterson, J. E., McKern, N. M., and Emery, D. L., Eds., CSIRO Division Animal Health/Australian Wool Corp., Glebe, N. S. W., 1986.

Chapter 4

FOOT ABSCESS OF CATTLE

B. L. Clark

TABLE OF CONTENTS

I. DEFINITION AND ETIOLOGY

Foot abscess is a name which could be legitimately applied to any suppurative deep-seated infection of the foot, but in cattle it is commonly used to describe the deep diffuse infection of the dermal and subdermal tissues of the hoof by *Fusobacterium necrophorum,* which may occasionally extend into the distal interphalangeal joint and which is usually associated with fever and lameness. There are many synonyms for the cattle disease such as footrot, fouls, foul-in-the-foot, panaritium, phlegmona interdigitalis, and interdigital necrobacillosis. The common name for the analogous disease of sheep is foot abscess (in order to distinguish it from ovine footrot). There is much merit for using the same descriptive term in both cattle and sheep, choosing either foot abscess or interdigital necrobacillosis.[1,2]

The primary etiological agent of foot abscess in cattle is considered to be *F. necrophorum.*[2,3] It is always present in large numbers in necrotic material from lesions. A second anaerobic microorganism, *Bacteriodes melaninogenicus* is also present in the lesions and Berg and Loan[4] have postulated that both microorganisms act together to cause the disease. They injected a mixture of pure cultures of *F. necrophorum* and *B. melaninogenicus* intradermally into the interdigital skin of 28 bovine feet and produced 15 swollen feet and ulceration of the interdigital skin in another 10 feet. They also injected another ten feet intradermally with a pure culture of only *F. necrophorum* and produced three swollen feet and ulceration of the interdigital skin in two others, but considered that *B. melaninogenicus* may also have contributed to these lesions because the latter organism was later recovered from them, as well as *F. necrophorum.* However, pure cultures of *B. melaninogenicus* alone given either by intradermal or s.c. injection caused only minor or no lesions in inoculated feet.[4,5] Clark et al.[5] found that isolates of *F. necrophorum* that had been recovered from lesions of foot abscess in cattle and sheep were able to produce typical lesions in calves when pure cultures were injected s.c. through the interdigital skin. The inclusion of *B. melaninogenicus* with *F. necrophorum* in the inoculum did not seem to augment or contribute to the development of lesions when given by either s.c. or intradermal injection and they concluded that *F. necrophorum* was the primary etiological agent of foot abscess in cattle.

The pathogenic isolates belong to either biotypes A or AB,[6] are characterized by flat, irregularly shaped, grayish colonies on solid media, grow as dense filamentous suspensions in liquid media, and produce a soluble leucocidin and a hemolysin.[7] Other isolates of *F. necrophorum,* belonging to biotype B, are commonly present on the interdigital skin of cattle and in their environment, produce convex yellow colonies on solid media, flocculate in liquid media, produce little or no leucocidin or hemolysin, and are relatively avirulent.[7] In the study of the etiology of the disease in cattle, it is important to distinguish between the pathogenic strains of biotypes A and AB and the relatively nonpathogenic strains of biotype B. The association between strains of biotype A and AB and lesions of foot abscess indicates that potential vaccines should be sought among these strains.

II. CLINICAL SIGNS

The first sign noticed is acute lameness, usually in one limb only. Lameness is usually accompanied by a moderate systemic reaction with a fever of 39 to 40°C.[8] The animal objects strongly to examination of the extremely painful affected foot. On inspection of the limb, spreading of the claws and swelling of tissues in the interdigital space and around the coronet are obvious. There is temporary depression of milk yield in cows, and affected bulls may show temporary infertility.[8] The animal is obviously in pain, walks as little as possible, prefers to lie down, grazes little, and loses body weight rapidly.[9] The disease is characterized by cellulitis that causes the swelling between the digits and around the coronet and that, in severe cases, extends around and above the fetlock. Liquefactive necrosis takes place within a few days and pus is then discharged, commonly through a fissure opening into the interdigital cleft, or through sinuses opening on the coronary margin.

Foot abscess has a typical external lesion in the interdigital space which generally commences as acute dermatitis in the posterior aspect, spreading to involve the whole space with outpouring of exudate from the skin. The skin commonly splits along the top of the interdigital space with formation of a deep fissure with swollen protruding edges which may extend the whole length of the space or be localized to either the anterior or posterior aspects.[8-10] Pus is never present in large quantities, but the edges of the fissure are covered with necrotic material and the lesion has a characteristic foul odor.[8-10] Occasionally, no external lesion develops and there is simply lameness with swelling of the coronet, and such cases have been termed "blind fouls".[1] Usually, recovery is rapid once the abscess has discharged, but commonly the disease process lasts 2 to 3 weeks from first signs to resolution.

A severe but infrequent complication is the extension of infection to involve the distal interphalangeal joint and associated ligaments of one of the two digits of the foot. The lateral digit is most commonly involved and usually a hind limb is affected.[1] The animal is febrile, very lame, carries the extremely painful foot, and loses weight rapidly. The whole foot becomes very swollen and painful and swellings extend up the limb. Medical treatment of this complication is rarely successful and surgical intervention may be necessary if the animal is to be returned to normal production.

III. DIFFERENTIAL DIAGNOSIS

The clinical diagnosis of foot abscess in cattle is based on lameness due to lesions in the hoof, usually of the one limb, the presence of swellings in the interdigital space and around the coronet, the development of typical fissured lesions in the interdigital space that discharge pus, and the generally acute and transitory nature of the disease. The isolation of *F. necrophorum* organisms belonging to either biotypes A or AB from typical lesions support such a clinical diagnosis.

Extensive surveys of lameness in cattle have shown that most cases are due to lesions of the feet rather than to those of the legs or the trunk. Russell et al.[11] conducted a survey of dairy cattle in 48 veterinary practices in Great Britain in 1977 and ascertained the cause of lameness in 7526 cases in 1821 herds containing 136,800 cattle, and established that 88.3% of the cattle were lame due to lesions in the feet. The survey found that the most common lesions of the hooves causing lameness were foot abscess 16.7%, white-line abscess 15.6%, sole ulcer 13.6%, punctured sole 10.4%, underrun heel 8.7%, aseptic laminitis 5.3%, interdigital dermatitis 4.8%, white-line separation 4.7%, foreign body in the sole 3.5%, deep sepsis 3.5%, interdigital foreign body 2.2%, overworn sole 2.2%, and sandcrack 1.3%. The differential diagnosis of foot abscess requires that all lesions be correctly identified and this requires that the animal be observed standing and in motion, and that the lame feet be thoroughly examined and trimmed.[12] Brief descriptions of the lesions that need to be differentiated from foot abscess are given below. While most are those described by Russell et al.,[11] it should be noted that the relative incidences of these conditions in other parts of the world could be affected by climate, systems of animal management, nutrition, and breed of cattle.

A. White-Line Abscess

White line abscess is caused by penetration of the white-line area on the abaxial border of the distal surface of the claw by sharp grit or slivers of stone which in turn initiate a septic laminitis of the abaxial wall, which travels upwards, eventually producing abscessation at the coronary band.[13] The lateral digit of the hind limb is generally the one involved and frequently both hind feet are affected simultaneously. It occurs most commonly in older cows.[14,15] Malformations of the foot due to overgrowth, chronic laminitis, and chronic interdigital dermatitis are predisposing factors. Diagnosis depends on careful paring of the sole. In severe or neglected cases, sepsis can extend to the navicular bursa, the tendon sheaths, and ultimately to the distal interphalangeal joint.[1]

B. Sole Ulcer

This lesion develops in the axial groove on the axial border of the sole, halfway between the heel and the toe, where the horn is naturally thin, and is associated with excessive wear of the sole on hard surfaces. It most commonly occurs in the outer hind claws and in abnormally shaped and overgrown feet. Chronic interdigital dermatitis appears to be a predisposing factor.[16] It is most common in older cows.[14,15] An open ulcer through the sole develops, becomes infected, and leads to either a localized or a widespread septic pododermatitis.

C. Punctured Sole

A traumatic penetration of the sole with infection of laminae and pus production, it appears to be most common in the hind feet.[11]

D. Underrun Heel

This condition involves erosion and separation of the soft horn on the bulbs of the heel commonly associated with abnormal stance of the foot due to overgrowth of the toe. There may be some infection of sensitive laminae. It is most common in the hind feet and is often associated with interdigital dermatitis.

E. Aseptic Laminitis

This is a diffuse aseptic inflammation of the sensitive laminae of the feet which may be chronic, subacute, or acute. There is lameness, usually involving more than one foot. The chronic form, with elongation of the toes, concentric transverse rings around the horn, and chronic lameness is the most usual. It is less common in pasture-fed animals. In housed cattle, it is frequently associated with nutritional disorders consequent upon feeding high levels of concentrates to newly calved cows.

F. Footrot

Interdigital dermatitis is caused by *B. nodosus* and is characterized by erosions and hyperkeratosis of the epidermis of the interdigital skin.[17,18] The lesion is essentially a diffuse necrosis of the epidermis of the interdigital skin, but will occasionally affect the sensitive laminae of the horn of the heels and cause some separation. The skin becomes extensively hyperkeratotic and fissured and exudes serous fluid, but there is little swelling of the foot.[2] It occurs under wet conditions, and during the winter and spring months in temperate areas, between 40% and 60% of the feet of cattle on pasture are likely to be affected with interdigital dermatitis.[19] It is a very prevalent infectious condition which has a pronounced tendency to recur in the same feet, producing chronic lesions and malformations. Although lameness is not a prominent feature, severe lameness can be seen in 2- to 4 week-old calves and in older, heavier cows. In the former, there is extensive ulceration of the interdigital skin, and in the latter, the lesion that seems to produce the most severe lameness is underrunning of the heels; the weight of the affected animal is possibly an aggravating factor. Interdigital dermatitis due to *B. nodosus* is considered to be a predisposing cause of foot abscess in cattle,[2,19] to sole ulcer, underrun heels and misshapen feet,[16] and to interdigital fibroma.[20]

G. White-Line Separation

White-line separation is due to breakdown of the integument of the white line, usually at the abaxial border of the lateral digit of the hind feet and lameness is due to impaction of foreign material into the crack.

H. Sandcrack

A vertical crack in the horn of the wall which may involve sensitive tissue and cause lameness is called sandcrack.

I. Interdigital Fibroma

This is a large outgrowth or proliferation of the anterior interdigital skin which is seen particularly in older cows and in Hereford cattle.[20] It is considered to be a tissue response to chronic interdigital dermatitis. The skin on the fibroma often becomes inflamed and ulcerated and produces chronic mild lameness. It appears to be a predisposing factor to foot abscess in cattle.

J. Verrucose Dermatitis

This is a proliferative inflammatory condition of the skin of the palmar or plantar surfaces of the feet extending from the heels to the fetlock joint. The condition is seen particularly in feedlot cattle which are crowded together in wet, muddy conditions and can occur in outbreaks.[8] Although the condition is found more commonly on hind feet, all four feet can be affected. There is considerable pain and lameness and *F. necrophorum* is present in large numbers.[8]

K. Viral Diseases

Several major viral diseases cause lesions of the feet and produce lameness. Foot and mouth disease and vesicular stomatitis cause vesicles on the skin of the feet, and blue tongue produces coronitis. Ephemeral fever and mucosal disease may also be responsible for lameness.

L. Other Conditions

Lameness can be due to ''fescue foot'' caused by grazing tall fescue grass. In this condition, there is peripheral dry necrosis of the hind feet which causes lameness similar to foot abscess. Sprains of the fetlock with swellings down to the coronet can be confused with foot abscess.[8,9] Other miscellaneous causes of lameness include overworn sole, foreign body in the sole and interdigital skin, and deep sepsis.

IV. EPIDEMIOLOGY

Foot abscess occurs following entry of pathogenic *F. necrophorum* organisms belonging to biotypes A or AB through lesions of the interdigital skin. They do not appear to be able to invade the intact normal skin. Injury is therefore a predisposing factor; this may be trivial such as a small abrasion between the claws derived from contact with stones, hardened mud, or sticks, or severe, following penetration by a nail or wire, or from pressure due to rubber washers becoming attached to claws.[9] Such abrasions or injuries are more likely to occur when the skin is soft and swollen due to continual wetting. Greenough[1] has drawn attention to the following factors that have been postulated as important for the development of foot abscess in cattle.

1. The effect of water, wet dung, and mud for softening and macerating the interdigital epidermis
2. The effect of dried or frozen mud, stones, and stubble in abrading the interdigital tissue, and possibly lowering its resistance to infection

3. The effect of thick mud adhering to the coronet in facilitating local proliferation of anaerobic pathogens such as *F. necrophorum*
4. The effect of other diseases of the interdigital skin for predisposing to entry of *F. necrophorum*

Thus, stony ground, lanes, and gateways filled with sharp stones or gravel, pasturing on stubble, and wet, muddy conditions, are all commonly considered to be associated with the occurrence of foot abscess in cattle. Several authors have observed that some cases of foot abscess in cattle are preceded by interdigital dermatitis,[2,19,21] and an analogous situation pertains in sheep.[22,23]

Despite the usual sporadic occurrence of cases of foot abscess in herds, the contagious nature of the disease is suggested by the occurrence of outbreaks which often affect 20 to 25% of groups of cattle within short periods of time[8,9] and by the introduction of infection to farms by transient cattle. The fact that foot abscess in cattle tends to be a continuing problem on some farms and yet rarely, if ever, occurs on others is also consistent with the contagious nature of the disease, but indicates that there are as yet unknown factors that limit transmission.

While the disease can affect cattle of all ages, prevalence is highest in lactating dairy cows and in cattle in feedlots. It is not so prevalent in calves, young cattle, nonlactating cows, and bulls on infected dairy farms.[9] It is not as common in beef as it is in dairy cattle. However, in feedlots the prevalence can be very high. The main explanation for the higher prevalence in feedlot cattle and in lactating dairy cattle, compared to others, is that in the two former categories, there is regular movement of larger numbers of animals in close formation through yards, gateways, and around troughs, usually with wet and muddy underfoot conditions. In feedlots, cattle frequently live continuously under such conditions where opportunities for foot abrasion and injury and for exposure to infection, are optimal. Discharges from diseased feet would have maximum opportunity to contaminate the interdigital skin of other cattle under such conditions.

In dairy cows, foot abscess is most prevalent during early lactation.[9,14,15] Eddy and Scott[14] found that 41% of all recorded cases occurred during the first 2 months after calving, and Rowlands et al.[15] in a larger survey, also in Great Britain, found 31% of all cases in the same period. The prevalence of lameness in dairy cows appears generally to follow the yearly calving pattern.[24] The greater prevalence of all lameness, including that due to foot abscess, during early lactation suggests that the foot tissues may be softer, or more susceptible to infection at that time, which in turn may reflect the stress of maximum lactation following calving.[14]

The effect of season on the prevalence of foot abscess differs between years and areas, and is probably related to differences in rainfall patterns. Thus, while some authors report the prevalence to be highest in winter, others find it to be highest in autumn,[14,15] and yet others in spring and early summer.[9] In making observations on the effect of season, other factors such as the effect of the stage of lactation, should be taken into account in order to avoid confusion.[15] Furthermore, observations on rainfall are of reduced value if much of the water in the cowyards and laneways originates from taps and hoses. Nevertheless, there is general, if not unanimous, agreement in the veterinary literature, that the prevalence is increased by higher rainfall and that the disease is more common in wetter seasons.

Differences in the type of management systems do not seem to be associated with any significant effect in the prevalence of foot abscess in dairy cattle. Thus, the prevalence of foot abscess in Great Britain was similar in dairy cattle kept on pastures, in cowsheds, in strawyards, or in cubicles with concrete yards.[25] However, the prevalence of conditions such as sole ulcer and white-line abscess was affected by the type of management, being generally higher in cattle kept on hard surfaces. The standard of management, as distinct from the type of management system, is of more importance in the epidemiology of bovine foot abscess. Failure to attend to the repair of damaged surfaces in laneways, yards, and gateways, to provide concrete pathways for cattle to walk to and from milking parlors, to eliminate obvious danger spots where cattle have to congregate, and failure to immediately treat clinical cases of foot abscess when seen are all likely to facilitate the occurrence of foot abscess in the herd.

The prevalence in cattle of different breeds appears to be similar. *Bos indicus* (Brahman and Sahiwal) cattle have been reported to be less susceptible to foot abscess than *Bos taurus* (Australian Illawarra Shorthorn and Charolais) cattle.[26]

Dairy cows of all ages appear to be equally susceptible to foot abscess, with only a slight increase being reported with increased age.[15] Despite this, foot abscess was found to be the most important cause of lameness in first- and second-lactation cows, reflecting the relatively low prevalence of lameness due to sole ulcer, white-line disease, punctured sole, underrun heel, and interdigital dermatitis in younger compared to older cows.[15] Foot abscess is a disease which tends to occur in normally shaped feet rather than in abnormally shaped or overgrown hooves, whereas some of the other conditions such as sole ulcer and underrun heel are most prevalent in misshapen feet.[11]

Most reports indicate that foot abscess is more prevalent in the hind than in the front feet,[11,14] although others have found them to be equally affected.[9]

Most surveys of lameness in dairy cows have used veterinary practice records which did not include cases which were treated by farmers, and estimates of the national prevalence of lameness in cattle based on such records probably would be far too low to be accurate.[14,24,27] However, it is likely that the relative importance of the various causes of lameness in cattle are fairly presented. Such records indicate that foot abscess caused 14.3% of all lameness in dairy cows in Somerset in 1974, 1975, and 1976,[14] and 16.7% of all lameness in dairy cows in the British Isles in 1977.[11] In individual practices, the relative proportion may be much higher and thus Greenough[1] reported that 40% of all lameness seen in cattle in his practice was due to foot abscess. To estimate the prevalence of foot abscess in dairy cattle regionally or nationally requires knowledge of the annual prevalence of lameness, which may be as high as 25%.[24,27] Data collected between April 1981 to March 1982 from 185 herds in England and Wales by the University of Edinburgh, Dalgety Spillers Dairy Herd Health and Productivity Service showed that 25% of the 21,000 dairy cows were actually treated for lameness, 6.3% by veterinary surgeons and 18.7% by farmers.[24] On individual farms, the incidence of lameness varied from 2% of cows in some herds to 55% in others. Thus, the actual prevalence of foot abscess in British dairy cattle in the period 1977 to 1981 may have been between 3% and 4% annually. The disease appears to be as prevalent in dairy cattle in the tropics as in the temperate zones.[28] The disease appears to occur in cattle in all countries.

Loss of production occurs in affected cattle from effects on milk production,[24] bodyweight, and fertility.[29] While these losses can be minimized by prompt treatment, under British conditions it has been estimated that each case of lameness caused a loss of 18 pounds sterling (in 1983) to the owner and this did not include the cost of treatment.[24]

V. PATHOGENESIS AND PATHOLOGY

The pathogenesis of bovine foot abscess is not well understood, although the primary cause is known to be *F. necrophorum* organisms belonging to biotypes A and AB.[6] However, the presence of these organisms in the environment or on the interdigital skin does not automatically cause foot abscess because they do not appear to be able to invade the normal intact skin. Furthermore, injection of *F. necrophorum* into the superficial layers of the interdigital skin did not produce foot abscess, although deep ulcers were produced.[5] Foot abscess has been produced by injecting *F. necrophorum* cultures, or *F. necrophorum* plus *B. melaninogenicus* cultures intradermally into the interdigital skin of cattle but it is not clear if the injection was given superficially or deep into the dermis.[4] The disease is readily produced if cells of an 18-h culture of *F. necrophorum* are injected through the interdigital skin,[5] but only when cell numbers are in excess of 10^8 cells.[30] Preliminary colonization of devitalized areas of the interdigital skin by pathogenic *F. necrophorum* organisms, in order to engender huge localized populations, may be a necessary prerequisite to invasion of the subdermal tissues in the natural disease. It is well known that the occurrence of foot abscess in cattle is greatly influenced by trauma and maceration of the skin of the feet through prolonged exposure to water. These predisposing factors favor the development of large populations of *F. necrophorum, B. melaninogenicus, B. nodosus,* and other anaerobic microorganisms in skin lesions.[4] The bacteriology of this early phase of invasion is obscure, but synergism between *F. necrophorum,* and other organisms such as *B. melaninogenicus,*[4] *B. nodosus,*[10] or microaerophilic cocci,[31] within skin lesions may be essential for *F. necrophorum* to penetrate the skin in the natural disease. Another interdigital disease of cattle, namely interdigital dermatitis (footrot) which is caused by *B. nodosus,* and which is considered to be a predisposing factor in bovine foot abscess,[2,19,21] is characterized by severe erosions and hyperkeratosis of the interdigital skin. *B. nodosus* possesses a protease which is involved in the productions of the skin lesions. *F. necrophorum* possesses an exotoxin or leucocidin, which is destructive for leucocytes and for erythrocytes and which is essential for the production of foot abscess in cattle after the invasion through the skin is completed.[7] Unlike *B. nodosus,* it does not appear to produce large quantities of a protease, and synergism with *B. nodosus* would appear to offer advantages to *F. necrophorum* during the phase of skin penetration.[32]

When *F. necrophorum* penetrates the interdigital skin in sufficient numbers to establish infection in deeper tissues, a diffuse infection of the dermal and subdermal tissues of the foot results within 12 h to 3 d. Both its endotoxin[33] and leucocidin[34] are implicated in the pathogenesis of foot abscess. There is no fluid within the swollen tissue initially, but following the destruction of leucocytes by the leucocidin of *F. necrophorum,* affected tissue liquifies and discharges as sticky, creamy pus, generally through the fissure opening into the interdigital space. The leucocidin is a protein with a molecular weight of around 250,000 to 300,000 Da, and is inactivated by heating to 56°C for 30 min.[34]

If there is no involvement of the distal interphalangeal joint, resolution of the infection tends to be rapid once pus is discharged. Involvement of the joint is said to be more common where considerable sloughing of tissue has taken place in the anterior region of the interdigital space close to the coronet, because at this point the joint cavity is very close to the surface and relatively unprotected. It is nearly always a lateral digit that is involved. Due to the deep location of the joint, drainage is often hindered, and extension of infection to tendons may take place. Sinus formation tends to occur on the abaxial surface of the claw at the coronet.

The pathology of the disease is briefly described below. The tissue in the interdigital space is characterized by coaggulative necrosis of the skin and connective tissue, which extends from the anterior to the posterior cleft and to depths of up to 2 cm.[35] Surface cells are often desquamated and in chronic cases replaced by granulation tissue. Proximal to the necrotic tissue, the connective tissue is edematous and hyperaemic. Histopathology reveals coagulative necrosis with abundance of cells of *F. necrophorum.* Few leucocytes are present in the necrotic tissue.[35]

VI. TREATMENT AND CONTROL

The lame foot should first be thoroughly washed with water, brushed clean, and examined for lesions and foreign bodies. Examination may only involve simple restraint of the animal and lifting of the foot by hand, but if necessary the veterinarian should be prepared to sling the affected foot to an overhead beam or to cast the animal. All necrotic tissue in the interdigital space should be scraped or curretted away in order to facilitate drainage.[27] Trimming of the foot should be regarded as necessary to aid effective differential diagnosis.[12]

In most cases of foot abscess, further local treatment is not generally necessary, unless there is extensive tissue damage in the anterior region of the interdigital cleft. In this case, the lesion should be cleansed and dressed, the area protected by a bandage or pad, and the animal stabled in a dry stall in order to reduce the risk of extension of infection into the distal interphalangeal joint.[1]

Parenteral treatment with antibiotics should always be carried out as early as possible in the course of the disease and is necessary for best results. Immediate treatment as soon as possible after the onset of swelling and lameness will give excellent recovery in 2 to 4 d in the majority of cases.[8] If treatment is delayed for several days after the onset of signs, severe lesions tend to occur and the whole period of recovery will be extended. Sodium sulphadimidine (150 to 250 mg/kg) solution given by i.v. or s.c. injection, or procaine penicillin 10,000 units/kg given by i.m. injection are usually highly effective. It is generally recommended that the parenteral treament be repeated on three successive days, but often one treatment is effective.[9]

In severe refractory cases where there is involvement of the distal interphalangeal joint, consideration should be given to amputation of the affected digit.[1] This provides rapid relief from pain and permits return to normal milk production. However, owners appear to regard such animals as unsound and they are often culled at the end of lactation.[1] The use of toe supports and shoes has been advocated as alternatives to amputation of the digit.[12,36] The support is fitted or stuck onto the sound digit, and affords the animal immediate relief from pain by removing the weight bearing from the affected digit. The principle employed is that of giving protracted rest in order to allow serious lesions to heal and for ankylosis to take place.

The control of bovine foot abscess by prevention of injuries to the interdigital skin of cattle and by reduction of wet, muddy conditions is dependent on sound management. In most cases, this requires the provision of concrete pathways and concrete aprons around water and feed troughs, smooth surfaced laneways, gateways, and yards, and the elimination of sharp metal objects, nails, wire, glass, and broken stone and concrete from areas where cows walk. The control of interdigital dermatitis by footbathing in 5% formalin may be of some benefit in the control of foot abscess. The footbath is installed in a doorway so that milking cows have to walk through it twice daily and is reported to reduce the incidence of interdigital dermatitis.[37] Ethylenediamine dihydriodide (EDDI) has been used as a feed additive to prevent foot abscess in feedlot cattle.[38,39] Experimental results suggest that cattle receiving EDDI have significantly less severe lesions and lameness than control cattle.

Commercial vaccines against bovine foot abscess are available but their efficacy has not been established in controlled comparative trials. A marked reduction in the prevalence of foot abscess was found in dairy herds in which a commercial vaccine was used.[9] The prevalence of liver abscess in grain-fed cattle, also caused by strains of *F. necrophorum* of biotypes A and AB was apparently reduced by immunization. In this case, an alum-precipitated vaccine prepared from the cytoplasmic fraction of *F. necrophorum* caused a reduction in the prevalence of liver abscess from 35% in control to 10% in vaccinated cattle.[40] The mechanisms involved in immunity against *F. necrophorum* infection in cattle have not been elucidated, although studies in rabbits suggested that acquired resistance to infection could be antibody mediated.[41]

In a series of experiments, calves have been immunized with whole cells or various fractions of *F. necrophorum* strain FnB1 incorporated in mineral-oil adjuvant and then each foot experimentally exposed to foot abscess.[30] The results of these experiments showed that immunity was associated with vaccines containing the supernatant fluid of 18 h-old liquid cultures. The degree of protection afforded by a vaccine containing concentrated, cell-free, culture supernatant was estimated to be 60% when the proportion of feet in vaccinated calves that were severely swollen was compared with that in control calves.[30] In comparison, vaccines containing either washed whole cells or the cytoplasmic fraction of the cells gave limited protection. The culture supernatant contained the membrane and secreted antigens such as the exotoxins, capsular antigens, lipopolysaccharides, and outer-membrane antigens. While the precise nature of the antigens concerned with immunity has still to be determined, it is possible that they are identical or closely associated with the exotoxins.[30,41]

The encouraging results obtained in these experiments indicate that continuing research to identify and purify the protective antigens of *F. necrophorum* and to evaluate them as immunogens in commercial vaccines is warranted.

Finally, the effect of immunization of a group of 32 cattle with an autogenous vaccine containing two benign strains of *B. nodosus* on the prevalence of interdigital dermatitis, and indirectly, on foot abscess is mentioned. It was found that immunization caused a significant reduction in the severity of interdigital dermatitis in the feet of the vaccinated cattle during the following 4 months and that during this period, the prevalence of foot abscess was somewhat less in vaccinated (4/32) compared to control (8/32) cattle.[42] However, a much larger trial to investigate the effect of vaccination against *B. nodosus* on the prevalence of foot abscess is required before any conclusions can be drawn.

REFERENCES

1. **Greenough, P. R.,** Observations on some of the diseases of the bovine foot. Part I, *Vet. Rec.,* 74, 1, 1962.
2. **Beveridge, W. I. B.,** *Bacterial Diseases of Cattle, Sheep and Goats, Animal Health in Australia,* Vol. 4, Australian Government Publishing Service, Canberra, 1983, 108.
3. **Flint, J. C. and Jensen, R.,** Pathology of necrobacillosis of the bovine foot, *Am. J. Vet. Res.,* 12, 5, 1951.
4. **Berg, J. N. and Loan, R. W.,** *Fusobacterium necrophorum* and *Bacteroides melaninogenicus* as etiologic agents of foot rot in cattle, *Am. J. Vet. Res.,* 36, 1115, 1975.
5. **Clark, B. L., Stewart, D. J., and Emery, D. L.,** The role of *Fusobacterium necrophorum* and *Bacteroides melaninogenicus* in the aetiology of interdigital necrobacillosis in cattle, *Aust. Vet. J.,* 62, 47, 1985.
6. **Fievez, L.,** *Etude comparee des souches de Sphaerophorus necrophorus isolees chez l'Homme et chez l'Animal,* Presses Academiques Europeennes, Bruxelles, 1963.
7. **Emery, D. L., Vaughan, J. A., Clark, B. L., Dufty, J. H., and Stewart, D. J.,** Cultural characteristics and virulence of strains of *Fusobacterium necrophorum* isolated from the feet of cattle and sheep, *Aust. Vet. J.,* 62, 43, 1985.
8. **Blood, D. C., Radostits, O. M., and Henderson, J. A.,** *Veterinary Medicine,* 6th ed., Bailliere Tindall, London, 1983, 662.
9. **Gilder, R. P.,** Foot diseases of cattle, *Aust. Vet. J.,* 36, 151, 1960.
10. **Gupta, R. B., Fincher, M. G., and Bruner, D. W.,** A study of the etiology of foot rot in cattle, *Cornell Vet.,* 54, 66, 1964.
11. **Russell, A. M., Rowlands, G. J., Shaw, S. R., and Weaver, A. D.,** Survey of lameness in British dairy cattle, *Vet. Rec.,* 111, 155, 1982.
12. **Allenstein, L. C.,** Lamenesses of cattle, *Can. Vet. J.,* 22, 65, 1981.
13. **Edwards, G. B.,** White line disease of the foot in cattle, *Vet. Ann.,* 20, 227, 1980.
14. **Eddy, R. G. and Scott, C. P.,** Some observations on the incidence of lameness in dairy cattle in Somerset, *Vet. Rec.,* 106, 140, 1980.
15. **Rowlands, G. J., Russell, A. M., and Williams, L. A.,** Effects of stage of lactation, month, age, origin and heart girth on lameness in dairy cattle, *Vet. Rec.,* 117, 576, 1985.
16. **Raven, E. T. and Cornelisse, J. L.,** The specific contagious inflammation of the interdigital skin in cattle, *Vet. Med. Rev.,* 2/3, 223, 1971.

17. **Egerton, J. R. and Parsonson, I. M.,** Isolation of *Fusiformis nodosus* from cattle, *Aust. Vet. J.,* 42, 425, 1966.
18. **Thorley, C. M., Calder, H. A. McC., and Harrison, W. J.,** Recognition in Great Britain of *Bacteroides nodosus* in foot lesions of cattle, *Vet. Rec.,* 100, 387, 1977.
19. **Laing, E. A. and Egerton, J. R.,** The occurrence, prevalence and transmission of *Bacteroides nodosus* infection in cattle, *Res. Vet. Sci.,* 24, 300, 1978.
20. **Egerton, J. R.,** Diseases of cattle causing lameness, in Cattle Diseases, Refresher Course for Veterinarians, Sydney University Post-Graduate Committee in Veterinary Science, No. 42, 202, 1979.
21. **Clark, B. L., Stewart, D. J., and Emery, D. L.,** A study of role of *Bacteroides nodosus, Fusobacterium necrophorum,* and *Bacteroides melaninogenicus,* Summaries of the 22nd World Vet. Congr., Perth, 1983, 145.
22. **Roberts, D. S., Graham, N. P. H., Egerton, J. R., and Parsonson, I. M.,** Infective bulbar necrosis (heel-abscess) of sheep, a mixed infection with *Fusiformis necrophorous* and *Corynebacterium pyogenes, J. Comp. Pathol.,* 78, 1, 1968.
23. **West, D. M.,** Observations on an outbreak of foot abscess in sheep, *N.Z. Vet. J.,* 31, 71, 1983.
24. **Whitaker, D. A., Kelly, J. M., and Smith, E. J.,** Incidence of lameness in dairy cows, *Vet. Rec.,* 113, 60, 1983.
25. **Rowlands, G. J., Russell, A. M., and Williams, L. A.,** Effects of season, herd size, management system and veterinary practice on the lameness incidence in dairy cattle, *Vet. Rec.,* 113, 441, 1983.
26. **Frisch, J. E.,** The comparative incidence of footrot in *Bos taurus* and *Bos indicus* cattle, *Aust. Vet. J.,* 52, 228, 1976.
27. **Arkins, S.,** Lameness in dairy cows, *Ir. Vet. J.,* 35, 135, 1981.
28. **Soewardi, B.,** Dairy development in Indonesia, *Extension Bull. Food Fertilizer Technol. Centre Taiwan,* No. 237, 13, 1986.
29. **Lucey, S., Rowlands, G. J., and Russell, A. M.,** The association between lameness and fertility in dairy cows, *Vet. Rec.,* 118, 628, 1986.
30. **Clark, B. L., Emery, D. L., Stewart, D. J., Dufty, J. H., and Anderson, D. A.,** Studies into immunisation of cattle against interdigital necrobacillosis, *Aust. Vet. J.,* 63, 107, 1986.
31. **Slee, K. J.,** A microaerophilic coccus in pyogenic infections of ruminants, *Aust. Vet. J.,* 62, 57, 1985.
32. **Roberts, D. S. and Egerton, J. R.,** The aetiology and pathogenesis of ovine foot-rot. II. The pathogenic association of *Fusiformis nodosus* and *F. necrophorum, J. Comp. Pathol.,* 79, 217, 1969.
33. **Warner, J. F., Fales, W. H., Sutherland, R. C., and Teresa, G. W.,** Endotoxin from *Fusobacterium necrophorum* of bovine hepatic abscess origin, *Am. J. Vet. Res.,* 36, 1015, 1975.
34. **Emery, D. L., Dufty, J. H., and Clark, B. L.,** Biochemical and functional properties of a leucocidin produced by several strains of *Fusobacterium necrophorum, Aust. Vet. J.,* 61, 382, 1984.
35. **Flint, J. C. and Jensen, R.,** Pathology of necrobacillosis of the bovine foot, *Am. J. Vet. Res.,* 12, 5, 1951.
36. **Afford, H.,** Treatment of bovine foot conditions, *N. Z. Vet. J.,* 30, 162, 1982.
37. **Davies, R. C.,** Effects of regular formalin footbaths on the incidence of foot lameness in dairy cattle, *Vet. Rec.,* 111, 394, 1982.
38. **Berg, J. N., Brown, L. N., Ennis, P. G., and Self, H. L.,** Experimentally induced foot rot in feedlot cattle fed rations containing organic iodide (ethylenediamine dihydriodide) and urea, *Am. J. Vet. Res.,* 37, 509, 1976.
39. **Berg, J. N., Maas, J. P., Paterson, J. A., Krause, G. F., and Davis, L. E.,** Efficacy of ethylenediamine dihydriodide as an agent to prevent experimentally induced bovine foot rot, *Am. J. Vet. Res.,* 45, 1073, 1984.
40. **Garcia, M. M., Dorward, W. J., Alexander, D. C., Magwood, S. E., and McKay, K. A.,** Results of a preliminary trial with *Sphaerophorus necrophorus* toxoids to control liver abscesses in feedlot cattle, *Can. J. Comp. Med.,* 38, 222, 1974.
41. **Roberts, D. S.,** Toxic, allergenic and immunogenic factors of *Fusiformis necrophorus, J. Comp. Pathol.,* 80, 247, 1970.
42. **Clark, B. L., Stewart, D. J., Emery, D. L., Dufty, J. H., and Jarrett, R. G.,** Immunisation of cattle against interdigital dermatitis (foot-rot) with an autogenous *Bacteroides nodosus* vaccine, *Aust. Vet. J.,* 63, 61, 1986.

Section II: Bacteriology of the Major Pathogens

Section II

BACTERIOLOGY OF THE MAJOR PATHOGENS

J. R. Egerton, W. K., Yong, and G. G. Riffkin

Two species of Gram-negative, nonsporing, nonmotile, anaerobic bacteria, *Fusobacterium necrophorum* and *Bacteroides nodosus* have long been associated with diseases of the feet of sheep and other ruminants. The identification of these species and the research which has led to establishment of their etiological roles in a number of different diseases has been a major challenge for veterinary microbiologists since *F. necrophorum* was identified in footrot in sheep in 1904. Almost 40 years elapsed before *B. nodosus* was isolated from cases of ovine footrot.

A number of factors have increased the difficulty of establishing the role of these two organisms in infections of the feet of ruminants. Isolation, identification, and taxonomy of the principal bacterial were problematical with both organisms. *F. necrophorum,* generally known as the organism of necrobacillosis, has had a total of 33 synonyms ascribed to it in successive editions of *Bergey's Manual of Determinative Bacteriology.* Simon and Harris* believe that the multiplicity of names for this bacterium are due, in part, to uncertainty of investigators concerning its taxonomic position, its pleomorphism, and the difficulty experienced by many in its isolation and maintenance in culture. It is, furthermore, a pathogen of many species and many apparently different disease processes. Experimental reproduction of naturally occurring diseases has been complicated further by the occurrence of biotypes which vary in virulence.** Nevertheless, the identity of the necrosis bacillus as a member of the genus *Fusobacterium,* organisms which produce butyric acid as an endproduct of peptone or carbohydrate metabolism, is well established.

B. nodosus is another Gram-negative anaerobe which has been particularly difficult to isolate and characterize. In contrast to *F. necrophorum,* the host range of *B. nodosus* is limited to the ruminants. Also, whereas *F. necrophorum* is associated with disease in a variety of organs, infections with *B. nodosus* have only been described in the epidermal tissues of the foot. It is more demanding in its growth requirements, and is always accompanied in lesions by a multiplicity of other species of bacteria.

The taxonomic place of *B. nodosus* is far less certain than that of the necrosis bacillus. While there would be general agreement that both organisms belong in the family Bacteroidaceae and that the necrosis bacillus has the characteristics of a fusobacterium, there is less certainty about the generic status of *B. nodosus.* In fact, a recent review of species currently placed in the genus *Bacteroides* concluded that the footrot organism was not a *Bacteroides, sensu stricto.*

The chapters which follow are comprehensive reviews of research on the isolation and characterization of both *B. nodosus* and *F. necrophorum.* The chapter on *B. nodosus* is the first full description of the bacteriology of that organism since Beveridge*** published his monograph in 1941. Much progress has been made and Dr. Skerman has been a major contributor to that progress. In the interval between Beveridge's pioneering work and this publication, a number of obstacles to study of *B. nodosus* were overcome. Media which provided, empirically, the requirements of *B. nodosus* were devised by Thomas. Subse-

* Simon, P. C. and Harris, M., in *Handbuch der bakteriellen Infektionen bei Tieren,* Band 3, Blobel, H. and Schliessec, T., Eds., Gustav Fischer, Stuttgart, 1981, 639.

** Fierez, L., *Etude comparee des souches de Sphaerophorus necrophorus isolees chez l'Homme et chez l'Animal,* Presses Academiques Europeennes, Bruxelles, 1963.

***Beveridge, W. I. B., *Bull. Counc. Sci. Ind. Res. (Melbourne),* No. 140, 1941.

quently, it was found by Thorley that addition of 4% agar to isolation plates made recovery of *B. nodosus* from lesion material very much easier. Better media and more isolates in turn made possible more detailed and comparative studies on antigenic structure of the footrot organism and factors which determine its virulence. Dr. Skerman was a major contributor to this research and to the work on the isolation of *B. nodosus* (Chapter 5).

Much has been published previously on the bacteriology of the necrosis bacillus. The material in Chapter 6 is, however, a review by research workers who have investigated the role of *F. necrophorum* in foot diseases of cattle and sheep. Much of the other work published on *F. necrophorum* isolation and characterization has been concerned with quite different disease processes, like bovine liver abscess or calf diphtheria.

Dr. Clark and his colleagues (Chapter 6) have therefore given special emphasis to overcoming the difficulties of isolating *F. necrophorum* from lesion material which, in the case of the foot diseases, is contaminated with other organisms.

Chapter 5

ISOLATION AND IDENTIFICATION OF *BACTEROIDES NODOSUS*

T. M. Skerman

TABLE OF CONTENTS

I. INTRODUCTION

A. Classification

Bacteroides nodosus is a Gram-negative, nonsporeforming, nonflagellated, strictly anaerobic bacterium classified in the family Bacteroidaceae.[1] In the first complete description of the organism, Beveridge[2] named it *Fusiformis nodosus,* the new combination, *Bacteroides nodosus* being subsequently applied by Mraz et al.[3]

B. nodosus is grouped among nonsaccharolytic species of the genus *Bacteroides* and shows a limited range of activity in standard biochemical tests. The DNA-base composition of one strain examined was 45 mol% of guanine and cytosine,[4] thus lying within the range 28% to 61% quoted for other species of *Bacteroides.*[1] A culture of *B. nodosus* deposited in the American Type Culture Collection as accession no. 25549 has been designated the neotype strain.[5]

B. Occurrence

B. nodosus is the essential transmitting agent of footrot in sheep[2] and goats[6,7] and of certain infectious interdigital conditions in cattle.[8] It has also been isolated from a foot lesion in a farmed deer.[9] It does not normally occur in healthy animals and appears to be capable of proliferation only within living epidermal tissues of the hoof in host species. Other Gram-negative fusiform and spirochaete-like bacteria are generally associated with *B. nodosus* in foot lesions and are considered to have contributory roles in the infective process.[2,10,11]

B. nodosus populations may remain associated with chronic footrot lesions for prolonged periods. In sheep formerly affected with footrot, the bacteria may continue to survive in small pockets within the horn of feet in the absence of overt clinical disease. These foci of infection may then act as reservoirs for the carryover of viable organisms until environmental conditions become suitable for the reactivation of hoof lesions. In the external environment of soil or pasture, bacteria deposited from affected feet can survive for brief periods, probably no longer than 3 to 4 d under the most favorable conditions of temperature and humidity,[12,13] thus allowing infection to be transmitted within flocks or herds.

C. Preliminary Identification

Provisional diagnosis is based on the characteristic cellular morphology of *B. nodosus* observed by microscopic examination of necrotic material taken from active hoof lesions. Smears prepared on glass slides are air-dried, fixed by gentle heat, and stained by Gram's method. The organisms may be recognized as strongly stained straight or slightly curved large rods, commonly having pronounced swollen ends (Figure 1). The terminal swellings may contain more deeply stained material.

Invariably, *B. nodosus* are outnumbered in stained smears by other components of the footrot microbiota and by environmental contaminants; consequently, it may be necessary to search several microscopic fields to detect them. Filamentous Gram-negative bacteria are usually conspicuously abundant in material from advanced footrot lesions.

Occasionally, single *B. nodosus* cells that are detached from detritus may be seen to be surrounded by smaller Gram-negative bacteria arranged in a characteristic radial pattern first noted by Beveridge[2] and sometimes referred to as "the Beveridge phenomenon".[14]

A fluorescent-antibody stain developed by Roberts and Walker[15] was applied to the detection of *B. nodosus* in smears from necrotized tissue in hoof lesions in sheep and cattle.[16] Although it appeared that the preparation contained antibody to cell-surface antigens present throughout the species, it did not discriminate between serogroups or between different biotypes that occur in natural infections. Because *B. nodosus* is so readily recognized by microscopic examination of Gram-stained specimens, the fluorescent-antibody stain has not been widely used.

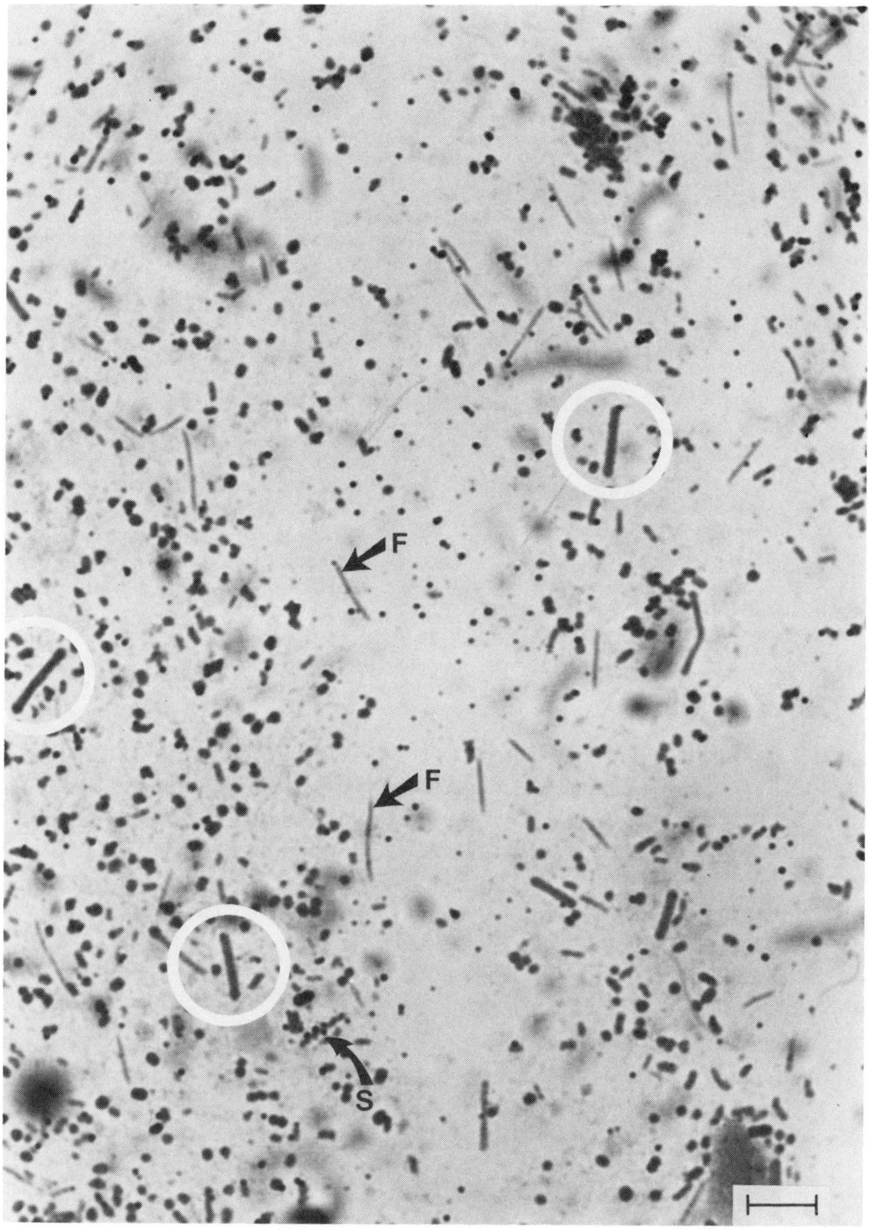

FIGURE 1. Gram-stained smear of lesion material from ovine footrot. Large rod-shaped cells of *B. nodosus* with characteristic terminal swellings are located within the circles. Associated Gram-negative organisms (arrows) include fusiform (F) and spirochaete-like bacteria (S). Scale bar equals 5 μm.

II. SPECIAL CONSIDERATIONS

A. Isolation

1. Field-Sampling Procedures

As hoof lesions are usually grossly contaminated with environmental detritus, *B. nodosus* must be isolated from dense populations of extraneous bacteria as well as from other components of the footrot microflora. Using a wooden probe, cotton swab, or point of a scalpel

Table 1
COMPOSITION OF AGAR ISOLATION AND MAINTENANCE MEDIA FOR *B. NODOSUS*

Medium	Basic composition (g/l)[a]	Supplements and use	Ref.
Hoof agar	Powdered ovine hoof, 20; meat extract,[b] 5; agar,[c] 20 or 40; pH = 6.8	The higher (4%) agar concentration is preferred for the isolation medium; 2% agar concentration allows development of agar-corroding characteristics of colonies	29, 32
Eugonagar®[d]	Eugonagar®, 45.4; pH = 7.0 (product contains Trypticase®,[d] 15; phytone-peptone,[d] 5; sodium chloride, 4; sodium sulfite, 0.2; L-cystine, 0.7; dextrose, 5.5; agar, 15)	Yeast extract[c] (0.2%) added plus	
		10% defibrinated equine blood, or	37
		10% defibrinated bovine blood	31
TAS agar[e]	Trypticase®,[d] 15; meat extract,[b] 5; proteose-peptose,[c] 5; yeast extract,[c] 2; L-arginine HCl, 5; DL-serine, 1.5; agar,[c] 40; pH = 7.8	Used for both isolation and maintenance, inclusion of 10% defibrinated equine or bovine blood optional	18

[a] Inoculated media incubated anaerobically under 90% H_2 + 10% CO_2.
[b] Lab-lemco, Oxoid Limited, Basingstoke, Hants, England, or beef extract, Difco® Laboratories, Detroit, Michigan, U.S.
[c] Difco® Laboratories, Detroit, Michigan, U.S.
[d] BBL Division of Becton Dickinson and Co., Cockeysville, Maryland, U.S.
[e] Modified from original formulation due to different source of agar.

blade, necrotic material should be carefully collected from the depth of the hoof lesion, preferably at the point of separation of the horn. Active lesions with a moist appearance rather than desiccated hoof tissue should be selected for sampling. Interdigital skin lesions can be sampled by swabbing or gentle scraping with a scalpel blade.

Samples should be promptly transferred to a vial of freshly prepared (deoxygenated) transport medium for processing in the laboratory, or inoculated directly on to plates of an isolation medium (see below).

2. *Transport of Specimens*

Transport media preferred by different investigators include standard (Difco®) Stuart's transport medium,[8,14] semisolid Stuart's transport medium,[4] and cysteine — HCl/phosphate buffered saline/agar.[17] Commercial anaerobic specimen collectors (Vacutainer Systems®, Becton Dickinson and Co., Rutherford, N.J., U.S.) may also be used.

3. *Isolation Media and Techniques*

Compositions of commonly used isolation media are shown in Table 1. A higher than normal concentration of agar is used (e.g., up to 4 to 5% w/v Difco® agar) which enhances development of *B. nodosus* colonies, as well as suppressing the spreading growth of contaminants.[17,18] As agars from different commercial suppliers vary in their gel strength, it is important that they are identified. Freshly poured plates, containing agar medium to a depth of 6 to 8 mm, should be well dried and stored under hydrogen in anaerobic jars. Plates of hoof agar (Table 1) may be stored aerobically in plastic bags in the refrigerator.[19]

For inoculation, portions of necrotic material in transport medium are streaked heavily on to plates in parallel lines 8 to 10 mm apart. The plates are then incubated under 90% H_2 + 10% CO_2 at 37°C for up to 5 d. Plates of isolation medium may also be inoculated directly from samples taken in the field and placed in airtight jars containing catalysts and a commercial gas-generating sachet (Gas-pak, BBL Microbiology Systems, Beckton Dickinson and Co., Maryland, U.S.). The sealed jars are then returned to the laboratory for

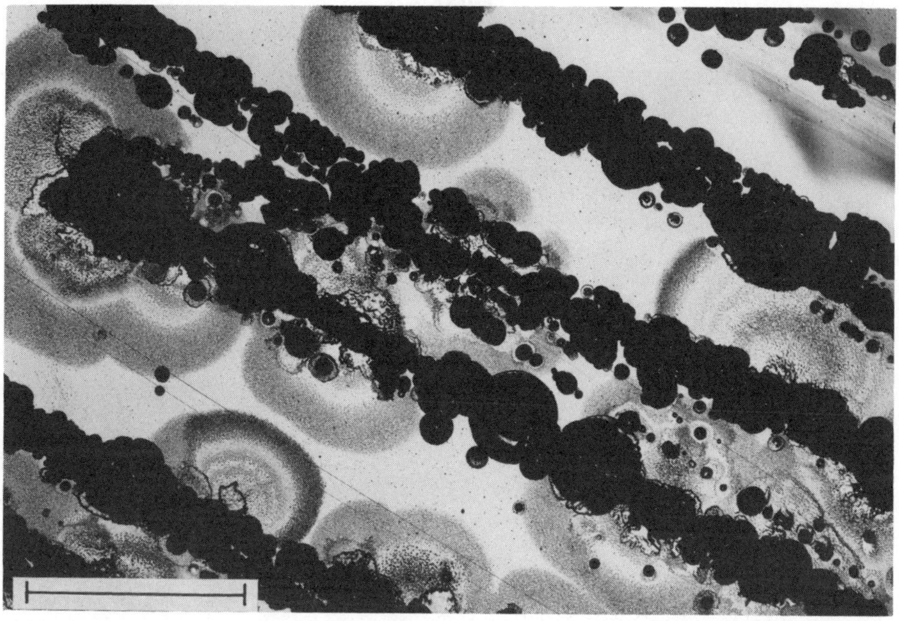

FIGURE 2. Colonies of *B. nodosus* growing on primary isolation plates inoculated with material from an ovine footrot lesion. Note the flat, spreading growth with a fine granular-fimbriate edge surrounding a clearer mid-zone. Scale bar equals 5 mm. (From Thorley, C. M., *J. Appl. Bacteriol.*, 40, 1976. With permission.)

incubation. In spite of the dense, mixed bacterial growth that results from the heavy inoculum, *B. nodosus* colonies or sectors thereof can usually be identified by their flat, spreading appearance and characteristic fimbriate outer areas with a ground-glass texture (Figure 2). Similar growth, sometimes as a narrow confluent band, may also be located along the outer margins of the streaks of mixed contaminants.

Examination of plates is carried out most satisfactorily under diffuse angled reflected light, using either a hand-held magnifying lens, or a stereomicroscope at a magnification of × 15. Growth from the peripheral areas of suspected *B. nodosus* colonies is picked off using either a flamed loop or the tip of a wire bent into a right angle and then streaked on plates of agar maintenance medium for purification.

B. Routine Cultivation

1. Agar Media

The composition of solidified media for the purification and maintenance of isolates is shown in Table 1. Some formulations specify the addition of defibrinated equine or bovine blood to enhance certain features of *B. nodosus* colonies. Conventional anaerobic jars containing an oxygen-scavenging catalyst are satisfactory for incubating inoculated plates at 37°C in an atmosphere of 90% H_2 + 10% CO_2.

Although *B. nodosus* is incapable of multiplication in an oxygenated environment, the organisms will tolerate limited exposure to air and fully grown cultures on plates may be left on the laboratory bench for up to about 12 h without significant viability being lost. This allows a good deal of latitude for subculturing cloned isolates without the necessity for strict anaerobic procedures. Although *B. nodosus* colonies may be macroscopically visible on plates after 48 h, agar cultures are usually incubated for 4 to 5 d. Such cultures will remain viable for 2 to 3 weeks if the jars are not opened and are stored in the dark at room temperature.

Table 2
COMPOSITION OF LIQUID MEDIA FOR *B. NODOSUS*

Medium	Basic composition (g/l)[a]	Supplements and modifications	Ref.
Eugonbroth®[b]	Eugonbroth®[c], 18; yeast extract,[d] 5; L-arginine HCl, 2.5; lysine, 2.5; final pH = 7.5.	L-cysteine HCl (added after autoclaving), 0.5; plus	
		Gelatin hydrolysate[e], 50	34
		Trypsin 1:250,[d] 10	45
TAS broth	Ingredients as for TAS agar, Table 1, minus agar plus $MgSO_4 \cdot 7H_2O$, 2; thioglycollic acid (95% w/v), 0.63 ml; 20% Na_2CO_3 solution (added after autoclaving), 10 ml; final pH = 6.8.	Nil, basic composition only	18
		$CaCl_2$, 1.5; thioglycollic acid replaced by L-cysteine $HCl \cdot H_2O$, 0.5;	27
		Na_2CO_3 buffer replaced by HEPES, Na salt, 13.05 g	50, 54
		Thioglycollic acid and Na_2CO_3 omitted for heavily inoculated medium	

[a] All media incubated in the presence of oxygen-free CO_2 after inoculation.
[b] BBL Division of Becton, Dickinson and Co., Cockeysville, Maryland, U.S.
[c] Composition as in Eugonagar®, Table 1, minus agar.
[d] Difco® Laboratories, Detroit, Michigan, U.S.
[e] Gibco Diagnostic, Madison, Wisconsin, U.S.

2. Liquid Media

Details of liquid media specifically designed or adapted for *B. nodosus* are shown in Table 2. All formulations include reducing agents and, after sterilization, the media must be protected from atmospheric oxygen. Many culture failures can be attributed to media having become oxidized or stored for too long. Anaerobic techniques based on those described by Holdeman and Moore[20] are generally applicable for dispensing and sterilizing the media, for aseptically adding heat-labile ingredients, and for inoculation.

A comparatively simple system followed in the author's laboratory utilizes as culture vessels, glass bottles of capacity ranging from 100 to 2000 ml and which are half filled with medium and closed either with soft-rubber bungs (American Sterilizer Co., PA, U.S.) or perforated screw caps fitted with rubber liners. The caps are fitted loosely on the bottles for autoclaving, after which they are immediately tightened. After the media have cooled, heat-labile constituents, CO_2, and, subsequently, inoculum are introduced with syringes via 22-gauge hypodermic needles. For subculturing, required amounts of liquid cultures are withdrawn similarly from the bottles with sterile syringes.

As it may be difficult to pick off with a loop sufficient growth from the characteristically flat *B. nodosus* colonies in order to inoculate liquid media directly, broth cultures can be initiated promptly as follows. Using a flamed stainless-steel spatula, a 1 cm × 1 cm square of agar containing one or more discrete *B. nodosus* colonies is cut out and aseptically transferred to a Universal bottle containing 1 to 2 ml of sterile water. The bottle is capped and shaken on a vortex mixer for ~10 sec. The dislodged organisms are then withdrawn in suspension with a syringe for seeding 50 ml volumes of medium in culture bottles. Providing the inoculum is taken from young agar cultures incubated no longer than 3 d, growth in these primary broth cultures is usually attained within 24 to 48 h at 37°C. When liquid cultures are transferred to fresh medium, inocula of 0.02 volumes result in optimum growth in up to 72 h, depending on the strain of *B. nodosus* concerned. Growth of *B. nodosus* in broth cultures is evenly dispersed, with a light, silky texture.

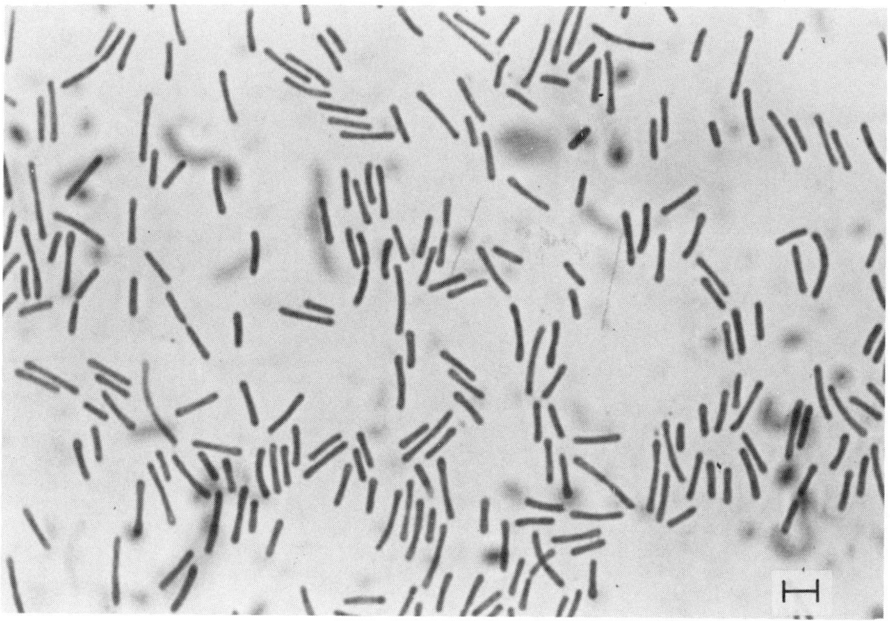

FIGURE 3. *B. nodosus* from a culture grown on TAS agar for 3 d and stained by Gram's method. Scale bar equals 5 μm. Note the predominance of slightly curved rods with swollen ends.

C. Conservation of Strains

Conventional freeze-drying techniques are satisfactory for long-term storage of cultures using a suspending medium of 7% glucose + 7% peptone. Cultures required more frequently may also be stored by injecting about 2 ml aliquots of a fully grown broth culture with a sterile syringe fitted with a 26-gauge needle into sterile 100 × 16 mm Vacutainer® tubes (No. 6430, additive-free, Becton Dickinson and Co., Rutherford, N. J., U.S.) which are then stored at −70°C. Subcultures are revived by thawing the contents of the tubes and withdrawing 0.2 ml volumes of the bacterial suspensions through the caps with a syringe and plating directly onto agar. Cultures stored under these conditions have withstood repeated thawing and refreezing for several months, while those left in the frozen state have remained viable for up to 5 years.

III. GENERAL PROPERTIES

Characteristics shown to be common to all strains of *B. nodosus* examined are described under the following headings.

A. Cellular Morphology

Cells are straight or slightly curved rods, 1 to 1.7 × 3 to 6 μm (Figure 3). They often have terminal swellings, although this feature is less pronounced in organisms subcultured in fluid media.

Bacteria from young agar cultures stained by Albert's method[21] or with Loeffler's methylene blue commonly show large polychromatic granules towards the poles and at intermediate sites within the organisms (Figure 4).[4,22] Cells do not bear flagella but may have large numbers of pili,[23] which have been reported to vary according to changes in colony morphology and with the virulence of the organisms.[24-26] A twitching pattern of motility was observed by Depiazzi and Richards.[27] The feature was observed in cells which adhered

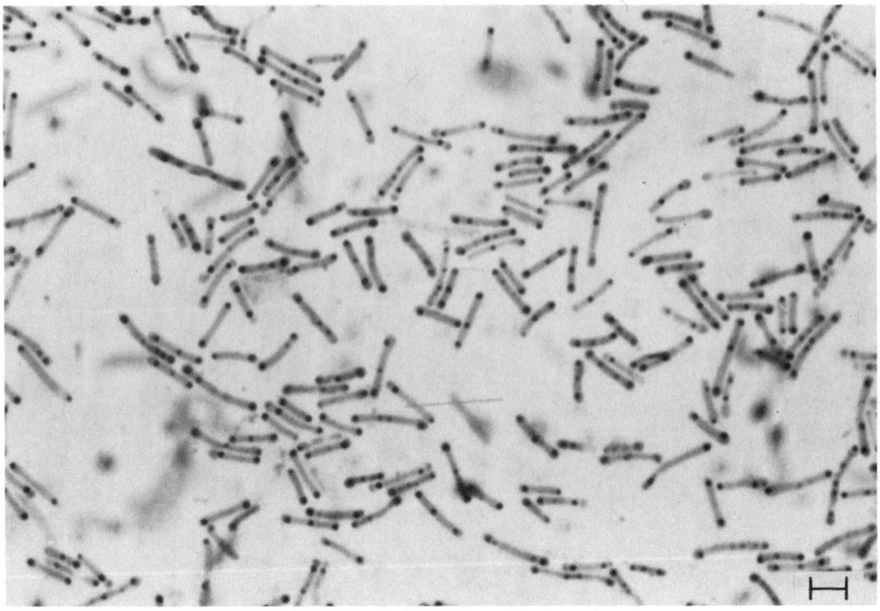

FIGURE 4. *B. nodosus* from a 3-d agar culture and stained with Loeffler's methylene blue, showing the presence of dark-staining intracellular granules located predominantly at the poles. Scale bar equals 5 μm.

to the glass coverslip in hanging-drop preparations from liquid cultures. The level of motility in liquid medium did not appear to be related to the number of pili borne by the cells.

Reports regarding the presence of a capsule are conflicting.[22,26,28] The chemical composition of the cell wall is unknown. Ultrastructural details of *B. nodosus* are described in Chapter 7.

B. Growth Requirements and Nutrition

The formulation of earlier isolation media with complex ingredients such as horse serum, meat-liver digests, or autolysed trypsin suggested that nutritionally, *B. nodosus* might be a highly fastidious organism. Improved anaerobic techniques, and the development of more reproducible media without natural enrichments have largely overcome the difficulties experienced by early investigators in isolating and propagating strains. However, the inclusion in agar media of finely ground ovine-hoof powder, originally introduced as a ''habitat-stimulating'' component by Thomas,[29] is still regarded of value for obtaining primary isolates of *B. nodosus* and for accentuating certain characteristics of their colonial growth.

Relatively little is known of the specific nutritional requirements of *B. nodosus*. Skerman[18] showed that in a liquid, anaerobically prepared, peptone yeast-extract basal medium, only four additional materials were necessary to obtain growth of cultures from small inocula. These were, CO_2, a tryptic digest of casein, a reducing agent, and L-arginine. Growth was further enhanced by serine, and, in the case of some strains, by lysine. In the complete medium (i.e., TAS medium, Table 2), no further growth stimulation occurred when it was supplemented with carbohydrates, fatty acids, lipids, Tween 80, hemin, hemoglobin, vitamin K, equine or bovine serum, rumen fluid, ox bile, or extracts and digests of liver, trypsin, and hoof. Relatively large amounts of arginine were required for optimum growth. Subsequent studies[30] indicated that arginine catabolism by *B. nodosus* ultimately results in ATP production, hence this amino acid may represent a significant energy source for the organism during cellular multiplication.

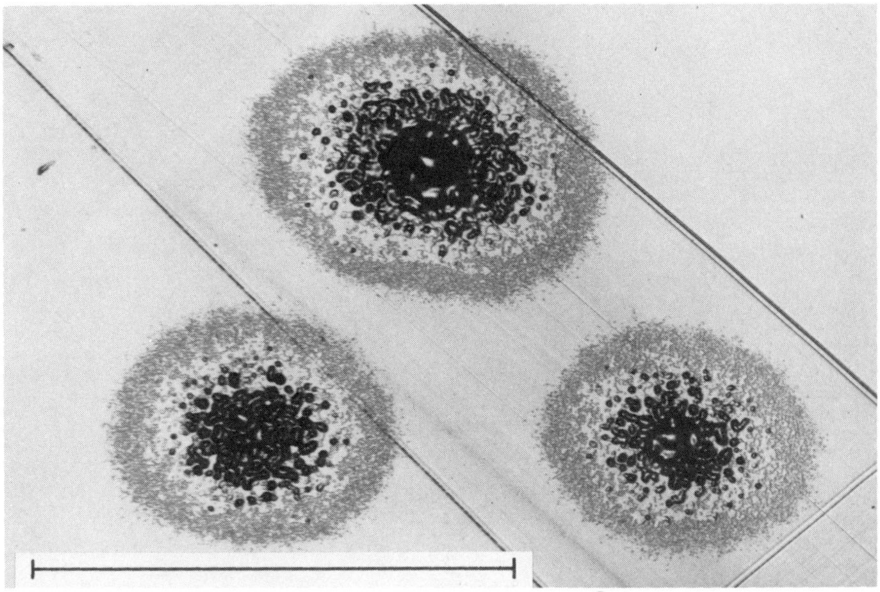

FIGURE 5. Colonies of *B. nodosus* with a central papillate zone, a clear mid-zone, and a wide, outer fimbriate zone. Scale bar equals 5 mm. (From Thorley, C. M., *J. Appl. Bacteriol.*, 40, 1976. With permission.)

Growth of all strains is most rapid at 37°C and within a pH range of 6.8 to 7.8. Multiplication does not occur below 20°C, and the organisms are killed by heating at 50°C for 5 min. Yields of bacteria in liquid culture media may range from 3 to 6 $\times$ 10^8 organisms per ml, depending on the strain concerned and the amount of inoculum.

C. Colony Morphology

Considerable diversity in the appearance of surface colonies produced by strains originating from different clinical conditions leads to some difficulty in categorizing forms that can be regarded as typical. In addition, colony morphology is markedly influenced by extrinsic cultural factors such as the medium composition, agar concentration, and depth of agar medium within plates. Furthermore, minor variations in the form of colony produced by the same strain can arise on plates prepared from different batches of medium, on plates of the same medium incubated in different anaerobic jars, and even between successive subcultures.

The appearance of colonies commonly developed on a 5% (Difco®) agar medium by isolates of *B. nodosus* from cases of virulent ovine footrot is shown in Figures 5 and 6. The main features of this colony form are a central zone occupied by small isolated papillae or beads, a pale, granular mid-zone, and a peripheral zone of varying width with a ground-glass texture and fimbriate edge. A fourth feature, not evident in Figure 5, is a small, pit-like depression etched into the agar surface directly underlying the central zone. The latter characteristic tends to be more readily observed in cultures grown on hoof medium containing 4% (Difco) agar (Figure 7).

Variations in these four basic features of colonies formed by different strains have been described and illustrated by Stewart et al.[31,32] and are summarized as follows:

1. A central zone occupied by a prominent dome that is either reticulate and slightly rough, or smooth (mucoid)
2. A mid-zone that is dense and granular, narrow and smooth, or absent

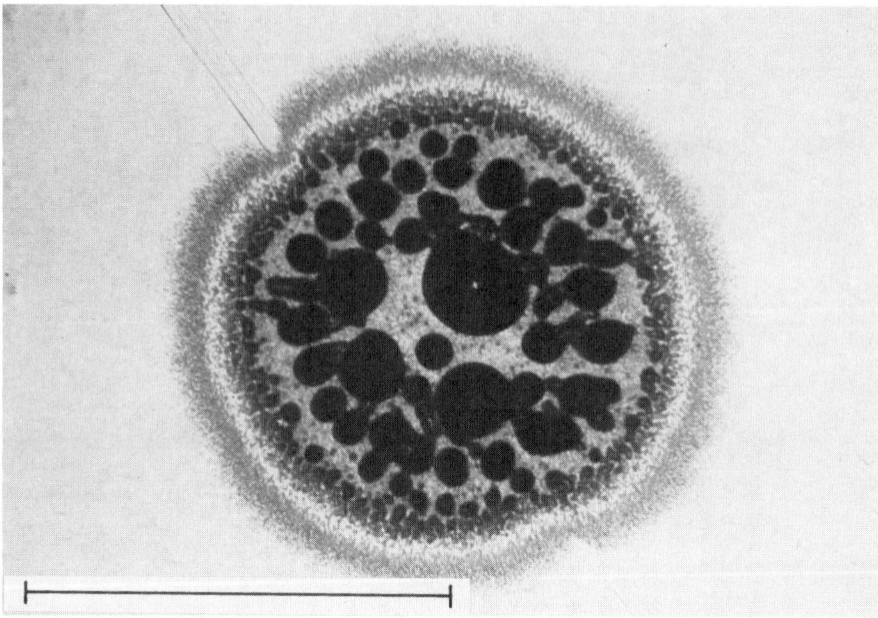

FIGURE 6. Colony formed by an isolate of *B. nodosus* from ovine virulent footrot featuring a more extensive central papillate zone than shown by colonies in Figure 5, but a more narrow fimbriate margin. Scale bar equals 5 mm. (From Skerman, T. M. et al., *Infect. Immun.*, 32, 1981. With permission.)

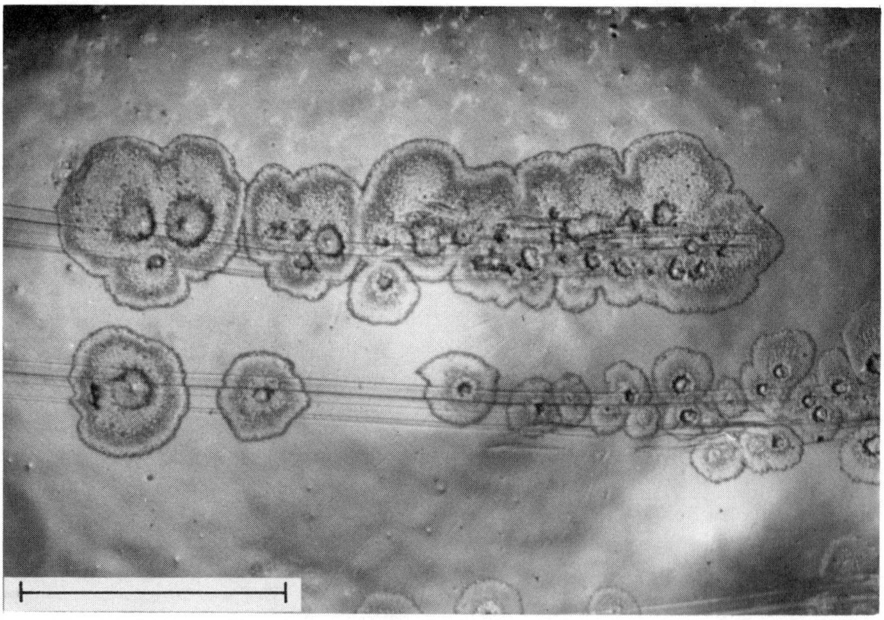

FIGURE 7. Colonies of *B. nodosus* grown on hoof agar, showing the corrosive pitting of the agar surface at their centers. (From Skerman, T. M., *J. Gen. Microbiol.*, 87, 1975. With permission.)

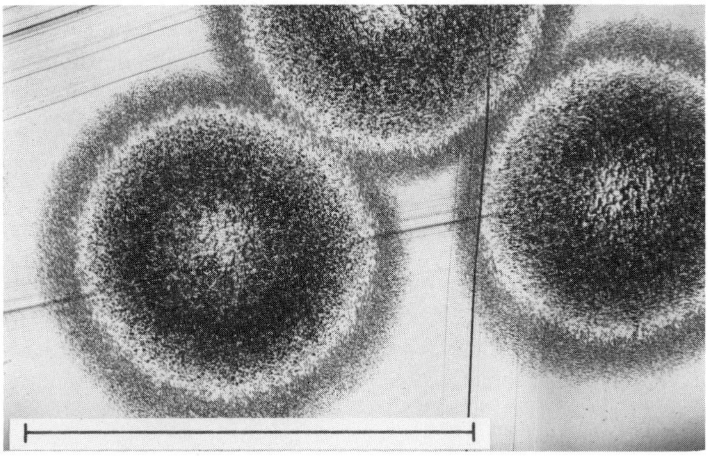

FIGURE 8. Colonies of a bovine isolate of *B. nodosus*. The central zone consists of a raised dome of coherent growth with a slightly rough surface. Scale bar equals 5 mm.

3. A peripheral finely granular zone that is either of intermediate width, narrow, or absent
4. A more shallow pitting and corrosion of the agar underlying the colony, or no agar corrosion

Figure 8 illustrates the type of colony formed by a bovine isolate of *B. nodosus* in which the entire area of the central zone is occupied by a raised dome with a coherent, slightly rough surface.

Changes in the type of colony formed by ovine *B. nodosus* isolates were noted after prolonged subculturing of strains in liquid media.[17,25] These changes were accompanied by certain alterations in the ultrastructure of the organisms and a loss of their virulence for sheep. However, the biochemical properties and extracellular enzyme activities of the wild-type form generally were retained. As far as is known, no strains resembling these culture variants occur in natural populations of *B. nodosus*.

The differentiation of strains on the basis of colony morphology is discussed in Section IV.A.

D. Biochemical Properties

From the results of several studies, the following characteristics appear to be common to all strains.[4,14,25,33-35] Procedures for carrying out the tests indicated were described by Skerman et al.[25]

Positive reactions — Ammonia production from arginine, asparagine, serine and threonine; ornithine decarboxylase; phosphatase (weak); H_2S production; selenite reduction; proteolytic activity on gelatin, casein and albumen.

Negative reactions — Acid or gas production from carbohydrates; ammonia production from phenylalamine, cystine, citrulline, ornithine; starch hydrolysis; aesculin hydrolysis; indole production; nitrate reduction; growth in 0.1% bile; hemolysis; arginine decarboxylase; catalase; oxidase; urease; DNase; coagulase; lipase; lecithinase; hyaluronidase.

E. Sensitivity to Antimicrobial Agents

B. nodosus is sensitive to a wide range of antimicrobial agents *in vitro*.[36,37] Gradin and

Schmitz[38] examined 18 strains and showed that penicillin was the most effective antibiotic tested. Others, in order of relative efficacy, were cefamandole, clindamycin, tetracycline, chloramphenicol, erythromycin, sodium cefoxitin, tylosin tartrate, nitrofurazone, tinidazole, and dihydrostreptomycin sulfate. Of the chemicals commonly used in footbaths for the treatment of footrot, copper sulfate was most effective, followed by zinc sulfate, then formalin.

F. Confirmatory Identification

The lack of activity by *B. nodosus* in many biochemical tests for differentiating nonhemolytic, nonsaccharoclastic, Gram-negative anaerobic bacteria limits the suitability of standard protocols for confirming the identity of organisms isolated from hoof lesions. The following supplementary tests are proposed for isolates that have been provisionally identified as *B. nodosus* on the basis of their characteristic cell morphology, staining reactions, and cultural features as described previously.

1. Gelatin Hydrolysis

Inoculate TAS agar or Eugonagar® (Table 1) containing 1% (Difco®) gelatin with a culture as a single streak along the diameter of a plate and incubate anaerobically for up to 3 d. Flood the surface with acid mercuric chloride;[21] the presence of clear zones bordering the growth of *B. nodosus* indicates gelatin hydrolysis.

2. Ammonia Production from Amino Acids

Prepare a suspension of bacteria from a fully grown agar culture in sterile 0.025 *M* phosphate buffer (pH 6.8) and adjust the density to about 5×10^8 cells per ml. Add 50 μl amounts of cell suspension to tubes containing equal volumes of 0.03 *M* solutions of the following amino acids: arginine, asparagine, citrulline, cystine, phenylalamine, threonine, and serine. Incubate tubes (including a control tube containing the cell suspension without amino acid) in a 37°C waterbath for 4 h. Add 1 drop of 20% NaOH plus 2 drops of Nessler's reagent and note the color change indicating ammonia production. Compare the profile of reactions with that typical of *B. nodosus*, as indicated in Section III.D.

3. Amino Acid Decarboxylase Activity

Prepare a dense aqueous suspension of bacteria from an agar culture (about 5×10^9 cells per ml). Prepare 0.03 *M* solutions of arginine, ornithine, and lysine in 0.0125 *M* phthalate buffer (pH 5.0) and add bromocresol purple indicator.[21] Mix equal volumes (50 μl) of the bacterial suspension and amino acid + indicator solution in tubes and incubate in a 37°C waterbath for 2 h. Each set of tests should include a bacterial suspension control in buffer without an amino acid. A positive reaction (blue coloration of the indicator) is given by *B. nodosus* for ornithine decarboxylase and negative reactions for arginine and lysine decarboxylases.

4. Animal Inoculation

Conclusive identification of *B. nodosus* may be achieved by experimentally infecting sheeps' feet with cultures and reisolating organisms with similar characteristics from interdigital or hoof lesions that are produced. Procedures for conditioning animals, infecting and bandaging their feet, and carrying out subsequent clinical examinations were described by Egerton et al.[10] It is desirable that a minimum of three sheep without prior experience of footrot and from a breed recognized as susceptible to the disease be used for testing each isolate. Each animal should be inoculated in one fore foot and one hind foot of the opposite side, the other feet being bandaged but not inoculated. Isolation and characterization of *B. nodosus* isolates from freshly collected clinical material may then be attempted as described previously.

IV. DIFFERENTIATION OF STRAINS

The recognition of clinically different forms of *B. nodosus* infection in sheep and cattle has led to many investigations of features that distinguish isolates from the various conditions, particularly those that may be related to the pathogenicity of the organisms or that may serve as markers in epidemiological studies. Experiments with artificially infected sheep have verified substantial differences between strains in their ability to reproduce the types of clinical infection that have been characterized. Thus, ovine virulent strains are responsible for the most invasive infections causing extensive damage to the matrix of the hoof in the majority of feet, whereas benign strains cause inflammation and necrosis of the interdigital skin without significant separation of the horn.[13,33] Strains of moderate invasiveness causing mild underrunning infections in a limited number of feet are considered to be of intermediate virulence.[39,40] Laboratory tests which so far have shown promise in the differentiation of such strains by their *in vitro* properties are described below.

A. Colony Type

Variation in one or more of the four principal features of *B. nodosus* colonies (see Section III.B.) has led to a wide range of colony morphotypes being described. Early reports indicated that cultures of virulent organisms produced spreading colonies with beaded centers and fimbriate peripheral areas, whereas benign isolates and avirulent culture variants produced nonfimbriate colonies of the smooth (mucoid) central-dome type.[17,24,25,41] In general, this also applied to the larger number of isolates compared by Stewart et al.[32] However, at least one of their virulent strains, and four strains of intermediate virulence, produced relatively smooth colonies similar to those of benign strains, indicating that the distinctions formerly noted were not clear cut. According to these authors, better discrimination was achieved when the agar-corroding tendency of strains was taken into account. When growth on hoof medium containing 2% agar was gently scraped off the surface, the appearance of crater-like depressions directly underlying the colonies was a characteristic feature of cultures of virulent strains, but was less marked or absent from those of intermediate and benign strains.

B. Degree of Piliation

Electron microscopy studies have revealed a wide variation in the number of pili borne by organisms from different strains.[31] Skerman et al.[25] and Every and Skerman[26] reported that bacteria from virulent strains were more highly piliated than those from benign strains, but these results were not confirmed for benign strains studied by Depiazzi and Richards.[27] Stewart et al.[32] also concluded that the degree of piliation of organisms was not closely correlated with their virulence for sheep.

Because pili play a dominant role in protective immunity against footrot,[26,42,43] as well as representing the principal antigenic structures by which different serogroups of *B. nodosus* may be recognized,[44,45] the selection of representative strains from those serogroups that consistently produce well-piliated organisms is an important consideration in the production of footrot vaccines.

The degree of piliation in bacteria from agar cultures has been reported to be considerably influenced by the medium composition.[32] Results of a similar comparison of cells from cultures in fluid media are not known.

C. Cell Motility

The phenomenon of twitching motility was first reported in *B. nodosus* by Depiazzi and Richards.[27] The level of motility was shown to be related to the size (diameter) of surface growth formed when cultures were stab-inoculated into plates of agar medium. Strains considered to have a high proportion of motile cells tended to form large, spreading colonies due to the surface translocation of organisms.

When 14 *B. nodosus* strains representing virulent, intermediate, and benign categories were compared by the above authors, it was found that virulent strains were highly motile, formed large, spreading colonies, and produced a thermostable protease. Benign strains had low motility, formed small colonies, and produced a thermolabile protease. Strains of intermediate virulence showed a low level of motility, produced small colonies, but a thermostable protease.

D. Caseinolytic Activity

Quantitative differences among isolates from field cases of virulent ovine footrot and those from benign ovine footrot in their hydrolytic activity on casein were used in a diagnostic test described by Egerton and Parsonson.[33] When organisms were heavily inoculated as streaks on plates of agar medium containing 1% w/v sodium caseinate, zones of precipitation and subsequently of clearing developed from the borders of growth on the medium. Cultures of the virulent isolates formed substantially larger zones of clearing than did those of benign strains, evidently corresponding to the greater amount of extracellular protease produced by the virulent organisms. A proteolytic index, calculated from measurements of these zones, showed a correlation with the virulence of the organisms. Because of their greater sensitivity, more recently developed quantitative and qualitative tests for *B. nodosus* proteolytic activity have superseded the caseinase-plate assay.

E. Elastase Activity

Comparison of a virulent wild-type *B. nodosus* strain with less virulent and avirulent variant organisms derived from it *in vitro* showed that whereas cultures of the prototype organisms hydrolyzed elastin particles incorporated in agar media, this activity was reduced or absent in cultures of the variants.[4,25,46] This apparent correlation between elastase activity and virulence was confirmed by an investigation of 87 ovine and bovine isolates, which showed that whereas cultures of ovine virulent isolates invariably produced elastase, ovine benign and bovine isolates were elastase negative.[47] Similar distinctions among three ovine virulent and three benign or avirulent strains were revealed in a later study,[32] which also showed that among ovine strains of intermediate virulence, elastase activity varied, and in the case of some strains was delayed.

These results indicate a reasonably close correspondence between elastase activity of cultures and the virulence of the organisms, as determined in sheep-challenge experiments. However, Claxton[48] has pointed out how confusing results can arise when the test is applied specifically to the diagnosis of field cases of virulent, mild, or benign footrot from which isolates have originated. This is due to the finding that elastase-positive and elastase-negative isolates can sometimes be obtained from the same flock, and even from the same infected foot.

Medium for the elastase test is prepared by adding 0.3 to 1.0 g powdered bovine elastin to each 100-ml volume of agar medium. By autoclaving the elastin separately in a small volume of water and adding this, after vigorous shaking, to the molten agar just before pouring the plates, an evenly dispersed suspension can be obtained. Addition of 0.15% $CaCl_2$ to the agar medium and extending the incubation of inoculated plates for up to 22 d has been shown to increase the sensitivity of the elastase test.[32,48] Results of a test comparing the elastase activities of different *B. nodosus* strains are illustrated in Figure 9.

F. Protease Thermostability

Depiazzi and Richards[49] showed that when extracellular proteases that had been produced in cultures of *B. nodosus* strains were further incubated at 37°C for periods of up to 11 d, the activities of those preparations from virulent strains were markedly more stable than those from benign strains. Protease activity was measured after growth had ceased, using

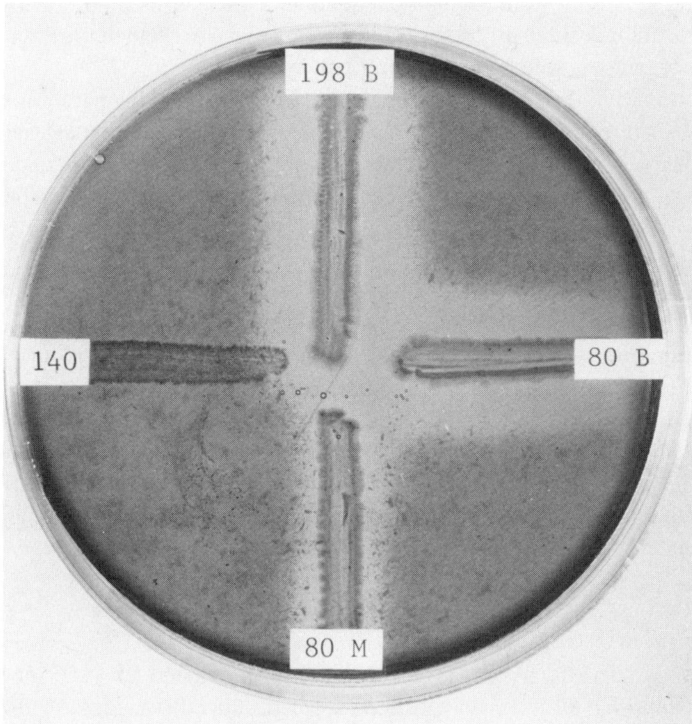

FIGURE 9. Comparison of cultures of four *B. nodosus* strains for their ability to solubilize particles of elastin incorporated in TAS agar medium after incubation for 8 d. Strains 198B and 80B, with the greatest elastase activity, produced virulent footrot in experimentally infected sheep; 80M, with diminished elastase activity, was a laboratory variant derived from 80B and produced only mild underrunning lesions in sheep; 140, with no apparent elastase activity, was isolated from ovine benign foot lesions.

hide powder azure as the substrate. Modifications of this test, originally described as the degrading-proteinase test, in which incubation of the enzyme preparations was carried out at 55°C,[50] or at 60°C,[51] enabled results to be obtained more rapidly. In the modification proposed by Green,[51] enzyme extracted from the agar medium upon which cultures had been grown was used to perform the test, which could be completed within 4 h.

G. Electrophoretic Zymogram Patterns

Qualitative distinctions among strains have been demonstrated in characteristic isozyme patterns and molecular weights determined for the proteinases after polyacrylamide gel electrophoresis.[52-54] Separate patterns typical of virulent and benign strains were recognized and confirmed in a later study, which also showed that patterns for intermediate strains varied.[55] In a comparison of the properties of purified preparations of five of the individual proteinase isozymes from virulent strains with those of three isozymes from benign strains, the "virulent" isozymes were found to have a greater thermal stability at 60°C and generally were most active in hydrolyzing substrates such as fibrinogen, α-elastin, collagen III, α-keratin, and hemoglobin.[56]

Immunoelectrophoretic techniques have also been applied to the characterization of antigenic components of extracellular products, including proteinases, of virulent and benign strains.[57,58] Antigens that were possibly unique to either virulent or benign strains were identified; some of these were proteases, others may have represented ultrastructural fractions of *B. nodosus* organisms released following cell lysis.

At present, no single laboratory test provides an unequivocal separation of strains associated with the different clinical forms of *B. nodosus* infection that have been recognized. However, Stewart et al.[32] showed that, considered together, the results of colony typing, determinations of protease thermostability, elastase activity, and zymogram patterns provided a broad discrimination among representative virulent, intermediate, and benign strains. It remains to be established whether this applies generally to the range of wild-type strains in the general population.

Stewart et al.[39] have pointed out that, from the evidence of pen challenge experiments, *B. nodosus* strains possess a spectrum of virulence for sheep. *In vitro* distinctions among strains may therefore be difficult to define precisely, the virulence of any individual isolate being due to a balance between such attributes as revealed by present tests, or to the possession of other as yet unidentified factors, or to an interaction between these. A clearer elucidation of these features and their interplay is therefore required so that more effective tests can be developed for the differentiation of isolates in relation to those clinical conditions in which they are implicated.

V. PATHOGENICITY

When sheep, rabbits, guinea pigs, and mice were inoculated s.c. and intradermally with cultures of *B. nodosus,* little or no reactions were produced.[2] Evidently, the pathogenicity of this organism is expressed restrictively in its association with the hoof epidermis of only a few ruminant host species, and/or in its interactions *in situ* with other microorganisms having ancillary etiological roles in the clinical disease. Although the incidence and course of these infections are substantially influenced by environmental factors, their severity essentially reflects the balance between the invasiveness of the organisms and the relative susceptibility of the host. Thus, organism and host factors that result in some *B. nodosus* infections being confined to the interdigital skin and soft horn of the hoof, and others leading to severe necrotic disruption of the hoof matrix are central to a complete understanding of the infective process.

Compared with other species of Gram-negative anaerobic bacteria that are pathogenic for animals, *in vitro* studies so far have revealed few properties of *B. nodosus* that might be assigned a primary role in the pathogenesis of natural infections. Thus, attention has been mainly focused on features such as the ultrastructure, colony morphology, and extracellular protease activity of strains which also serve to differentiate virulent from less virulent organisms.

The possibility that degree of piliation might have some role in the pathogenicity of *B. nodosus* arose from reports that virulent wild-type organisms appeared to be more richly piliated than those of benign strains[25,26] and from the fact that pilus antigens have a dominant role in protective immunity against virulent footrot.[26,42,43] However, as some benign strains also produce abundant pili and some virulent isolates are poorly piliated, the significance of pili *per se* as virulence determinants now seems doubtful.[27,55] Thus, in experiments described by Stewart et al.,[32] virulent footrot was produced in sheep by both a poorly piliated strain (forming small, relatively smooth colonies) and a heavily piliated strain (forming beaded, fimbriate colonies) when the interdigital skin of the animals was scarified prior to challenge. In unscarified feet, the highly piliated strain proved to be more virulent. This result suggested that either heavy piliation, or an unknown factor associated with beaded colonies, or both, may enhance the virulence of *B. nodosus* for sheep where the interdigital epidermis has not been damaged.

Although pili have been shown to have an adhesive function in some other bacteria,[59] this could not be demonstrated in *B. nodosus*.[41]

A possible link between the degree of twitching motility of *B. nodosus* bacteria and their virulence was proposed by Depiazzi and Richards.[27] The formation of flat, spreading colonies on agar cultures by wild-type strains was attributed to the pronounced surface translocation of organisms displaying this characteristic motility pattern, a feature that was most marked in virulent strains. The authors considered that, in combination with thermostable protease activity, a high level of motility in these organisms might facilitate their invasion of proteinaceous epidermal tissue.

Because of its prominence as an extracellular enzyme, the protease produced by *B. nodosus* has long been thought to have a role in the pathogenicity of these organisms.[2] Thomas[60] considered that the pathological changes in sheeps' feet affected with virulent footrot was due to a direct attack by keratinolytic proteases secreted by the bacteria. However, histological studies of active footrot lesions provided no evidence for this.[10] The production of elastase by more virulent strains of *B. nodosus* has also been considered as possibly having some pathogenic role, perhaps in contributing to the extensive separation of the horn from the underlying epidermal tissues that occurs in advanced ovine footrot.[47] Another *in vitro* characteristic shared by virulent strains studied by Stewart et al.[32] was the capacity to corrode agar immediately underlying colonies on plates. It was suggested that if this was due to production of an agarolytic enzyme by these organisms, such an activity *in vivo* might assist them to colonize the interdigital skin by the degeneration of complex polysaccharides present in the intercellular cement between epithelial cells.

Although some strong associations have now been observed between certain *in vitro* properties or products of *B. nodosus* strains and their virulence, the significance of these features to exact pathogenic mechanisms attributable to the organisms in clinical infections remains to be substantiated. Other factors may also await discovery, and it must be accepted that under *in vitro* conditions, selection of genotypes or phenotypic changes could result in the loss of sought-for determinants of pathogenicity. This subject is further complicated by the fact that the diseases concerned typically depend on the simultaneous or sequential presence of *B. nodosus* and *Fusobacterium necrophorum,* (see also Chapters 1 and 9).[10,11] Furthermore, Roberts and Egerton[11] presented evidence for a synergistic association between these two species, each contributing directly to the pathogenic process as well as promoting the survival and proliferation of the other in the infected tissues. While this *B. nodosus-F. necrophorum* relationship might involve aspects such as the interchange of essential nutrients, the generation of low oxidation-reduction potentials, or the interference with host nonspecific and immunospecific defense processes, the actual basis of the synergistic association has not been resolved.

In the past 5 years, evidence has accumulated indicating that in sheep that have been managed alike and not vaccinated against footrot, genetic factors may account for a significant amount of the variation observed among the animals in the occurrence of the disease.[61-63] Presumably, animals having a greater resistance to infection possess mechanisms that intercept or counter those by which the pathogenicity of potentially invading microorganisms is expressed. Thus, an elucidation of factors determining the virulence or pathogenicity of *B. nodosus* may ultimately provide a valuable clue to the nature of innate resistance to footrot in sheep and to means of identifying genetically superior animals for the breeding of less disease-prone flocks.

ACKNOWLEDGMENTS

The author is sincerely grateful to C. Michael Thorley, who kindly supplied Figures 2, 5, and 8. Thanks are also extended to D. R. Ris and G. W. de Lisle for their helpful comments on the text, to A. W. Barkus and D. C. Thurley for photographic assistance, and to J. I. Coulson, J. Tibble, and S. J. Dolden for typing the manuscript.

REFERENCES

1. **Holdeman, L. V., Kelley, R. W., and Moore, W. E. C.,** Family 1. Bacteroidaceae, in *Bergey's Manual of Systematic Bacteriology,* Vol. 1, Krieg, N. R. and Holt, J. G., Eds., Williams & Wilkins, Baltimore, 1984, 628.
2. **Beveridge, W. I. B.,** Foot-rot in sheep: a transmissible disease due to infection with *Fusiformis nodosus* (n. sp.), *Bull. Counc. Sci. Ind. Res. (Melbourne),* No. 140, 1941.
3. **Mraz, O., Tesarcik, J., and Varejka, F., Eds.,** *Nomina und Synonyma der Pathogenen und Saprophytaren Microben,* VEB Gustav Fischer Verlag, Stuttgart, 1963, 85.
4. **Skerman, T. M.,** *Bacteroides nodosus,* the causative agent of contagious footrot in sheep, in *Actual Data on the Biology and Pathology of Anaerobic Bacteria,* Bittner, J., Dumitriu, C., and Zamfirescu, M., Eds., Medical Publishing House, Bucharest, 1977, 327.
5. **Cato, E. P., Holdeman, L. V., and Moore, W. E. C.,** Proposal of neotype strains for 7 nonsaccharolytic *Bacteroides* species, *Int. J. Syst. Bacteriol.,* 29, 427, 1979.
6. **Mohler, J. R. and Washburn, H. J.,** Foot-rot of sheep, *U.S. Dept. Agric. Bur. Anim. Ind. Bull.,* No. 63, 1904.
7. **Claxton, P. D. and O'Grady, K. C.,** Footrot in goats and characterisation of caprine isolates of *Bacteroides nodosus,* in *Footrot in Ruminants,* Stewart, D. J., Peterson, J. E., McKern, N. M., and Emery, D. L., Eds., Commonwealth Scientific and Industrial Research Organisation, Australia, 1986, 119.
8. **Egerton, J. R. and Parsonson, I. M.,** Isolation of *Fusiformis nodosus* from cattle, *Aust. Vet. J.,* 42, 425, 1966.
9. **Skerman, T. M.,** Isolation of *Bacteroides nodosus* from hoof lesions in a farmed red deer *(Cervus elaphus)* (correspondence), *N. Z. Vet. J.,* 31, 102, 1983.
10. **Egerton, J. R., Roberts, D. S., and Parsonson, I. M.,** The aetiology and pathogenesis of ovine foot-rot. I. A histological study of the bacterial invasion, *J. Comp. Pathol.,* 79, 207, 1969.
11. **Roberts, D. S. and Egerton, J. R.,** The aetiology and pathogenesis of ovine foot-rot. II. The pathogenic association of *Fusiformis nodosus* and *F. necrophorus, J. Comp. Pathol.,* 79, 217, 1969.
12. **Graham, N. P. H. and Egerton, J. R.,** Pathogenesis of ovine foot-rot: the role of some environmental factors, *Aust. Vet. J.,* 44, 235, 1968.
13. **Egerton, J. R.,** *Bacteroides nodosus,* in *Handbuch der bakteriellen Infektionen bei Tieren,* Band III, Blobel, H. and Schliesser, T., Eds., Gustav Fischer, Stuttgart, 1981, 627.
14. **Richards, R. B., Depiazzi, L. J., Edwards, J. R., and Wilkinson, F. C.,** Isolation and characterisation of *Bacteroides nodosus* from foot lesions of cattle in Western Australia, *Aust. Vet. J.,* 56, 517, 1980.
15. **Roberts, D. S. and Walker, P. D.,** Fluorescein labelled antibody for the diagnosis of foot-rot, *Vet. Rec.,* 92, 70, 1973.
16. **Thorley, C. M., Calder, H. A. M., and Harrison, W. J.,** Recognition in Great Britain of *Bacteroides nodosus* in foot lesions of cattle, *Vet. Rec.,* 100, 387, 1977.
17. **Thorley, C. M.,** A simplified method for the isolation of *Bacteroides nodosus* from ovine foot-rot and studies on its colony morphology and serology, *J. Appl. Bacteriol.,* 40, 301, 1976.
18. **Skerman, T. M.,** Determination of some *in vitro* growth requirements of *Bacteroides nodosus, J. Gen. Microbiol.,* 87, 107, 1975.
19. **Claxton, P. D.,** Personal communication, 1981.
20. **Holdeman, L. V. and Moore, W. E. C., Eds.,** *Anaerobe Laboratory Manual,* 3rd ed., Virginia Polytechnic Institute and State University, Blacksburg, VA, 1975.
21. **Cowan, S. T.,** *Cowan and Steel's Manual for the Identification of Medical Bacteria,* 2nd ed., revised, Cambridge University Press, London, 1974.
22. **Cooper, B. S.,** Differences in morphology of *Bacteroides nodosus* attributable to culture media, *N.Z. Vet. J.,* 25, 16, 1977.
23. **Stewart, D. J.,** An electron microscopic study of *Fusiformis nodosus, Res. Vet. Sci.,* 14, 132, 1973.
24. **Short, J. A., Thorley, C. M., and Walker, P. D.,** An electron microscope study of *Bacteroides nodosus:* ultrastructure of organisms from primary isolates and different colony types, *J. Appl. Bacteriol.,* 40, 311, 1976.
25. **Skerman, T. M., Erasmuson, S. K., and Every, D.,** Differentiation of *Bacteroides nodosus* biotypes and colony variants in relation to their virulence and immunoprotective properties in sheep, *Infect. Immun.,* 32, 788, 1981.

26. **Every, D. and Skerman, T. M.,** Surface structure of *Bacteroides nodosus* in relation to virulence and immunoprotection in sheep, *J. Gen. Microbiol.,* 129, 225, 1983.
27. **Depiazzi, L. J. and Richards, R. B.,** Motility in relation to virulence of *Bacteroides nodosus, Vet. Microbiol.,* 10, 107, 1985.
28. **Stewart, D. J. and Egerton, J. R.,** Studies on the ultrastructural morphology of *Bacteroides nodosus, Res. Vet. Sci.,* 26, 227, 1979.
29. **Thomas, J. H.,** A simple medium for the isolation and cultivation of *Fusiformis nodosus, Aust. Vet. J.,* 34, 411, 1958.
30. **Millar, K. R., Skerman, T. M., Erasmuson, S. K., and Bond, G. C.,** Unpublished data, 1986.
31. **Stewart, D. J., Clark, B. L., Peterson, J. E., Emery, D. L., Smith, E. F., Griffiths, D. A., and O'Donnell, I. J.,** The protection given by pilus and whole cell vaccines of *Bacteroides nodosus* strain 198 against ovine foot-rot induced by strains of different serogroups, *Aust. Vet. J.,* 62, 153, 1985.
32. **Stewart, D. J., Peterson, J. E., Vaughan, J. A., Clark, B. L., Emery, D. L., Caldwell, J. B., and Kortt, A. A.,** The pathogenicity and cultural characteristics of virulent, intermediate and benign strains of *Bacteroides nodosus* causing ovine foot-rot, *Aust. Vet. J.,* 63, 317, 1986.
33. **Egerton, J. R. and Parsonson, I. M.,** Benign foot-rot — a specific interdigital dermatitis of sheep associated with infection by less proteolytic strains of *Fusiformis nodosus, Aust. Vet. J.,* 45, 345, 1969.
34. **Schmitz, J. A. and Gradin, J. L.,** Serotypic and biochemical characterisation of *Bacteroides nodosus* isolates from Oregon, *Can. J. Comp. Med.,* 44, 440, 1980.
35. **Lee, S. W., Alexander, B., and McGowan, B.,** Purification, characterisation, and serologic characteristics of *Bacteroides nodosus* pili and use of a purified pili vaccine in sheep, *Am. J. Vet. Res.,* 44, 1676, 1983.
36. **Egerton, J. R., Parsonson, I. M., and Graham, N. P. H.,** Parenteral chemotherapy of ovine foot-rot, *Aust. Vet. J.,* 44, 275, 1968.
37. **Gradin, J. L. and Schmitz, J. A.,** Selective medium for isolation of *Bacteroides nodosus, J. Clin. Microbiol.,* 6, 298, 1977.
38. **Gradin, J. L. and Schmitz, J. A.,** Susceptibility of *Bacteroides nodosus* to various antimicrobial agents, *J. Am. Vet. Med. Assoc.,* 183, 434, 1983.
39. **Stewart, D. J., Clark, B. L., and Jarrett, R. G.,** Observations on strains of *Bacteroides nodosus* of intermediate virulence to sheep, *Aust. Adv. Vet. Sci.,* 219, 1982.
40. **Stewart, D. J., Clark, B. L., and Jarrett, R. G.,** Differences between strains of *Bacteroides nodosus* in their effects on the severity of foot-rot, bodyweight and wool growth in Merino sheep, *Aust. Vet. J.,* 61, 348, 1984.
41. **Stewart, D. J.,** Studies on the Morphology and Antigenic Structure of *Fusiformis nodosus,* Ph.D. thesis, University of Sydney, Sydney, 1975.
42. **Stewart, D. J., Clark, B. L., Peterson, J. E., Griffiths, D. A., and Smith, E. F.,** Importance of pilus-associated antigen in *Bacteroides nodosus* vaccines, *Res. Vet. Sci.,* 32, 140, 1982.
43. **Walker, P. D., Short, J. A., Thomson, R. D., and Roberts, D. S.,** The fine structure of *Fusiformis nodosus* with special reference to the location of antigens associated with immunogenicity, *J. Gen. Microbiol.,* 77, 351, 1973.
44. **Egerton, J. R.,** Surface and somatic antigens of *Fusiformis nodosus, J. Comp. Pathol.,* 83, 151, 1973.
45. **Claxton, P. D., Ribeiro, L. A., and Egerton, J. R.,** Classification of *Bacteroides nodosus* by agglutination tests, *Aust. Vet. J.,* 60, 331, 1983.
46. New Zealand Ministry of Agriculture and Fisheries, Annual Report of Research Division 1976-77, 149.
47. **Stewart, D. J.,** The role of elastase in the differentiation of *Bacteroides nodosus* infections in sheep and cattle, *Res. Vet. Sci.,* 27, 99, 1979.
48. **Claxton, P. D.,** Elastase activities of ovine *Bacteroides nodosus* isolates, in *Footrot in Ruminants,* Stewart, D. J., Peterson, J. E., McKern, N. M., and Emery, D. L., Eds., Commonwealth Scientific and Industrial Research Organization, Australia, 1986, 99.
49. **Depiazzi, L. J. and Richards, R. B.,** A degrading proteinase test to distinguish benign and virulent ovine isolates of *Bacteroides nodosus, Aust. Vet. J.,* 55, 25, 1979.
50. **Kortt, A. A., O'Donnell, I. J., Stewart, D. J., and Clark, B. L.,** Activities and partial purification of extracellular proteases of *Bacteroides nodosus* from virulent and benign footrot, *Aust. J. Biol. Sci.,* 35, 481, 1982.
51. **Green, R. S.,** A method to differentiate between virulent and benign isolates of *Bacteroides nodosus* based on the thermal stability of their extracellular proteinases, *N. Z. Vet. J.,* 33, 11, 1985.
52. **Every, D.,** Proteinase isoenzyme patterns of *Bacteroides nodosus:* distinction between ovine virulent isolates, ovine benign isolates and bovine isolates, *J. Gen. Microbiol.,* 128, 809, 1982.
53. **Kortt, A. A., Burns, J. E., and Stewart, D. J.,** Detection of the extracellular proteases of *Bacteroides nodosus* in polyacrylamide gels: a rapid method of distinguishing virulent and benign ovine isolates, *Res. Vet. Sci.,* 35, 171, 1983.

54. **Gordon, L. M., Yong, W. K., and Woodward, C. A. M.,** Temporal relationships and characterisation of extracellular proteases from benign and virulent strains of *Bacteroides nodosus* as detected in zymogram gels, *Res. Vet. Sci.*, 39, 165, 1985.
55. **Stewart, D. J., Peterson, J. E., Vaugham, J. A., Clark, B. L., Emery, D. L., Caldwell, J. B., and Kortt, A. A.,** Clinical and laboratory diagnosis of benign, intermediate and virulent strains of *Bacteroides nodosus,* in *Footrot in Ruminants,* Stewart, D. J., Peterson, J.E., McKern, N. M., and Emery, D. L., Eds., Commonwealth Scientific and Industrial Research Organization, Australia, 1986, 81.
56. **Green, R. S.,** Characterisation of certain proteinase isoenzymes produced by benign and virulent strains of *Bacteroides nodosus, J. Gen. Microbiol.*, 131, 2871, 1985.
57. **Yong, W. K., Moses, E. K., and Riffkin, G. G.,** Characterisation of virulence associated antigens for differentiating virulent and benign strains of *Bacteroides nodosus,* in *Footrot in Ruminants,* Stewart, D. J., Peterson, J. E., McKern, N. M., and Emery, D. L., Eds., Commonwealth Scientific and Industrial Research Organisation, Australia, 1986, 109.
58. **Yong, W. K. and Gordon, L. M.,** Comparative antigenic analysis of extracellular proteins of *Bacteroides nodosus* isolated from virulent and benign ovine footrot, *Vet. Microbiol.*, 12, 135, 1986.
59. **Ottow, J. C. G.,** Ecology, physiology and genetics of fimbriae and pili, *Annu. Rev. Microbiol.*, 29, 79, 1975.
60. **Thomas, J. H.,** The bacteriology and histopathology of footrot in sheep, *Aust. J. Agric. Res.*, 13, 725, 1962.
61. **Egerton, J. R., Ribeiro, L. A., Kieran, P. J., and Thorley, C. M.,** Onset and remission of ovine footrot, *Aust. Vet. J.*, 60, 334, 1983.
62. **Emery, D. L., Stewart, D. J., and Clark, B. L.,** The comparative susceptibility of five breeds of sheep to foot-rot, *Aust. Vet. J.*, 61, 85, 1984.
63. **Skerman, T. M.,** Genetic variation and inheritance of susceptibility to footrot in sheep, in *Footrot in Ruminants,* Stewart, D. L., Peterson, J. E., McKern, N. M., and Emery, D. L., Eds., Commonwealth Scientific and Industrial Research Organisation, Australia, 1986, 73.

Chapter 6

THE ISOLATION AND IDENTIFICATION OF *FUSOBACTERIUM NECROPHORUM*

B. L. Clark, J. A. Vaughan, and D. L. Emery

TABLE OF CONTENTS

I. INTRODUCTION

The importance of the nonsporing, obligate, anaerobic bacteria as a cause of disease in man and in animals has been recognized for many decades, yet they are often neglected in routine field and laboratory investigation of infections. This is due to failure of clinicians to collaborate with microbiologists to achieve isolation of these organisms, the necessity of providing special conditions and procedures, and to limitation in equipment, particularly in smaller laboratories. While the most fastidious anaerobic bacteria certainly require the use of specialized anaerobic equipment, such as the anaerobic glove box, and sophisticated methods for isolation, culture, and identification, simpler, more rapid techniques are suitable for isolation and identification of most of the known anaerobic pathogenic bacteria, including *Fusobacterium necrophorum.*

In this chapter, emphasis will be placed on methods and media that are suitable for the isolation and identification of *F. necrophorum* in smaller laboratories, although reference will be made to the use of sophisticated instruments as an aid to identification.

Necrobacillosis is the general term which is applied to lesions in various parts of the body in which *F. necrophorum* plays the principal pathological role, although it is not necessarily the initiating agent nor the only pathogen present.[1] In most cases, necrobacillosis is a localized infection of skin or mucous membrane following lowering of resistance or trauma, but deeper tissues and internal organs, especially the liver, may be involved. Thus, *F. necrophorum* is the principal pathogen in calf diphtheria affecting the pharynx and larynx, in foot abscess and liver abscess of cattle and sheep,[2] and in fatal purulent conditions in wallabies.[3] It is a secondary invader in other conditions, such as necrotic enteritis in swine and virulent footrot in sheep.[1] In mixed infections involving *F. necrophorum, Bacteroides nodosus,* and *Corynebacterium pyogenes,* synergic mechanisms where one microorganism compensates for deficiencies of growth and virulence factors in the others, have been described.[4]

F. necrophorum is widespread in nature and is a common inhabitant of the gut in herbivora and, hence, found in their feces. Herbivorous animals, especially ruminants, are susceptible to infection whereas carnivores and humans are relatively resistant.[1] Experimental infections can be produced in almost all species of animals, although cats, dogs, rats, fowls, pigeons, and guinea pigs are less susceptible. The species most commonly used to study experimental infections are rabbits, mice, and guinea pigs.[5]

F. necrophorum, the causative agent of necrobacillosis, is a Gram-negative, nonmotile, nonsporing, obligate anaerobe. It forms long filaments 0.5 to 1.0 μm wide in young cultures and in active lesions, but later, these break down into short rods of 1 to 4 μm, or to coccoid elements. In culture media, the filaments break down into rods after 12 h to 5 d incubation. These filaments often stain unevenly, giving them a beaded appearance that is helpful in identification.[1] The ends of the cells may be round or tapered. The beaded filamentous forms are more common in liquid media and rods are more common in growth on solid media.[6] The organism produces a potent endotoxin and a soluble exotoxin that has leucocidal and hemolytic activities. It also causes hemagglutination of erythrocytes from several animal species.

The nomenclature for *F. necrophorum* has varied considerably and it has been called *Bacillus necrophorus, Fusiformis necrophorus,* and *Sphaerophorus necrophorus,* among other names, at various times.[6] The multiplicity of synonyms reflect the original uncertainty of the taxonomic position of the organism, its pleomorphism, and early difficulties in culturing anaerobes.[5] The present genus *Fusobacterium* has been assembled from organisms that are similar in that they are all obligate anaerobes, nonmotile, Gram-negative, nonsporing, pleomorphic rods that produce butyric acid as an endproduct of the metabolism of carbohydrate or peptone.[6] *Fusobacterium necrophorum* grows easily and rapidly, and identification is well within the capabilities of the smaller veterinary laboratory. We believe that the enthusiasm of clinicians and microbiologists interested in the control of diseases caused by *F. necrophorum* will be greatly stimulated by collaboration in the routine isolation and identification of the organism for epidemiological and laboratory studies.

II. SPECIAL REQUIREMENTS FOR ISOLATION

A. General Considerations

Obligate anaerobes are more exacting in their physical and nutritional requirements for isolation and growth than aerobic bacteria. The *in vivo* conditions where obligate anaerobes are actively metabolizing are often more reduced than can be readily achieved in transport and culture media. Furthermore, isolation requires more exacting conditions than subsequent growth in pure culture.[7]

It is important to collect specimens for culture from the depths of necrobacillosis lesions, endeavoring to avoid all environmental contamination. This may entail removal of loose or surface horn or tissue in cases of foot abscess. In the case of closed abscesses, wash the surface of the overlying skin and obtain specimens of pus using a sterile, disposable syringe and needle. Contamination from bowel contents and flora should be avoided when collecting specimens from liver abscesses.

It is essential to protect specimens from atmospheric oxygen as much as possible during collection and forwarding to the laboratory, and in practice this means always using either an anaerobic specimen collector (Vacutainer® Anaerobic Transporter, Becton-Dickinson) or prereduced transport media. Large quantities of necrotic tissue or pus should be put on swabs because small quantities frequently dry out during transit with loss of viability of the anaerobes in the specimen. Immediately put the swab either into an anaerobic specimen collector or into transport media. If considerable time must elapse prior to culture, keep the tube at 4°C both during transit and after arrival at the laboratory. In practice, properly collected specimens on swabs in such tubes can be held at 4°C for a week prior to culturing, if necessary. Failure to maintain the tube at 4°C will result in overgrowth of the specimen with facultative anaerobic organisms if they are also present. Cultural enrichment in order to increase the numbers of *F. necrophorum* cells prior to plating onto solid media is contraindicated due to the existing large count of these cells in specimens (10^6 to 5×10^9/g),[8,9] and to the risk of encouraging overgrowth by facultative anaerobes.

Gram-stained smears should be prepared in the laboratory from each specimen prior to commencing isolation as a general aid to recognizing the presence of *F. necrophorum* and of other organisms. Typically, *F. necrophorum* cells are pale, irregular-staining, Gram-negative, pleomorphic rods and filaments with swollen areas, in smears from abscesses.[10,11] Other organisms which are recognizable in the smear, often do not exceed 10^5 cells per gram in number.[8] Findings from direct microscopic examination of specimens are an important indicator for quality control. Subsequent failure to recover suspected *F. necrophorum* cells in culture indicates deficiencies in isolation and culture techniques.[10]

B. Special Needs for Isolation of *F. necrophorum*

The primary requirement is for anaerobic conditions, which in the small laboratory is best achieved by the use of media that has been freshly prepared, and by the use of anaerobic jars for culture. Other anaerobic cultural methods which are suitable for large specialized laboratories are the use of the anaerobic chamber or glove box and prereduced anaerobically sterilized media.[12] It is always preferable to use freshly prepared media for isolation of *F. necrophorum* whenever possible, and this is particularly the case with liquid media. In the case of solid media, such as brain heart infusion blood agar,[8] it is possible to use plates stored aerobically in cellophane bags in the refrigerator (4°C) for up to 2 weeks after preparation without loss of efficiency in the isolation of *F. necrophorum*.[10] Suitable anaerobic jars are those which are vented by Schrader valves and used with "cold" catalysts composed of palladium-coated alumina pellets enclosed in wire mesh, making them both safe and convenient to use. Following introduction of gas mixtures containing hydrogen into the jar, there is catalytic combination of all oxygen with hydrogen to form water, thus providing anaerobic conditions. It is advisable to include three to four packets of catalysts in each jar. These catalysts operate successfully at room temperature but become inactivated by H_2S and excessive moisture. To regenerate them, heat them in a dry oven at 160 to 170°C for 2 h on a regular basis. Various systems are available for providing suitable gas mixtures. The easiest method is to use the commercially available Gas Pak (BBL) H_2-, and CO_2-generator envelope which is activated by adding 10 ml of water.[13] In this method, the opened envelope is placed in the jar with inoculated media, the water then added, and the jar immediately closed. Anaerobic conditions are produced within 2 h. Another method involves the replacement of air in the jar with oxygen-free gas mixtures containing N_2, H_2, and CO_2 by the evacuation-replacement system. In this system, a vacuum pump is used to withdraw air from the jar repeatedly, three to five times, with gas being introduced between each "evacuation" in order to create a final oxygen-free atmosphere containing 80 to 90% N_2, 5 to 10% H_2, and 5 to 10% CO_2.[13]

F. necrophorum has an absolute requirement for at least 1% CO_2 in the gas mixture in order to grow.[14] In practice, it is usual to introduce 5 to 10% CO_2 to achieve optimal growth. Growth declines in atmospheres containing more than 20% CO_2.[14]

The nutritional requirements of *F. necrophorum* during isolation are adequately met by either brain heart infusion agar (Difco®), or Eugonagar® (BBL), when both bases are supplemented with yeast extract, hemin, vitamin K_1, L-cysteine HCl, and sheep, ox, or rabbit blood.[7] Other media have been recommended, but in our hands, these two have given excellent results. We routinely use supplemented brain heart infusion blood agar (BHIBA) as the solid medium in plates, and Eugonbroth (BBL), supplemented with yeast extract (EY) as the liquid medium for making dilutions during isolation (see Appendix).

The range of pH over which the organism is able to grow is governed by the nutritional adequacy of the medium. In simpler medium, the range was given as pH 5.5 to 8.6,[15] whereas in enriched medium it was pH 5.3 to 9.3.[16] The optimum pH for maximum growth in Simon's medium 156 was 6.84 while the optimum pH for acid production was 7.7.[16]

The optimum temperature for growth is 37°C, although good growth has been observed between 30 and 40°C,[5] and the range for growth has been reported to be from 22 to 43°C.[17] *F. necrophorum* is killed by exposure to 55°C for 15 min.[18]

C. Inoculation of Media for Isolation

Remove swab containing the specimen from the anaerobic specimen collector, or from the transport medium (see Appendix) and immediately immerse in 10 ml of freshly prepared EY broth. Thoroughly wash the swab so as to distribute the specimen evenly into the broth and assume that the initial dilution factor is 1:100. Prepare further decimal dilutions to 10^{-6} in EY and then distribute 0.1 ml volumes of the 10^{-4} and 10^{-6} dilutions over the surface of BHIBA plates using two plates for each dilution. Incubate the plates under anaerobic conditions without delay. In practice, this is a very successful and satisfactory method which depends on the fact that numbers of *F. necrophorum* cells in specimens usually exceed those of other organisms by at least a factor of 10^2. Success of the method can be jeopardized by failure to maintain inoculated transport media or anaerobic specimen collectors at refrigerator temperatures (2 to 4°C) when the time between collection of specimens and processing in the laboratory exceeds 2 to 3 h.

Selective solid media has been prepared for the isolation of *F. necrophorum* and other pathogenic, nonsporing, obligate anaerobes (see Appendix). A medium which is selective for *Fusobacterium* species and differential for *F. necrophorum* is the *Fusobacterium* egg-yolk agar prepared by Morgenstein et al.[19] This medium, which has a Brucella agar base (Difco) contains vancomycin, neomycin, josamycin, and egg yolk.[19] *F. necrophorum* is relatively resistant to vancomycin, josamycin, and neomycin, although susceptible to kanamycin and colistin.[13] On this medium, *F. necrophorum* typically shows a strong lipase reaction which enables colonies to be easily recognized. Another selective medium, phenylethyl alcohol agar, which has a trypticase soy agar (BBL) base and contains 0.25% phenylethyl alcohol, crystal violet, and egg yolk, has been used for isolation of the organism from bovine liver abscesses.[20] On this medium, *F. necrophorum* grows as blue colonies surrounded by a zone of clearing.[20]

D. Incubation and Examination

Plates are held for 3 d at 37°C under anaerobic conditions, prior to primary examination, and then reexamined again after a further 2 to 5 d incubation. It is essential to avoid unnecessary exposure of plates to atmospheric oxygen during examinations. Colonies of *F. necrophorum* can generally be recognized by their typical appearance but Gram-stained smears should be made and examined for presumptive identification of the colonies. Typical colonies should be subcultured onto a fresh BHIBA plate in order to obtain a pure culture prior to commencing identification tests. It should be noted that *F. necrophorum* has a shorter life on solid and in liquid media than in semisolid media, and subculture will be necessary at weekly intervals. The addition of 0.5% Bacto agar to Simon's medium 156,[16] which may be incubated aerobically, will keep *F. necrophorum* viable for up to 6 months.[42] When liquid media is being inoculated from a plate, use a heavy inoculum. In practice, when subculturing liquid media it is advisable to use a 10% inoculum v/v.[5,21]

III. MORPHOLOGY AND GROWTH CHARACTERISTICS

F. necrophorum is usually described as being a pleomorphic organism which may be present in culture as either coccoid, short or long rods, or as long filaments. A few organisms in young cultures of certain strains show fusiform or round swellings that can be 1.8 μm in diameter. The cellular morphology tends to be dependent on the age of the culture, the strain of the organism, and upon the medium. In glucose broth, cells are 0.5 to 0.7 μm in diameter with swellings up to 1.8 μm.[6] The cell length ranges from coccoid bodies to filaments of over 100 μm in length and the ends of the cells are either round or tapered. Filamentous forms with metachromatic granular inclusions tend to predominate in young cultures, particularly in liquid media, whereas coccoid bodies and rods are more common in older cultures grown on solid media.[6] In enriched liquid media, filaments appear at about 16 h and break down to rods at about 24 to 36 h.[5] The metachromatic granules that are a characteristic of this organism and which are responsible for the beaded appearance in stained smears, start to appear at about 12 h in actively growing cells, but only become prominent in cultures more than 24 h old. The distribution of metachromatic granules in cells is uneven and their shape varies from spherical to irregular.[5] *F. necrophorum* does not possess flagella or spores.

Ultrastructural studies have shown that the outer cell wall is convoluted, appearing to contain channels that connect the cytoplasm with the exterior of the cell wall and which may be used for excretion of waste products or exotoxin.[5] The cell wall has an inner and outer membrane and ribosomes are visible in the cytoplasm.[5,22]

F. necrophorum is a Gram-negative, nonacid-fast organism which is stained easily with all the ordinary aniline dyes. In 16- to 20-h old cultures, nearly all organisms stain evenly, but after 24 h, deeply staining areas alternating with faintly staining areas commence to appear giving a beaded or "string of beads" appearance.[18] Methylene blue and Giemsa stains reveal these metachromatic granules better than the Gram stain. In 3-day-old cultures, many cells will stain poorly and short cells may appear faint except for a deeply staining granule at each end which gives a bipolar appearance.

Three different types of colonies are produced on solid media by various strains of *F. necrophorum* and form the basis for the biotyping scheme of Fièvez, in which two distinct biotypes designated A and B and an intermediate biotype AB have been described.[15] On blood agar after 2 d incubation at 37°C, type-A colonies are 2 to 4 mm in diameter, flat, smooth, irregular in outline, metallic gray in color, and are surrounded by a β-type (clear) zone of hemolysis which usually has a diameter equal to, or twice that of the colony. Type-B colonies are 1 to 2 mm in diameter, convex, regular in outline, yellow in color, and rough in appearance. They are slower in development and are less hemolytic than type-A colonies, usually producing an α-type (greenish in color with partial lysis) zone of hemolysis which may not extend far beyond the colony and may only be immediately underneath the colony. Colonies of the type AB have a intermediate appearance quite similar to that which would be produced by superimposing a colony of type B on a colony of type A. Thus, it is a colony of 2 to 4 mm in diameter, has a raised, convex, yellowish central section surrounded by gray, flat growth, which is irregular in outline, and looks like a "fried egg" colony. The degree of hemolysis is usually the same as for type A colonies.

Study of the patterns of hemolysis produced by colonies of type A, AB, and B on blood agar prepared using blood from six different species, or by action of the hemolysin on erythrocyte suspensions suggests that chicken, ox, and horse erythrocytes are the most sensitive, with those of sheep, rabbit, and pig following in order of increasing resistance.[23]

In enriched liquid medium, the cells of types A and AB tend to produce cultures of homogenous turbidity whereas the cells of type B tend to produce flocculent growth which sediments, leaving the broth clear. This is because type-A and -AB strains produce suspensions of filaments in liquid media, whereas type-B strains grow as short cells which sediment.[15]

Examination of isolates of *F. necrophorum* obtained from foot abscess lesions in cattle and sheep,[23] and from liver abscess in cattle[9,15] indicates that the majority belong to biotypes A and AB, whereas those isolated from the interdigital lesions of bovine feet,[23] and from rumenal contents[9] belong to biotype B. The isolates in biotypes A and AB are virulent and are pathogenic for cattle, sheep, and mice whereas isolates in biotype B are nonvirulent and of low pathogenicity. Thus, presumptive identification, and differentiation between virulent and nonvirulent strains can be achieved by assessing the color, cellular and colonial morphology, and degree of hemolysis of colonies grown on blood agar.[23]

IV. BIOCHEMICAL REACTIONS AND IMMUNOLOGICAL TESTS

The biochemical reactions characteristic of *F. necrophorum* are production of indole from tryptophane, deamination of threonine to propionate, conversion of lactate to propionate, and production of butyric acid and traces of propionic and acetic acid during growth in media containing peptone, yeast extract, and glucose.[6] It usually, but not always, hydrolyzes gelatin and digests meat. It produces desoxyribonuclease and lipase. It does not produce catalase, lecithinase, or phosphatase and does not reduce nitrate or hydrolyze esculin. Growth is not enhanced by bile or desoxycholate, and with exception of limited acid production from glucose and fructose, it does not ferment sugars.[6]

Human, chicken, and pigeon red blood cells are strongly hemagglutinated by strains belonging to biotypes A and AB but not by strains belonging to biotype B.[15] Red blood cells from ox, sheep, rabbit, horse, and guinea pig are not as sensitive.

Antibiotic susceptibility testing of *F. necrophorum* has shown that resistance to erythromycin 3 μg/ml is variable and that the organism is susceptible to chloramphenicol 12 μg/ml, clindamycin 1.6 μg/ml, penicillin 2 μ/ml, tetracycline 6 μg/ml, and to cephalothin 1 μg/ml.[6] It is resistant to vancomycin (5 μg disc) but susceptible to kanamycin (1000 μg disc) and to colistin (10 μg disc).[11]

Various rapid micromethods have been developed and are available commercially, for the identification of anaerobic isolates. The RapID ANA system (Innovative Diagnostic Systems, Inc., Atlanta, GA, U.S.) is a rapid system that is growth independent and requires no anaerobic incubation. This system requires a minimal cell number and measures the activity of preformed enzymes in reactions that are completed after incubation in air for only 4 h. This system has been used successfully in identifying *F. necrophorum* isolates without the need of additional biochemical tests or gas-liquid chromatography.[42]

A large number of investigators have attempted to devise satisfactory serological procedures for aiding the identification and classification of *F. necrophorum*. The study of cross-agglutination reactions using freshly prepared or formalinized cell suspensions, and antisera prepared in rabbits has shown that while there appears to be considerable serological heterogeneity in the species, there is also partial sharing of antigens.[15,18,24-26] While several authors have been able to separate isolates of *F. necrophorum* into different antigenic groups on the basis of cross-agglutination patterns,[18,25,26] others have experienced difficulty. The organism has a notorious tendency to agglutinate spontaneously, which in practice is a considerable handicap. Complement fixation, flocculation, and precipitation tests have not given reliable results when applied to antigens prepared from *F. necrophorum.*[15] Passive hemagglutination tests using red blood cells sensitized with heat-extracted carbohydrate antigens prepared from isolates of *F. necrophorum* and specific rabbit antisera have also demonstrated the serological heterogeneity of the species.[9,27] A typing scheme that will allow differentiation of strains of *F. necrophorum* on the basis of cell-wall antigens has not yet been devised. A hemagglutination-inhibition test which depends on use of specific rabbit antisera to prevent *F. necrophorum* from hemagglutinating human red blood cells can be used for rapid species identification of *F. necrophorum.*[28]

While it is possible in reference laboratories to carry out a large number of biochemical and antibiotic sensitivity tests in order to fully identify presumptive isolates of *F. necrophorum,* in smaller laboratories, only a limited number of tests can usually be conducted. The essential identification tests to conduct on organisms presumed to be *F. necrophorum* on the basis of growth characteristics, and colonial and cellular morphology are those for the production of indole, lipase, and catalase, and antibiotic-disc susceptibility to kanamycin, vancomycin, and colistin. A rapid test for indole production is the spot test. This test is conducted by smearing cells of the isolate grown on a solid medium containing tryptophane onto a filter paper saturated with 1% paradimethylaminocinnamaldehyde in 10% concentrated HCl (v/v). A positive reaction is the rapid development of a blue color. Lipase production is determined from observation of the growth of the isolate on egg-yolk agar (see Appendix). The development of an irridescent sheen or "oil on water" appearance on the surface of growth indicates lipase activity.[10] Sensitivity to discs containing either 10 μg colistin, 1000 μg kanamycin, or 5 μ vancomycin is determined on solid media inoculated evenly with the isolate and incubated anaerobically at 37°C for 48 h. Inhibition-zone diameters of 10 mm or more indicate susceptibility, while smaller zones of inhibition indicate resistance.[13]

Although analysis of volatile fatty acid byproducts by gas-liquid chromatography (GLC) is not essential for the identification of *F. necrophorum,* GLC is the most practical method of rapidly identifying the presence of short-chain fatty acids in culture. GLC analysis can be performed on any broth which supports good growth of the isolate.[11] Production of the volatile butyric acid in the absence of iso-acids by *F. necrophorum* is determined by identification of its presence in an ether extract of culture broth when compared with an ether extract of uninoculated medium.

V. PATHOGENICITY

The pathogenicity of a microorganism is defined as the ability to cause damage or disease in the host or the relative capacity to surmount host defenses.[29] It is difficult to quantify, being dependent on both host and parasite factors. The factors that might contribute to the pathogenicity of *F. necrophorum* include an exotoxin (leucocidin),[30,32] a hemolysin,[15-30] a potent endotoxin,[33,34] and a polysaccharide capsule. Strains of biotypes A and AB are pathogenic and are isolated predominantly from infected lesions, whereas those of biotype B are nonpathogenic and are isolated from the gastrointestinal tract and from the skin.[9,15,23,35] Strains of biotypes A and AB produce relatively large amounts of leucocidin and hemolysin compared to strains of biotype B that produce little or no leucocidin and hemolysin.[15,23,36] The hemolysin most readily lyses ox, horse, and chicken red blood cells; those from sheep are less susceptible, while those from rabbit and pig are most resistant.[23] The relative susceptibility of red cells to lysis by cultural supernatant from *F. necrophorum* parallels the relative sensitivities of leucocytes from the same species to cytolysis by leucocidins produced by the same strains.[32]

All biotypes possess a toxic lipopolysaccharide of equivalent biological potency for which the LD_{50} for mice is 584 μg.[33] Only toxigenic strains appear to have capsules, antigens of which can be readily demonstrated in the supernatant from liquid cultures.[43] Although the antigenic structure of the capsular polysaccharides of *F. necrophorum* has not been examined, capsular material from *Bacteroides fragilis* produced abscesses after i.p. inoculation into mice.[37] Strains of either biotypes AB or B are engulfed with equal facility by bovine polymorphonuclear leucocytes in the absence of serum.[43]

The principal factor elaborated by pathogenic strains of *F. necrophorum* is the exotoxin, a cytolysin which is destructive for leucocytes, erythrocytes, and a variety of cultured cells including fibroblasts, macrophages, and kidney cells,[32,38] and which initiated degenerative changes in the skin of rabbits.[30] The importance of the leucocidin to the infectivity of *F. necrophorum* is emphasized by correlations between toxin production and the capacity to produce intra-abdominal abscesses in rodents,[9,31,39] and the inability of nontoxigenic strains to cause foot abscess in cattle following inoculation through the interdigital skin.[23] The associations among the titers of leucocidin, hemolysin, and cytotoxicity for cultured cells,[32] and among respective neutralization titers in serum from immunized animals, suggest that the cytolytic effects are mediated by a common molecule. The activity of the hemolysin is comparable to that of a phospholipase A and lysophospholipase.[40]

The most commonly used experimental animals are rabbits, mice, and guinea pigs.[5] Rabbits are particularly susceptible.[18] S.c. and i.m. inoculations of culture produce areas of progressive edema and necrosis and the animals usually die within 4 to 70 d.[18,24] I.v. inoculation produces disseminated infection in rabbits, with a special tendency for involvement of liver, lungs, joints, and tendon sheaths. Intradermal inoculation in rabbits produces a local abscess which tends to burst and heal. The average minimal infective dose for rabbits and mice (i.p.), appears to be about 2×10^7 cells of a young virulent culture.[5,41] In mice, s.c. inoculation tends to produce abscesses which rapidly discharge their contents, or the inoculum is lost through puncture wounds.[5,41] I.p. inoculation is commonly used in mice. Necrosis of the liver is commonly seen. Guinea pigs are much less susceptible than rabbits or mice and s.c. inoculation commonly produces abscesses which either burst or heal, or spread into adjacent s.c. tissue and musculature.[18]

APPENDIX

A. Media

1. Transport Media

a. Cary and Blair Transport Medium

Cary and Blair medium (BBL)	12.5 g
Calcium chloride, 1% aqueous	9 ml
Resazurin 0.05% aqueous	5 ml
L-Cysteine HCl	0.5 g
Distilled water	980 ml

Put all ingredients except cysteine into a flask and heat until dissolved. Allow to cool while gassing flask with CO_2 using a glass tube hooked over the rim of the flask. Add cysteine, dissolve, and adjust pH to 8.4 and then gas with N_2 in the same way. Dispense 10 ml into screw-capped bottles (25 ml) while gassing each with N_2 and then fasten screw caps down firmly. Steam for 15 min on 3 successive days. Store in refrigerator.

b. SBL Transport Medium

Bacto agar	10 g
Thiogycollic acid	0.5 ml
Sodium glycero-phosphate	10 g
$CaCl_2$ 1% aqueous	10 ml
L-Cysteine HCl	250 mg
Methylene blue 0.1% aqueous	2 ml
Distilled water	950 ml

Prepare in the same way as for Cary and Blair transport medium except that the pH is adjusted to 7.2. Store in refrigerator.

2. Culture Media

a. Brain Heart Infusion Blood Agar[8]

Use Brain Heart Infusion Agar (Difco®) as the base and supplement with 5% bovine blood, 0.5% yeast extract (Difco®), 0.5% beef extract (Difco®), 5 μg/ml hemin (Sigma), 0.5 μg/ml vitamin K_1 (Sigma), and 0.05% cysteine HCl (Sigma).

b. Eugon Blood Agar

Use Eugonagar® (BBL) as the base and supplement with 5% bovine blood, 0.2% yeast extract (Difco®), 5 μg/ml hemin (Sigma), 0.5 μg/ml vitamin K_1 (Sigma), and 0.05% cysteine HCl (Sigma).

c. Brucella Blood Agar

Use Brucella agar (Difco®) as the base and supplement with 5% bovine blood, 5 μg/ml hemin (Sigma), 0.5 μg/ml vitamin K_1, and 0.05% cysteine HCl (Sigma).

d. Eugon Broth Supplemented with Yeast Extract (EY)

Eugonbroth (BBL) pH 8.2 was supplemented with 0.2% yeast extract (Difco).

e. Egg-Yolk Agar

Peptone or trypticase	40 g
Na_2HPO_4	5.0 g
KH_2PO_4	1.0 g
NaCl	2.0 g
$MgSO_4$	0.1 g
Glucose	2.0 g

Haemin (Sigma)	5 mg
Agar	20.0 g
Distilled water	1000 ml

Adjust pH to 7.5, autoclave at 121°C for 15 min, cool to 50°C, add yolk of two eggs, mix, and pour plates. (It is advisable to use egg yolk that has been obtained in advance and tested for sterility.)

3. Selective Medium

a. Fusobacterium Egg Yolk Agar[19]

The medium is prepared from Brucella Agar base (Difco®) supplemented with 0.5% Na_2HPO_4, 0.1% KH_2PO_4, 0.01% $MgSO_4$, 5 μg/ml haemin (Sigma), and 1 ml/l polysorbate 80 (BBL). The pH is adjusted to 7.6 and the medium autoclaved at 121°C for 15 min, then cooled to 50°C, and 5 μg/ml vancomycin (Eli Lilly & Co), 100 μg/ml neomycin (Upjohn), and 3 μg/ml josamycin (Yamanouchi Pharmaceutical Co., Tokyo) added. Add sterile egg yolk to a final concentration of 2.5% v/v, and pour plates.

REFERENCES

1. **Beveridge, W. I. B.,** *Bacterial Diseases of Cattle, Sheep and Goats, Animal Health in Australia,* Vol. 4, Australian Government Publishing Service, Canberra, 1983, 178.
2. **Langworth, B. F.,** *Fusobacterium necrophorum:* its characteristic and role as an animal pathogen, *Bacteriol. Rev.,* 41, 373, 1977.
3. **Smith, G. R., Oliphant, J. C., and Parsons, R.,** The pathogenic properties of *Fusobacterium* and *Bacteroides* species from wallabies and other sources, *J. Hyg.,* 92, 165, 1984.
4. **Roberts, D. S.,** Synergic mechanisms in certain mixed infections, *J. Infect. Dis.,* 120, 720, 1969.
5. **Simon, P. C. and Harris, M.,** *Fusobacterium necrophorum (Sphaerophorus necrophorus),* in *Handbuch der bakteriellen Infektionen bei Tieren,* Band 3, Blobel, H. and Schliesser, T., Eds., Gustav Fischer, Stuttgart, 1981, 639.
6. **Moore, W. E. C., Holdeman, L. V., and Kelley, R. W.,** Genus 11 *Fusobacterium* Knorr 1922, in *Bergey's Manual of Systematic Bacteriology,* Vol. 1, Krieg, N. R. and Holt, J. G., Eds., Williams & Wilkins, Baltimore, 1984, 631.
7. **Holdeman, L. V., Cato, E. P., and Moore, W. E. C., Eds.,** *Anaerobe Laboratory Manual,* 4th ed., V. P. I. Anaerobe Laboratory, Virginia Polytechnic Institute and State University, Blacksburg, 1977.
8. **Berg, J. N. and Loan, R. W.,** *Fusobacterium necrophorum* and *Bacteroides melaninogenicus* as etiologic agents of foot rot in cattle, *Am. J. Vet. Res.,* 36, 1115, 1975.
9. **Berg, J. N. and Scanlan, C. M.,** Studies of *Fusobacterium necrophorum* from bovine hepatic abscesses: biotypes, quantitation, virulence, and antibiotic susceptibility, *Am. J. Vet. Res.,* 43, 1580, 1982.
10. **Allen, S. D. and Siders, J. A.,** Procedures for isolation and characterization of anaerobic bacteria, in *Manual of Clinical Microbiology,* 3rd ed., Lennette, E. H., Editor-in-Chief, American Society for Microbiology, Washington, D.C., 1980, 397.
11. **Finegold, S. M. and Citron, D. M.,** Gram-negative, nonsporeforming anaerobic bacilli, in *Manual of Clinical Microbiology,* 3rd ed., Lennette, E. H., Editor-in-Chief, American Society for Microbiology, Washington, D.C., 1980, 431.
12. **Hungate, R. E.,** A roll tube method for cultivation of strict anaerobes, in *Methods in Microbiology,* Vol. 3B, Norris, J. R. and Ribbons, D. W., Eds., Academic Press, London, 1969, 117.
13. **Sutter, V. L., Citron, D. M., and Finegold, S. M.,** *Wadsworth Anaerobic Bacteriology Manual,* 3rd ed., C. V. Mosby Co., St. Louis, 1980, 1.
14. **Reilly, S.,** The carbon dioxide requirements of anaerobic bacteria, *J. Med. Microbiol.,* 13, 573, 1980.
15. **Fièvez, L.,** *Etude Compareè des souches de Sphaerophorus necrophorus isolees chez l'Homme et chez l'Animal,* Presses Academiques Europeennes, Brussels, 1963.
16. **Simon, P. C.,** Cultivation and maintenance of *Sphaerophorus necrophorus.* II. Influence of pH on maximum growth and acid production in medium 156, *Can. J. Comp. Med.,* 39, 89, 1975.
17. **Simon, P. C.,** The effect of temperature on growth and survival of *Fusobacterium necrophorum* isolated from bovine liver abscesses, *Can. J. Comp. Med.,* 41, 169, 1977.

18. **Beveridge, W. I. B.,** A study of twelve strains of *Bacillus necrophorus* with observations on the oxygen intolerance of the organism, *J. Pathol. Bacteriol.*, 38, 467, 1934.
19. **Morgenstein, A. A., Citron, D. M., and Finegold, S. M.,** New medium selective for *Fusobacterium* species and differential for *Fusobacterium necrophorum, J. Clin. Microbiol.*, 13, 666, 1981.
20. **Fales, W. H. and Teresa, G. W.,** A selective medium for the isolation of *Sphaerophorus necrophorus, Am. J. Vet. Res.*, 33, 2317, 1972.
21. **Garcia, M. M. and McKay, K. A.,** A satisfactory medium and technique for the mass cultivation of *Sphaerophorus necrophorus, Can. J. Microbiol.*, 19, 296, 1973.
22. **Berg, J. N. and Evans, J. W.,** Identification of common antigens in ribosome-rich extracts from *Fusobacterium necrophorum, Am. J. Vet. Res.*, 46, 127, 1985.
23. **Emery, D. L., Vaughan, J. A., Clark, B. L., Dufty, J. H., and Stewart, D. J.,** Cultural characteristics and virulence of strains of *Fusobacterium necrophorum* isolated from the feet of cattle and sheep, *Aust. Vet. J.*, 62, 43, 1985.
24. **Orcutt, M. L.,** A study of *Bacillus necrophorus* obtained from cows, *J. Bacteriol.*, 20, 343, 1930.
25. **Feldman, W. H., Hester, H. R., and Wherry, F. P.,** The occurrence of *Bacillus necrophorus* agglutinins in different species of animals, *J. Infect. Dis.*, 59, 159, 1936.
26. **Fales, W. H. and Teresa, G. W.,** Fluorescent antibody technique for identifying isolates of *Sphaerophorus necrophorus* of bovine hepatic abscess origin, *Am. J. Vet. Res.*, 33, 2323, 1972.
27. **Warner, J. F., Fales, W. H., and Teresa, G. W.,** Passive hemagglutination test for determining the immune response of rabbits to *Sphaerophorus necrophorus* of bovine hepatic abscess origin, *Am. J. Vet. Res.*, 35, 551, 1974.
28. **Simon, P. C.,** A simple method for rapid identification of *Sphaerophorus necrophorus* isolates, *Can. J. Comp. Med.*, 39, 349, 1975.
29. **Sparling, P. F.,** Bacterial virulence and pathogenesis: an overview, *Rev. Infect. Dis.*, 5 (Suppl. 4), 637, 1983.
30. **Roberts, D. S.,** Toxic, allergenic and immunogenic factors of *Fusiformis necrophorus, J. Comp. Pathol.*, 80, 247, 1970.
31. **Coyle-Dennis, J. E. and Lauerman, L. H.,** Biological and biochemical characteristics of *Fusobacterium necrophorum* leukocidin, *Am. J. Vet. Res.*, 39, 1790, 1978.
32. **Emery, D. L., Dufty, J. H., and Clark, B. L.,** Biochemical and functional properties of a leucocidin produced by several strains of *Fusobacterium necrophorum, Aust. Vet. J.*, 61, 382, 1984.
33. **Garcia, M. M., Charlton, K. M., and McKay, K. A.,** Characterization of endotoxin from *Fusobacterium necrophorum, Infect. Immun.*, 11, 371, 1975.
34. **Warner, J. F., Fales, W. H., Sutherland, R. C., and Teresa, G. W.,** Endotoxin from *Fusobacterium necrophorum* of bovine hepatic abscess origin, *Am. J. Vet. Res.*, 36, 1015, 1975.
35. **Scanlan, C. M. and Hathcock, T. L.,** Bovine rumenitis-liver abscess complex: a bacteriological review, *Cornell Vet.*, 73, 288, 1983.
36. **Scanlan, C. M., Berg, J. N., and Fales, W. H.,** Comparative *in vitro* leukotoxin production of three bovine strains of *Fusobacterium necrophorum, Am. J. Vet. Res.*, 43, 1329, 1982.
37. **Weinstein, W. M., Onderdonk, A. B., Bartlett, J. G., and Gorbach, S. L.,** Experimental intra-abdominal abscesses in rats: development of an experimental model, *Infect. Immun.*, 10, 1250, 1974.
38. **Garcia, M. M., Alexander, D. C., and McKay, K. A.,** Biological characterization of *Fusobacterium necrophorum* cell fractions in preparation for toxin and immunization studies, *Infect. Immun.*, 11, 609, 1975.
39. **Emery, D. L., Vaughan, J. A., Clark, B. L., and Stewart, D. J.,** Virulence determinants of *Fusobacterium necrophorum* and their prophylactic potential in animals, in *Footrot in Ruminants,* Proc. Workshop, Melbourne, 1985, Stewart, D. J., Peterson, J. E., McKern, N. M., and Emery, D. L., Eds., CSIRO Division Animal Health/Australian Wool Corp. Glebe, N. S. W., 1986, 267.
40. **Abe, P. M., Kendall, C. J., Stauffer, L. R., and Holland, J. W.,** Hemolytic activity of *Fusobacterium necrophorum* culture supernatants due to presence of phospholipase A and lysophospholipase, *Am. J. Vet. Res.*, 40, 92, 1979.
41. **Emery, D. L. and Vaughan, J. A.,** Generation of immunity against *Fusobacterium necrophorum* in mice inoculated with extracts containing leucocidin, *Vet. Microbiol.*, 12, 255, 1986.
42. **Vaughan, J. A.,** Unpublished data.
43. **Emery, D. L.,** Unpublished data.

Section III: Immunology

Section III

IMMUNOLOGY

W. K. Yong, G. G. Riffkin, and J. R. Egerton

I. INTRODUCTION

The science of immunology had its beginnings about 200 years ago when Edward Jenner first observed that immunity to small pox could be induced in people by injecting them with lymph from the blisters of vaccinia lesions on the teats of cows. Although nobody knew the exact nature of infection at that time, this observation led to the discovery of the first useful *antigen* that could evoke an immune response and give rise to resistance against a highly infectious disease. From this audacious beginning, immunology has grown and is now recognized as a field of biomedical endeavor in its own right, by no means restricted to the area of infection, and the term immunity no longer necessarily implies a response that is protective to the individual. This preamble describes in a didactic way the nature of antigens and the consequences when animals are exposed to them.

II. ANTIGENS

Antigens are defined as those biochemical substances capable of inducing a specific immune response. Usually, antigens induce the formation of antibody, as well as cellular types of immune reactions, although there are instances in which one or the other type is stimulated preferentially.

An antigenic determinant (or epitope) requires consistent shape and charge pattern if it is to be recognized by a specific antibody. Since the antibody-combining site is relatively small, most large organic molecules and almost all biological molecules, with the exception of lipids, fulfill this criterion. The most extensively studied antigens are proteins and polysaccharides, which can be strongly antigenic when administered either in soluble form or as part of a more complex structure, for example, a bacterial cell wall.

Most natural antigens are multivalent. They are either large protein molecules with a large number of different epitopes on their surfaces, or are polymers, presenting the same few epitopes in a large array. Examples of such multivalent but unideterminant antigens are *Escherichia coli* endotoxin, capsular polysaccharides from streptococci and pneumococci, and polymerized flagellin. Chapters 7, 8, and 9 discuss these natural antigens of bacteria which cause footrot and foot abscess.

III. ANTIGEN RECOGNITION AND IMMUNITY

When antigen is presented to and recognized by the immune system the result will be an immunological response effected by the T lymphocytes (cell mediated) and/or the B lymphocytes (humoral or antibody). The two classes of lymphocytes are named after the tissues in which they originate. B lymphocytes are produced by the bone marrow, while T lymphocytes are produced by the thymus. Although both cell types are specifically involved in antigen recognition, only the B lymphocytes differentiate into antibody-secreting cells. On the other hand, T lymphocytes are involved in regulating the activities of B lymphocytes, expression of delayed-hypersensitivity reactions, resistance to viruses, fungi and certain facultative intracellular bacteria, rejection of allograft in transplantation immunity, and killing of other cells such as tumors. With the present knowledge of T-lymphocyte functions, the division of immune responses into cell mediated and humoral, although convenient conceptually, is therefore artificial and should now be regarded as of historical interest.

As the initial bacterial invasion in footrot and foot abscess is restricted to the avascular epidermis, many people have been skeptical of the capacity of a naturally infected animal to mount an immune response against the respective causal microorganism. Chapter 9 considers this problem and establishes without doubt that invasion of the hoof structure by bacteria is an adequate stimulus to the immune response. Chapters 10 and 11 explore the practical application of the immune response in diagnosis and immunotherapy, respectively.

Readers should note that the terms "fimbriae" and "pili" are accepted at present by the editors as synonyms which describe the same nonflagellar, filamentous appendages found on the surface of *Bacteroides nodosus*. The preferred usage of either term is therefore left to the respective author of each chapter.

Chapter 7

ANTIGENS OF *BACTEROIDES NODOSUS*

E. K. Moses and W. K. Yong

TABLE OF CONTENTS

I. INTRODUCTION

Studies on the antigens of *Bacteroides nodosus* have been primarily aimed at (1) the identification of immunogens for use in vaccines and (2) the identification and characterization of factors associated with virulence, with the objective of developing improved diagnostic reagents. Of particular interest have been the fimbriae, the extracellular proteases, and outer-membrane components of the cell wall. Therefore, for the convenience of classification, the antigens of *B. nodosus* can be divided into three categories: the surface antigens, the extracellular antigens, and the somatic antigens.

II. SURFACE ANTIGENS

A. Fimbriae

The terms "fimbriae" and "pili" are both commonly employed to describe the nonflagellar filamentous appendages found on the surface of bacteria. There is general disagreement as to which of these two terms is more correct for the description of such surface appendages.[1] In order to avoid alternatives generally accepted in the literature, such as "*fim*" (genetic marker), "Fim" (genotypically fimbriate), or "Fimo" (irreversibly nonfimbriate), Ottow[1] proposed that the notations "fimbriae", "fimbriate", or "fimbriated" be used to describe all nonflagellar bacterial appendages other than those clearly involved in the transfer of bacterial and viral nucleic acids. Mattick et al.[2] have adopted the convention suggested by Ottow[1] and, in the absence of other evidence or argument, the term "fimbriae" will also be used exclusively in this chapter.

1. Electron Microscopy

Stewart[3] and Walker et al.[4] demonstrated that *B. nodosus* possess fimbriae. These authors also demonstrated that the fimbriae are antigenic and that they represent the antigenic structure responsible for mediating the serological K-agglutination reaction.[5] Stewart[3] reported a mean diameter of 4.6 nm and a length often exceeding 10 μm, whereas Walker et al.[4] reported a diameter of 6 nm and a length of up to 5 μm. A polar localization was observed by Walker et al.[4]

2. Serological Analysis

Egerton[5] reported that *B. nodosus* produce at least two agglutinating antigens during growth on hoof-agar plates. They could be distinguished by the type of agglutination observed in the presence of antiserum, by differences in their heat lability, and by the ability of one of them to cause hemagglutination of sensitized sheep erythrocytes. By analogy with types of agglutination described for envelope and somatic antigens in *Escherichia coli* and *Pasteurella multocida,* Egerton[5] referred to the relatively heat-labile antigen of *B. nodosus* as K antigen, and that which resists boiling for 1 h as O antigen. However, unlike the K antigen of other Gram-negative bacteria, the K antigen of *B. nodosus* can be removed from the cells by washing. It has since been shown that this K antigen is a constituent of *B. nodosus* fimbriae.[3,4]

On the basis of agglutination reactions involving the K antigen, Egerton[5] classified 33 of 46 isolates derived from outbreaks of benign and virulent ovine footrot into three serotypes A, B, and C. Thorley,[6] using cross-tube agglutination tests, found that sera against English isolates gave characteristic agglutinations essentially the same as those described by Egerton.[5] Thorley[6] was able to extend the K antigen A, B, and C system of Egerton[5] by nine more serologically distinct groups among English isolates. Similarly, Schmitz and Gradin[7] characterized 97 North American *B. nodosus* isolates by the tube-agglutination test. Fourteen serotypes were identified, including isolates that were serologically similar to Australian serotypes A, B, and C.

Claxton et al.[8] examined 1267 Australian *B. nodosus* isolates by slide- and cross-tube agglutination tests. Of these, 1260 could be classified by slide agglutination into eight serogroups designated A to H. However, consistent with the findings of Schmitz and Gradin,[7] the results of cross-tube agglutination tests suggested that 16 or more subgroups or serotypes might exist.

3. Molecular Analysis

The fimbrial strand of *B. nodosus* consists of a single, repeated polypeptide subunit of approximate molecular weight 17 kDa.[9-11] However, Anderson et al.[12] have found that there is substantial variation in the electrophoretic mobility of the fimbrial subunit proteins from the prototypes of each of the eight serogroups. The apparent molecular weights of the fimbrial subunit proteins were found to range from about 16.5 kDa in serogroup D to 19 kDa in serogroup F, with the exception of serogroup H where fimbriae appeared to consist of two smaller polypeptides which in the prototype (H1) had apparent molecular weights of about 6 and 10 kDa. In serogroups, A, B, C, and E, the apparent molecular weights were clustered in the range of 17 to 17.5 kDa, whereas in serogroup G the subunit protein was about 18.5 kDa. Electrophoretic variation in the fimbrial subunit protein was also observed among different serotypes (i.e., within a serogroup), although in most cases this was not as pronounced as between the serogroups.[12]

Mattick et al.[10] found that there were two polypeptide antigens associated with fimbrial preparations. The 17 kDa structural subunit protein of fimbriae was the major component (approximately 98%) but there were also trace amounts of a polypeptide component with a molecular weight of approximately 80 kDa. The observation that this protein could not be separated from the fimbrial strand (17 kDa subunit) by a variety of techniques usually employed in the purification of fimbriae led to the suggestion that the 80 kDa polypeptide may be physically associated with the fimbrial strand.[10] Consequently, fimbrial fractions containing this protein were examined by electron microscopy and were shown to contain a significant number of "cap-shaped" structures, attached to one end of truncated fimbrial strands. Mattick et al.[10] have suggested that the 80 kDa polypeptide may actually constitute the basal protein which anchors the fimbria to the outer membrane, analogous to a similar protein recently described in enterotoxigenic strains of *E. coli.*[13] Electrophoretic variation in the basal protein (with an apparent molecular weight range from about 75 to 84 kDa) has been shown to occur at all levels of the serotypic heirarchy, in a manner which appears unrelated to the serological classification system of *B. nodosus.*[12]

Western transfer analyses have shown that fimbrial subunit proteins are strongly recognized by antiserum raised against the homologous serogroup, but not by antisera raised against heterologous groups.[14] However, the basal protein cross-reacts randomly among different isolates from various serogroups, depending on the antiserum employed, and appears to form separate groups independent of the primary system.[14]

McKern et al.[11] have reported the complete amino acid sequence of the fimbrial subunit protein from *B. nodosus* strain 198 (serogroup A) and compared it with those of some other bacteria whose fimbrial subunit proteins have been completely or partially sequenced. The subunit protein was shown to contain 151 amino acid residues corresponding to a molecular weight of 16.3 kDa in the unreduced form. The N-terminal amino acid was identified as *N*-methylphenylalanine and the C-terminal amino acid as asparagine. The hydrophobic N-terminal 24-residue sequence was shown to be highly homologous with that of fimbrial subunit proteins from *Neisseria gonorrhoeae, Moraxella nonliquefaciens,* and *Pseudomonas aeruginosa.* However, it was also pointed out that partial or complete amino acid sequences of fimbrial subunit proteins from several strains of *E. coli* show no significant homology with any regions of the fimbrial subunit protein from these organisms and conversely the two known C-terminal sequences of *E. coli* fimbrial subunit protein are quite hydrophobic, whereas in the fimbrial subunit protein from *B. nodosus* it is the N-terminal sequence which is particularly hydrophobic. Therefore, it appears that the fimbriae can consist of quite different amino acid sequences for different bacteria, but with one end necessarily being hydrophobic. It was suggested[11] that this highly conserved hydrophobic region may facilitate transport of the protein through cell membranes and promote interactions between the subunits to form filamentous structures.

In a later study, the amino acid sequence of the fimbrial subunit protein from *B. nodosus* strain 216 (serogroup E) was determined and compared with the previously determined sequence of fimbrial subunit protein from *B. nodosus* strain 198 (serogroup A).[15] The protein had a calculated molecular weight of 15.96 kDa and contained 151 amino acid residues, the same as the previously sequenced fimbrial subunit protein from strain 198. The sequence of the first 44 residues was common to both strains, including the unusual N-terminal amino acid, *N*-methylphenylalanine. Of the remaining 107 residues, 37% of them differed between the two strains. A comparison of profiles of hydrophilicity constructed from the sequence data revealed a similarity of patterns. Of particular interest was the conservation of the largest peak, with its apex at residues 71 to 72, since the data of Hopp and Woods[16] suggested that the point of highest local hydrophilicity was associated with an antigenic determinant of the protein.

4. Role in Infection

The role of *B. nodosus* fimbriae in infection is not known. The fimbriae of *B. nodosus* do not appear to have an adhesive function,[17] as in some other bacteria.[1] Every[9] has suggested that *B. nodosus* fimbriae are most like the group-4-type fimbriae defined by Ottow.[1] Fimbriae of this type have a polar location and appear to be involved in surface translocation by a phenomenon known as twitching motility.[18] Recent studies[19] have shown that although both virulent and benign strains of *B. nodosus* contain equivalent numbers of polar fimbriae, virulent isolates clearly exhibit a higher degree of twitching motility than do benign isolates. The fimbriae of *B. nodosus* may therefore act to promote the penetration by the bacteria of the epidermal matrix in the hoof. This is consistent with the pathogenesis of footrot, a disease characterized by a steady progression of *B. nodosus* through the proteinaceous epidermal tissue.

5. Role in Protection

The first indication that the humoral response to fimbriae of *B. nodosus* can confer protection against footrot in sheep was reported by Walker et al.[4] They vaccinated sheep with a purified preparation of fimbriae and found that the agglutinin responses elicited were identical to those given by injection of a commercial vaccine prepared from formalized whole cells. As it was known[20] that there was a significant correlation between the high agglutination titers found in animals vaccinated with whole cells, and subsequent protection against footrot, Walker et al.[4] concluded that fimbriae contained the protective antigen.

Further demonstrations that the humoral response to fimbriae confers protection have come from other experimental vaccine trials. Vaccination with preparations of isolated fimbriae confers protection that is as good as, if not superior to, that obtained with whole cells.[21-23] However, vaccination with nonfimbriate cells (generated in culture) is much less effective.[22,24,25]

The immunity conferred by vaccination with fimbriae is related to the serogroup involved.[21,22,26,27] However, the possible existence of cross-protective antigens associated with, but distinct from, the fimbriae has been reported.[26,27] and sera from sheep with footrot recognize a variety of *B. nodosus* proteins distinct from the fimbriae.[28]

B. Other Surface Structures

Several structures other than the fimbriae have been observed to project from the surface of *B. nodosus*. Blebs have been occasionally observed on most parts of the bacterial surface[4] and "fuzzy" projections or thin pili have been seen emerging either from the poles only, or sometimes from the entire surface.[29] Every and Skerman[30] have also observed large, rodlike structures, diffuse polar material (unlikely to be fimbriae or fuzzy projections reported by Short et al.[29]), and an outermost additional layer. In a later study, Every and Skerman[31] went on to show that in all strains and culture variants of *B. nodosus,* the structure of the plasma membrane, peptidoglycan layer, and outer membrane were similar. However, those strains and variants of low immunoprotective activity and virulence were distinguished from strains of high virulence and immunoprotective activity by their lack, or reduction, of one or two of the three surface structures: additional layer, diffuse polar material, and fimbriae. Therefore, the possible involvement of these three structures as virulence factors or protective immunogens should be considered.

Every and Skerman[31] have suggested that the additional layer may have a role in adhesion of *B. nodosus* to the epidermis of sheep and cattle. The basis for this suggestion was the demonstration that an additional layer was involved in the adhesion of *Aeromonas salmonicida* to fish tissues[32] and the possible involvement of an additional layer in the adhesion of Gram-negative rumen bacteria to substrates.[33] Confirmation of a similar role for the additional layer of *B. nodosus* must await demonstration that it is equivalent to those from bacteria of other genera.

III. EXTRACELLULAR ANTIGENS

A number of bacterial species elaborate extracellular enzymes to which their invasive properties have been attributed.[34] Proposed mechanisms whereby such enzymes could enhance virulence include destruction of (1) host-protective macromolecules in fluid such as mucus, (2) lipoprotein membranes that surround organelles and cytoplasm of defense cells, (3) ground substances such as hyaluronic acid, and (4) immunoglobulins such as secretory IgA.

A. Proteases

The first evidence indicating that *B. nodosus* produced proteases was the digestion of meat particles with the subsequent formation of tyrosine crystals observed by Beveridge[35] when *B. nodosus* was grown in cooked-heart broth.

Thomas[36] related the progress of experimental infections with pure cultures of *B. nodosus* in the sheep foot to the histopathological changes observed. He suggested that, "In the lesion of footrot, the separation of the stratum corneum, the detachment and disintegration of partially cornified cells, and the liquefaction of the stratum corneum with the formation of fissures and sinuses in the hoof, appeared to be the result of a digestive process and was probably caused by the activity of a keratinase elaborated by *B. nodosus.*" These early observations encouraged an examination of the enzymes produced by *B. nodosus.*

1. Substrate Specificity

Several substrate-specificity studies have been made on extracellular enzymes from various *B. nodosus* biotypes and culture variants to elucidate the role of these enzymes in the pathology of ovine footrot.

Thomas[37] tested partially purified preparations of *B. nodosus* extracellular enzymes for their ability to digest substrates of dermal and epidermal tissues. The preparations did not hydrolyze hyaluronic acid, chondroitin sulfate, or collagen; however, weak elastase activity was demonstrated. It was concluded that the absence of hyaluronidase, chrondroitinase, and collagenase activity, and the presence of weak elastase activity were consistent with the site of infection being in the epidermis of the foot, since hyaluronic acid, chondroitin sulphate, collagen, and elastin are constituents of dermal tissue. This was consistent with the results of Thomas,[36] who showed that there were no degenerative changes in the dermis in sections of sheeps' feet that had been infected with pure cultures of *B. nodosus*. However, Thomas[37] did find that the enzyme preparations liberated epithelial cells from frozen sections of sheep hoof, digested powdered sheep hoof in the presence of cysteine (a sulfhydryl compound), and produced ulcers when injected in the epidermis of the skin of sheep. Once again, the results implied that proteases of *B. nodosus* might have an activity analogous to that of keratinases.

Broad and Skerman,[38] therefore, decided to reexamine the protease activity of cultures grown under conditions described by Skerman,[39] which supported more rapid growth and higher yields of organisms than obtained by Beveridge[35] and Thomas,[37,40] and to study in greater detail its activity on keratin and other substrates. They demonstrated that a purified-protease preparation from a virulent strain of *B. nodosus* had some degradative action on fibrin, hoof and hide powders, and on gelatin, although on each the activity was less than on casein. They were not able to demonstrate any activity on bovine hemoglobin, elastin, human or bovine gammaglobulin, ovalbumin, or sheep, rabbit, or human albumin. However, in conflict with statements made by Thomas[40,41] to the effect that earlier work[42] had shown that hoof particles were digested when incorporated into agar media inoculated with *B. nodosus*, they found no evidence that *B. nodosus* was capable of attacking ovine hoof constituents during growth *in vitro*. They indicated that the effects, on pulverized hoof, of *B. nodosus* enzyme preparations, or indeed of the organisms in culture, may have little relevance to their effects on intact hoof keratin. They suggested that this was because the process of dry grinding reduces keratins to low-molecular-weight peptides,[43] so that the proportion of true keratin in the ball-milled hoof powder cannot be determined.[44] For these reasons, gamma-irradiated wool was used as substrate for keratinase. They found that the protease preparation did not possess any keratinase or pseudokeratinase activity when incubated for up to 90 h with native or denatured wool.

Kortt et al.[45] found that *B. nodosus* proteases digested casein, hide-powder azure, ^{125}I-labeled elastin, and denatured hemoglobin. However, there was no digestion of native bovine serum albumin, or keratin (as finely ground wool), and *B. nodosus* proteases did not hydrolyze the synthetic substrates α-*N*-benzoyl-L-arginine ethyl ester or *N*-benzoyl-L-tyrosine ethyl ester.

In addition, Green[46] found that electrophoretically purified *B. nodosus* proteases hydrolyzed collagen I, collagen III, and α-keratin. It should be noted that in previous studies,[38,45] a hard keratin (wool) was used as a substrate, whereas the observations of Green[46] were made on a soft keratin isolated from the stratum corneum, an epidermal component perhaps more allied to the site of *B. nodosus* infection.

2. Inhibitor Studies

Broad and Skerman[38] demonstrated that *B. nodosus* protease activity was inhibited by the metal ion chelator ethylene diamine tetraacetic acid (EDTA) but not by 1,10-phenanthroline, ovomucoid, soybean trypsin inhibitor or *N*-α-tosyl-L-lysyl chloromethane and *N*-tosyl-L-phenylalanyl chloromethane (TPCK) (specific trypsin and chymotrypsin inhibitors, respectively).

Kortt et al.[45] found that both virulent and benign protease activity were completely inhibited by phenylmethylsulfonyl fluoride (PMSF) and EDTA but not by 1,10-phenanthroline. TPCK inhibited approximately 30% of the activity of both *B. nodosus* proteases, and iodoacetate inhibited approximately 20 and 30% of the virulent and benign protease activity, respectively.

Depiazzi and Rood[47] confirmed that both virulent and benign protease activity were inhibited by PMSF, EDTA, and ethyleneglycol-*bis*-aminoethylether-*N,N*-tetraacetic acid (EGTA) but were not inhibited by 1,10-phenanthroline. In a later study on electrophoretically purified *B. nodosus* proteases, Green[46] demonstrated that the enzymes were inhibited by PMSF, diphenylcarbamyl chloride, TPCK, EGTA, and EDTA.

These results indicate that *B. nodosus* proteases are chymotrypsin-like serine proteases that require a divalent metal ion such as Ca^{2+}, Mg^{2+}, or Mn^{2+} for activity but not the heavy-metal ions such as Zn^{2+} and Fe^{2+}. The serine nature of *B. nodosus* proteases was further demonstrated by Gordon et al.[48] who reported their affinity for the seryl-specific ligand, phenyl boronate.[49]

3. *Molecular Weight Determinations*

There is considerable controversy in the literature over the molecular weights of the extracellular proteases produced by *B. nodosus*. Every,[50] using gradient polyacrylamide gels, reported that the molecular weights of the proteases ranged from 70 to 129 kDa and, more specifically, Every and Green[51] reported that five individual proteases purified from one strain had molecular weights of 78, 82, 88, 96, and 107 kDa when they used a nondenaturing gradient polyacrylamide gel electrophoresis system.

However, using gel-filtration techniques, Depiazzi and Rood[47] reported a molecular weight of 17 to 20 kDa and Gordon et al.,[48] 8 to 43 kDa for virulent and benign *B. nodosus* proteases. The work of Kortt et al.[52] does not support the contention that the apparent anomalies in molecular weight are a result of the method of determination. They have reported that the molecular weights of all the major *B. nodosus* proteases are in the range of 40 to 42 kDa based on gel-filtration data, nondenaturing gradient polyacrylamide gel electrophoresis, and SDS-polyacrylamide gel electrophoresis. Kortt et al.[52] also noted that although the proteases are relatively stable, there was evidence of limited autolysis during purification. Significant discrepancies in the molecular weight of the neutral protease of *Pseudomonas aeruginosa* have also been reported.[53] However, in the latter case, it was suggested that these discrepancies may be due to the different methods used to purify the enzymes, and that the autocatalytic activity of this protease may also contribute to the observed differences.

4. *Role in Laboratory Diagnosis*

Since Beveridge[35] first suggested that *B. nodosus* protease may be associated with the pathogenesis of footrot in sheep, there have been several reports of a correlation between the virulence of *B. nodosus* and proteolytic activity on casein,[54,55] hide-powder azure,[56] and elastin.[25,27]

Thomas[54] tested the production of proteases in broth cultures by strains of *B. nodosus* isolated from sheep with footrot and organisms isolated from sheep with foot scald, by measuring the degree of digestion of azocasein by samples of the cultures. The strains of *B. nodosus* isolated from sheep with footrot produced approximately twice as much protease as the organisms from sheep with foot scald, although the latter grew slightly better.

Egerton and Parsonson[55] suggested that the more serious disease with extensive separation of horn be called virulent footrot, and the scald associated with less proteolytic strains of *B. nodosus* be called benign footrot. Furthermore, they compared the proteolytic activity on casein-agar plates of 12 isolates of *B. nodosus* from flocks with benign footrot to that of 13 isolates from flocks with virulent footrot. They found that after 3 to 4 d growth on casein-agar plates, the streak of *B. nodosus* was surrounded by an inner clear zone and an outer ring of precipitate. The ratio of the width of the clear zone to that of the remaining band of precipitate was determined for each isolate at 4 d and was termed the proteolytic index. They found a highly significant difference between the mean proteolytic indices from virulent and benign strains.

Depiazzi and Richards[56] described an *in vitro* test for the differentiation of virulent and benign *B. nodosus* strains using a degrading-proteinase (= protease) technique which measured the rate of change of the enzyme activity in broth culture following the termination of bacterial growth. The spectrophotometric measurement of the release of a dye from a hide-powder-azure conjugate by the action of proteolytic enzymes provided an objective analysis of enzyme activity. The relative thermostability of proteolytic enzymes produced by virulent and benign *B. nodosus* isolates appears to form the basis of the degrading-proteinase test.[45,47] According to the authors,[56] the degrading-proteinase test provided a far greater degree of specificity than the modified casein-agar-plate method.[55]

Stewart[57] reported the development of a highly specific diagnostic test based on the production of elastase. He demonstrated a definite association between the virulence of field isolates of *B. nodosus* and elastase production. All isolates derived from virulent ovine footrot produced elastase, whereas isolates cultured from benign footrot were elastase negative. He also demonstrated a close correlation between results of the elastin-agar test and the degrading-proteinase test of Depiazzi and Richards.[56] He indicated that it would be important to determine in future studies whether the difference in behavior of *B. nodosus* in the degrading-proteinase test is attributable to elastase activity *per se* or to another stable enzyme or group of enzymes which were also associated with virulence.

Skerman et al.[25] demonstrated that *B. nodosus* cultures of all strains and colony variants (representing ovine and bovine clinical isolates as well as laboratory variants) had proteolytic activity on gelatin, casein, albumin, and milk. Quantitative differences in proteolytic activity, among strains, could not be categorized by either plate or gel diffusion methods. Pronounced zones of clearing in elastin plates were produced by all cultures forming what they termed "B-type" colonies (i.e., papillate or beaded [B]-type colonies which were produced by fresh isolates from advanced ovine footrot). However, reduced elastase activity was shown by cultures of M-variants (mucoid [M]-type colonies which were formed by isolates from noninvasive *B. nodosus* infections of the interdigital skin of sheep and cattle, and also by cultures of some primary B-type colonies passaged nonselectively *in vitro*). All other cultures were elastase negative.

Recently, further differentiation of the proteases produced by virulent and benign isolates has been achieved by the use of electrophoretic separation on polyacrylamide gel followed by detection of the proteases using a gelatin-agarose substrate.[45,48,58] This procedure has been successfully used to distinguish among *B. nodosus* isolates that cause virulent ovine footrot, benign ovine footrot, and bovine dermatitis based on the common and unique electrophoretic protease-isoenzyme bands detected for the various isolates.[48,50,59]

In an attempt to develop a rapid immunologically based diagnostic test, Yong et al.[60] successfully raised antibodies in rabbits by injecting them with various electrophoretically separated individual proteases from a virulent and a benign *B. nodosus* isolate. However, when these antibodies were conjugated with horseradish peroxidase and used in an enzyme-linked immunosorbent assay to differentiate whole cells from virulent and benign strains, extensive cross-reactions were demonstrated with each antiprotease antibody tested.

5. *Role in Protection*

Egerton and Merritt[61] demonstrated that sheep with footrot and those vaccinated with *B. nodosus* develop specific inhibitors of proteolytic activity, later shown to be IgG_1 and IgG_2 immunoglobulins.[62] The serum from normal sheep, rabbits, horses, guinea pigs, and mice also inhibits the proteolytic activity of *B. nodosus* culture filtrates. The active fraction in sheep, rabbit, and horse serum was a 19S component distinct from IgM, and identified by immunoelectrophoresis and gel electrophoresis as an α-macroglobulin.[63]

The potential usefulness of proteases as protective antigens seemed unlikely as it was observed that active infections persisted concurrently with high antiprotease titers.[62] Furthermore, a vaccine made from crude-protease preparations did not protect sheep against challenge with *B. nodosus*.[23] However, more recently it has been reported that the purified proteases are promising cross-protective antigens for providing broad-spectrum cover against a wide range of serotypes.[64] These conflicting results may be related to the purity of the protease antigen used.

B. Other Extracellular Antigens

Yong and Gordon[65] made a comparative immunoelectrophoretic study of the extracellular proteins excreted/secreted by virulent (strain 198) and benign (strain 305) *B. nodosus* using antisera from rabbits immunized with these proteins. Four precipitin peaks unique to the virulent strain, and five peaks unique to the benign strain were identified in homologous and heterologous crossed immunoelectrophoresis. In an attempt to characterize the different extracellular protein antigens, rabbit antisera were raised against an outer-membrane protein (OMP; molecular weight 35 kDa) and fimbriae from virulent *B. nodosus* (strain 198) and various proteases from virulent (strain 198) and benign (strain 305) *B. nodosus*. No precipitin band was observed when extracellular protein from both *B. nodosus* strains were reacted against anti-OMP and antifimbriae sera. However, single precipitin bands unique to one protease from the benign strain and one protease from the virulent strain were identified. These results suggested that specific antigens other than proteases or fimbriae are important in determining whether a *B. nodosus* isolate is virulent or benign.

IV. SOMATIC ANTIGENS

Gram-negative bacteria are characterized by a complex cell wall (envelope) consisting of an inner (cytoplasmic or plasma) membrane and an outer membrane. The two are separated by a relatively thin peptidoglycan layer and a periplasmic space. The outer membrane is normally composed of phospholipids, lipopolysaccharide (endotoxin), and a limited number of characteristic and major proteins.

B. nodosus has the typical cell-envelope layers of a Gram-negative bacterium: plasma membrane, peptidoglycan layer, and outer membrane.[4,66,67] In addition, Stewart and Egerton[67] reported the presence of a capsule which was observed by light and electron microscopy. However, Cooper,[66] using light microscope staining techniques, could not detect a capsule.

A. Lipopolysaccharide

The outer membrane of the Gram-negative bacterial cell wall incorporates the lipopolysaccharide (LPS) layer which has three important constituents, namely lipid A, R-specific polysaccharide core, and repeating O-specific units.[68] These three constituents, together with the underlying lipoprotein layer, constitute the endotoxin complex.

Stewart[69] examined LPS extracted from three *B. nodosus* isolates representative of serotypes A, B, and C. He found that *B. nodosus* LPS showed physiochemical and biological characteristics attributable to LPS from other Gram-negative bacteria. Electron microscopic examination also indicated a similar morphology to LPS derived from other Gram-negative bacteria. It was further demonstrated that *B. nodosus* LPS exhibit endotoxic activities characteristic of endotoxins from other Gram-negative bacteria including pyrogenicity, production of leucopaenia followed by leucocytosis, the primary inflammatory response in rabbit skin, and the Schwartzman reaction. However, by comparison with the enterobacterial endotoxins, the toxicity of *B. nodosus* LPS for rabbits and mice was low.

In a later immunological study,[70] using comparative double immunodiffusion and immunoelectrophoretic analyses of LPS from the same three *B. nodosus* isolates examined previously,[69] the presence of common heat-stable, high-molecular-weight, carbohydrate antigens was revealed. The formation of two precipitin lines indicated that the LPS was at least bi-specific, being composed of two antigenically distinct types of molecules. Indirect hemagglutination and tube-agglutination tests confirmed that the three isolates shared common LPS antigens and that there were no isolate-specific O antigens. The LPS preparations were capable of sensitizing sheep red blood cells only after alkali or heat treatment. Periodate oxidation suggested that the carbohydrate component of *B. nodosus* LPS was essential for its serological activity and insensitivity of LPS to pronase made it unlikely that protein was responsible for the major part of its serological activity. It was therefore concluded[70] that the serological properties of *B. nodosus* LPS resembled those of LPS from other Gram-negative bacteria.[71]

B. Outer-Membrane Proteins

Emery et al.[72] examined the outer-membrane proteins of several strains of defimbriated *B. nodosus* after lithium acetate extraction and SDS-polyacrylamide gel electrophoresis. In addition to the fimbrial subunit protein, five proteins were consistently identified with approximate molecular weights of 75, 50, 38, 34.5, and 26.5 kDa. Surface labeling with radioactive iodine revealed that all these proteins were exposed on the external surface of the bacterial membrane.

Based on the premise that cross-protection against *B. nodosus* infection depended on antibody to common bacterial determinants, a pattern of reactivity, which reflected cross-protective immunity *in vivo* was sought by Emery[73] among the outer-membrane proteins of *B. nodosus* from different serotypes. Consequently, antisera were produced in sheep against the individual outer-membrane proteins of *B. nodosus* strain 198. However, none of the spectra of serological reactions against these outer-membrane proteins matched the pattern of cross-protection observed *in vivo*.

V. CONCLUSIONS

Despite the fact that *B. nodosus* was identified as the essential contagious agent of ovine footrot in the early forties,[35] the first study devoted specifically to investigating antigens of *B. nodosus* was not published until 20 years later.[54] However, the last 10 years have witnessed an explosive interest in the identification and characterization of *B. nodosus* antigens associated with immunity and virulence.

Because of the economic significance of footrot, the majority of studies to date have been of an applied nature aimed at identifying potential antigens for vaccine development and, to a lesser extent, for improved diagnosis. From the foregoing, it is evident that two particular antigens, namely the fimbriae and the extracellular proteases have received the most attention.

The identification of fimbriae as the major serogroup-specific protective antigen has led to exhaustive immunological, molecular, ultrastructural, and physiological studies on them. The complete amino acid sequence of fimbrial subunit proteins from certain serogroups is known, the genes encoding these fimbrial subunit proteins have been cloned and sequenced, and morphogenetic expression has been achieved in *P. aeruginosa,* resulting in the development of a recombinant-DNA-based fimbrial vaccine.[74-76] Furthermore, the increased twitching motility of virulent compared with benign *B. nodosus* strains may be used for laboratory diagnosis.[19]

The extracellular proteases secreted by *B. nodosus* have undergone extensive biochemical characterization and a number of their physico-chemical properties form the basis of current laboratory diagnosis. They have recently been shown to be cross-protective antigens providing broad-spectrum cover against a wide range of serogroups,[64] and in time will undoubtedly be subjected to the detailed molecular analysis already afforded the fimbriae.

It would appear that there remain unidentified and uncharacterized antigens of *B. nodosus* that may also be important in providing cross-protection in vaccines and for the development of improved laboratory diagnosis.[77] Perhaps it is now appropriate, with the advent of modern technologies, for more rigorous studies to be undertaken on some of the other antigens already discussed in this chapter, for example, the outer-membrane proteins, to resolve these issues and possibly provide an insight into the molecular interactions involved in the pathogenesis of footrot.

ACKNOWLEDGMENTS

The authors' work referred to in this chapter has been supported by grants from the Wool Research Trust Fund and the Rural Credits Development Fund. E. K. Moses wishes to further acknowledge the award of a post-graduate scholarship from the Australian Wool Corporation.

REFERENCES

1. **Ottow, J. C. G.,** Ecology, physiology, and genetics of fimbriae and pili, *Annu. Rev. Microbiol.*, 29, 79, 1975.
2. **Mattick, J. S., Anderson, B. J., and Egerton, J. R.,** Molecular biology and footrot of sheep, in *Biotechnology and Recombinant DNA Technology in the Animal Production Industries,* Reviews in Rural Science 6, Leng, R. A., Barker, J. S. F., Adams, D. B., and Hutchinson, K. J., Eds., University of New England, Armidale, 1985, 79.
3. **Stewart, D. J.,** An electron microscopic study of *Fusiformis nodusus, Res. Vet. Sci.*, 14, 132, 1973.
4. **Walker, P. D., Short, J., Thompson, R. O., and Roberts, D. S.,** The fine structure of *Fusiformis nodosus* with special reference to the location of antigens associated with immunogenicity, *J. Gen. Microbiol.*, 77, 351, 1973.
5. **Egerton, J. R.,** Surface and somatic antigens of *Fusiformis nodosus, J. Comp. Pathol.*, 83, 151, 1973.
6. **Thorley, C. M.,** A simplified method for the isolation of *Bacteroides nodosus* from ovine foot-rot and studies on its colony morphology and serology, *J. Appl. Bacteriol.*, 40, 301, 1976.
7. **Schmitz, J. A. and Gradin, J. L.,** Serotypic and biochemical characterization of *Bacteroides nodosus* isolates from Oregon, *Can. J. Comp. Med.*, 44, 440, 1980.
8. **Claxton, P. D., Ribeiro, L. A., and Egerton, J. R.,** Classification of *Bacteroides nodosus* by agglutination tests, *Aust. Vet. J.*, 60, 331, 1983.
9. **Every, D.,** Purification of pili from *Bacteroides nodosus* and an examination of their chemical, physical and serological properties, *J. Gen. Microbiol.*, 115, 309, 1979.
10. **Mattick, J. S., Anderson, B. J., Mott, M. R., and Egerton, J. R.,** Isolation and characterization of *Bacteroides nodosus* fimbriae: structural subunit and basal protein antigens, *J. Bacteriol.*, 160, 740, 1984.
11. **McKern, N. M., O'Donnell, I. J., Inglis, A. S., Stewart, D. J., and Clark, B. L.,** Amino acid sequence of pilin from *Bacteroides nodosus* (strain 198), the causative organism of ovine footrot, *FEBS Lett.*, 164, 149, 1983.
12. **Anderson, B. J., Kristo, C. L., Egerton, J. R., and Mattick, J. S.,** Variation in the structural subunit and basal protein antigens of *Bacteroides nodosus* fimbriae, *J. Bacteriol.*, 166, 453, 1986.
13. **Gaastra, W. and de Graaf, F. K.,** Host-specific fimbrial adhesins of non-invasive enterotoxigenic *Escherichia coli* strains, *Microbiol. Rev.*, 46, 129, 1982.
14. **Anderson, B. J., Egerton, J. R., and Mattick, J. S.,** Electrophoretic and antigenic variation in the structural subunit and basal protein of *Bacteroides nodosus* fimbriae, in *Footrot in Ruminants,* Proc. Workshop, Melbourne, 1985, Stewart, D. J., Peterson, J. E., McKern, N. M., and Emery, D. L., Eds., CSIRO Division of Animal Health/Australian Wool Corp., Glebe, N.S.W., 1986, 219.
15. **McKern, N. M., O'Donnell, I. J., Stewart, D. J., and Clark, B. L.,** Primary structure of pilin protein from *Bacteroides nodosus* strain 216: comparison with the corresponding protein from strain 198, *J. Gen. Microbiol.*, 131, 1, 1985.
16. **Hopp, T. P. and Woods, K. R.,** Prediction of protein antigenic determinants from amino acid sequences, *Proc. Natl. Acad. Sci. U.S.A.*, 78, 3824, 1981.

17. **Stewart, D. J.,** Studies on the morphology and antigenic structure of *Fusiformis nodosus,* Ph.D. thesis, University of Sydney, 1975.
18. **Heinrichsen, J.,** Twitching motility, *Annu. Rev. Microbiol.,* 37, 81, 1983.
19. **Depiazzi, L. J. and Richards, R. B.,** Motility in relation to virulence of *Bacteroides nodosus, Vet. Microbiol.,* 10, 107, 1985.
20. **Egerton, J. R. and Roberts, D. S.,** Vaccination against ovine foot-rot, *J. Comp. Pathol.,* 81, 179, 1971.
21. **Every, D. and Skerman, T. M.,** Protection of sheep against experimental footrot by vaccination with pili purified from *Bacteroides nodosus, N. Z. Vet. J.,* 30, 156, 1982.
22. **Stewart, D. J., Clark, B. L., Peterson, J. E., Griffiths, D. A., and Smith, E. F.,** Importance of pilus-associated antigen in *Bacteroides nodosus* vaccines, *Res. Vet. Sci.,* 32, 140, 1982.
23. **Stewart, D. J.,** The role of various antigenic fractions of *Bacteroides nodosus* in eliciting protection against footrot in vaccinated sheep, *Res. Vet. Sci.,* 24, 14, 1978.
24. **Thorley, C. M. and Egerton, J. R.,** Comparison of alum-absorbed or non-alum-absorbed oil emulsion vaccines containing either pilate or non-pilate *Bacteroides nodosus* cells in inducing and maintaining resistance of sheep to experimental foot rot, *Res. Vet. Sci.,* 30, 32, 1981.
25. **Skerman, T. M., Erasmuson, S. K., and Every, D.,** Differentiation of *Bacteroides nodosus* biotypes and colony variants in relation to their virulence and immunoprotective properties in sheep, *Infect. Immun.,* 32, 788, 1981.
26. **Stewart, D. J., Clark, B. L., Emery, D. L., Peterson, J. E., and Fahey, K. J.,** A *Bacteroides nodosus* immunogen, distinct from the pilus, which induces cross-protective immunity in sheep vaccinated against footrot, *Aust. Vet. J.,* 60, 83, 1983.
27. **Stewart, D. J., Clark, B. L., Peterson, J. E., Emery, D. L., Smith, E. F., Griffiths, D. A., and O'Donnell, I. J.,** The protection given by pilus and whole cell vaccines of *Bacteroides nodosus* strain 198 against ovine footrot induced by strains of different serogroups, *Aust. Vet. J.,* 62, 153, 1985.
28. **O'Donnell, I. J., Stewart, D. J., and Clark, B. L.,** Serological identification of pilus antigen and other protein antigens of *Bacteroides nodosus* using electro-blot radioimmunoassay after electrophoretic fractionation of the proteins on sodium dodecyl sulfate polyacrylamide gels, *Aust. J. Biol. Sci.,* 36, 15, 1983.
29. **Short, J. A., Thorley, C. M., and Walker, P. D.,** An electron microscope study of *Bacteroides nodosus:* ultrastructure of organisms from primary isolates and different colony types, *J. Appl. Bacteriol.,* 40, 311, 1976.
30. **Every, D. and Skerman, T. M.,** Ultrastructure of the *Bacteroides nodosus* cell envelope layers and surface, *J. Bacteriol.,* 141, 845, 1980.
31. **Every, D. and Skerman, T. M.,** Surface structure of *Bacteroides nodosus* in relation to virulence and immunoprotection in sheep, *J. Gen. Microbiol.,* 129, 225, 1983.
32. **Udey, L. R.,** An additional cell wall layer associated with virulence of the fish pathogen, *Aeromonas salmonicida,* Abstr. Annu. Meet. Am. Soc. Microbiol., Washington, D.C., 1978, 24, (B66).
33. **Costerton, J. W., Damgaard, H. N., and Cheng, K. J.,** Cell envelope morphology of rumen bacteria, *J. Bacteriol.,* 118, 1132, 1974.
34. **Holder, I. A., Ed.,** *Bacterial Enzymes and Virulence,* CRC Press, Boca Raton, FL, 1982.
35. **Beveridge, W. I. B.,** Foot-rot in sheep: a transmissible disease due to infection with *Fusiformis nodosus* (n. sp.). Studies on its cause, epidemiology, and control, *Aust. Counc. Sci. Ind. Res. Melbourne,* Bull. No. 140, 1941, 1.
36. **Thomas, J. H.,** The bacteriology and histopathology of footrot in sheep, *Aust. J. Agric. Res.,* 13, 725, 1962.
37. **Thomas, J. H.,** The pathogenesis of footrot in sheep with reference to proteases of *Fusiformis nodosus, Aust. J. Agric. Res.,* 15, 1001, 1964.
38. **Broad, T. E. and Skerman, T. M.,** Partial purification and properties of extracellular proteolytic activity of *Bacteroides nodosus, N.Z. J. Agric. Res.,* 19, 317, 1976.
39. **Skerman, T. M.,** Determination of some *in vitro* growth requirements of *Bacteroides nodosus, J. Gen Microbiol.,* 87, 107, 1975.
40. **Thomas, J. H.,** Proteolytic enzymes produced in liquid media by *Fusiformis nodosus, Aust. J. Agric. Res.,* 15, 417, 1964.
41. **Thomas, J. H.,** A liquid medium for the growth of *Fusiformis nodosus, Aust. Vet. J.,* 39, 434, 1963.
42. **Thomas, J. H.,** A simple medium for the isolation and cultivation of *Fusiformis nodosus, Aust. Vet. J.,* 34, 411, 1958.
43. **Cohen, H. R.,** The effects of dry grinding on the properties of proteins. 5. Keratins, *Arch. Biochem.,* 4, 145, 1944.
44. **Mathison, G. E.,** The microbiological decomposition of keratin, *Ann. Soc. Belge Med. Trop.,* 44, 767, 1964.
45. **Kortt, A. A., O'Donnell, I. J., Stewart, D. J., and Clark, B. L.,** Activities and partial purification of extracellular proteases of *Bacteroides nodosus* from virulent and benign footrot, *Aust. J. Biol. Sci.,* 35, 481, 1982.

46. **Green, R. S.,** Characterization of certain proteinase isoenzymes produced by benign and virulent strains of *Bacteroides nodosus, J. Gen. Microbiol.*, 131, 2871, 1985.
47. **Depiazzi, L. J. and Rood, J. I.,** The thermostability of proteases from virulent and benign strains of *Bacteroides nodosus, Vet. Microbiol.*, 9, 227, 1984.
48. **Gordon, L. M., Yong, W. K., and Woodward, C. A. M.,** Temporal relationships and characterisation of extracellular proteases from benign and virulent strains of *Bacteroides nodosus* as detected in zymogram gels, *Res. Vet. Sci.*, 39, 165, 1985.
49. **Akparov, V. K. H. and Stepanov, V. M.,** Phenylboronic acid as a ligand for biospecific chromatography of serine proteinases, *J. Chromatogr.*, 155, 329, 1978.
50. **Every, D.,** Proteinase isoenzyme patterns of *Bacteroides nodosus:* distinction between ovine virulent isolates, ovine benign isolates and bovine isolates, *J. Gen. Microbiol.*, 128, 809, 1982.
51. **Every, D. and Green, R. S.,** Purification of individual proteinase isozymes from *Bacteroides nodosus* by use of polyacrylamide gel electrophoresis, a fluorogenic substrate detection system, and a simple electroelution apparatus, *Anal. Biochem.*, 119, 82, 1982.
52. **Kortt, A. A., Caldwell, J. B., Burns, J. E., Guthrie, R. E., and Stewart, D. J.,** Characterisation and substrate specificity of the extracellular proteases of *Bacteroides nodosus* from virulent and benign ovine isolates, in *Footrot in Ruminants, Proc. Workshop, Melbourne, 1985,* Stewart, D. J., Peterson, J. E., McKern, N. M., and Emery, D. L., Eds., CSIRO Division of Animal Health/Australian Wool Corp., Glebe, N.S.W., 1986, 237.
53. **Lory, S. and Tai, P. C.,** Biochemical and genetic aspects of *Pseudomonas aeruginosa* virulence, in *Genetic Approaches to Microbial Pathogenicity,* Current Topics in Microbiology and Immunology, Vol. 118, Goebel, W., Ed., Springer-Verlag, Berlin, 1985, 53.
54. **Thomas, J. H.,** The differential diagnosis of footrot in sheep, *Aust. Vet. J.*, 38, 159, 1962.
55. **Egerton, J. R. and Parsonson, I. M.,** Benign footrot — a specific interdigital dermatitis of sheep associated with infection by less proteolytic strains of *Fusiformis nodosus, Aust. Vet. J.*, 45, 345, 1969.
56. **Depiazzi, L. J. and Richards, R. B.,** A degrading proteinase test to distinguish benign and virulent ovine isolates of *Bacteroides nodosus, Aust. Vet. J.*, 55, 25, 1979.
57. **Stewart, D. J.,** The role of elastase in the differentiation of *Bacteroides nodosus* infections in sheep and cattle, *Res. Vet. Sci.*, 27, 99, 1979.
58. **Every, D.,** Quantitative measurement of protease activities in slab polyacrylamide gel electrophoresis, *Anal. Biochem.*, 116, 519, 1981.
59. **Kortt, A. A., Burns, J. E., and Stewart, D. J.,** Detection of the extracellular proteases of *Bacteroides nodosus* in polyacrylamide gels: a rapid method of distinguishing virulent and benign ovine isolates, *Res. Vet. Sci.*, 35, 171, 1983.
60. **Yong, W. K., Moses, E. K., and Riffkin, G. G.,** Characterisation of virulence associated antigens for differentiating virulent and benign strains of *Bacteroides nodosus,* in *Footrot in Ruminants,* Proc. Workshop, Melbourne, 1985, Stewart, D. J., Peterson, J. E., McKern, N. M., and Emery, D. L., Eds., CSIRO Division of Animal Health/Australian Wool Corp., Glebe, N.S.W., 1986, 109.
61. **Egerton, J. R. and Merritt, G. C.,** Serology of foot-rot: antibodies against *Fusiforims nodosus* in normal, affected, vaccinated and passively immunised sheep, *Aust. Vet. J.*, 49, 139, 1973.
62. **Merritt, G. C. and Egerton, J. R.,** IgG_1 and IgG_2 immunoglobulins to *Bacteroides (Fusiformis) nodosus* protease in infected and immunized sheep, *Infect. Immun.*, 22, 1, 1978.
63. **Merritt, G. C., Egerton, J. R., and Loi, J. S.,** Inhibition of *Fusiformis nodosus* protease and bovine trypsin by serum α-macroglobulin, *J. Comp. Pathol.*, 81, 353, 1971.
64. **Stewart, D. J., Clark, B. L., Emery, D. L., Peterson, J. E., and Kortt, A. A.,** The phenomenon of cross protection against footrot induced by vaccination of sheep with *Bacteroides nodosus,* in *Footrot in Ruminants,* Proc. Workshop, Melbourne, 1985, Stewart, D. J., Peterson, J. E., McKern, N. M., and Emery, D. L., Eds., CSIRO Division of Animal Health/Australian Wool Corp., Glebe, N.S.W., 1986, 185.
65. **Yong, W. K. and Gordon, L. M.,** Comparative antigenic analysis of extracellular proteins of *Bacteroides nodosus* isolated from virulent and benign ovine footrot, *Vet. Microbiol.*, 12, 135, 1986.
66. **Cooper, B. S.,** Differences in morphology of *Bacteroides nodosus* attributable to culture media, *N.Z. Vet. J.*, 25, 16, 1977.
67. **Stewart, D. J. and Egerton, J. R.,** Studies on the ultrastructural morphology of *Bacteroides nodosus, Res. Vet. Sci.*, 26, 227, 1979.
68. **Woolcock, J. B.,** Pathogenesis of bacterial infections: some determinants of virulence in Gram-negative bacteria, *Aust. Vet. J.*, 62, 177, 1985.
69. **Stewart, D. J.,** Biochemical and biological studies on the lipopolysaccharide of *Bacteroides nodosus, Res. Vet. Sci.*, 23, 319, 1977.
70. **Stewart, D. J.,** Studies on the antigenic structure of *Bacteroides nodosus, Res. Vet. Sci.*, 24, 293, 1978.
71. **Hofstad, T.,** Serological properties of lipopolysaccharide from oral strains of *Bacteroides melaninogenicus, J. Bacteriol.*, 97, 1078, 1969.

72. **Emery, D. L., Clark, B. L., Stewart, D. J., O'Donnell, I. J., and Hewish, D. R.,** Analysis of the outer membrane proteins of *Bacteroides nodosus,* the causal organism of ovine footrot, *Vet. Microbiol.,* 9, 155, 1984.
73. **Emery, D. L.,** Reactivity of sera from sheep immunised with individual outer membrane proteins of *Bacteroides nodosus* against heterologous bacterial strains, *Vet. Microbiol.,* 9, 453, 1984.
74. **Mattick, J. S., Anderson, B. J., Bills, M. M., Cox, P. T., Dalrymple, B., Egerton, J. R., Elleman, T. C., Finney, K., Hoyne, P. A., Lehrbach, P., Mckern, N. M., McNicholas, P., Schwartzkoff, C. L., Stewart, D. J., and Wright, D.,** Anaerobe to aerobe: the development of a recombinant DNA-based vaccine against ovine footrot, 7th Aust. Biotechnol. Conf. Proc., U. Melbourne, August 25—28, 1986, 91.
75. **Egerton, J. R., Cox, P. T., Anderson, B. J., Kristo, C., Norman, M., and Mattick, J. S.,** Protection of sheep against footrot with a recombinant DNA-based fimbrial vaccine, *Vet. Microbiol.,* 14(4), 393, 1987.
76. **Stewart, D. J. and Elleman, T. C.,** A *Bacteroides nodosus* pili vaccine produced by recombinant DNA for the prevention and treatment of footrot in sheep, *Aust. Vet. J.,* 64, 79, 1987.
77. **Rood, J. I. and Yong, W. K.,** Application of biotechnology to the diagnosis of footrot, this volume, chap. 14.

Chapter 8

ANTIGENS OF *FUSOBACTERIUM NECROPHORUM*

David L. Emery

TABLE OF CONTENTS

I. INTRODUCTION

The Gram-negative, anaerobic bacterium, *Fusobacterium necrophorum* is readily isolated from a variety of necrotic infectious conditions of domestic livestock. These include liver abscess and foot abscess of ruminants, calf diphtheria, and epithelial ulcers of pigs and horses.[1,2] It has also been implicated in the pathogenesis of ovine footrot.[3] The organism is considered an opportunist pathogen and can be recovered from the environment, normal skin (especially the feet), and from the gastrointestinal and urogenital microflora.[2] *F. necrophorum* is considered to colonize and invade devitalized tissue in which the redox potential of the tissue has been lowered by injury or deprivation of blood supply.[2] Since the generation of protective immunity against *F. necrophorum* has been inconsistent and rarely in excess of 60 to 70% efficacy, the identification of protective antigens of this bacterium has been largely empirical[4] (see also Chapter 10). To appreciate the variety of virulence determinants which have been examined either serologically or for prophylaxis, this chapter discusses the antigenic structure of *F. necrophorum* and its products. These will be discussed as cellular and extracellular antigens.

Strains of *F. necrophorum* which have been isolated from normal tissues or lesions have been classified historically on the basis of cultural characteristics. Two biovars, A and B, and an intermediate type AB (which closely resembles biotype A in virulence and cultural activity) have been identified.[5,6] A- and AB-type biovars are virulent for ruminants and rodents, produce leucocidin, and capsules, and are hemolytic for a range of erythrocytes.[6] B-biovars are avirulent for ruminants and less virulent for mice, nonhemolytic, and nontoxigenic, and do not possess capsules.[4,6] Therefore, it is against the toxigenic strains that immunity needs to be directed, but identification of specific molecules on toxigenic strains of *F. necrophorum* is complicated by cross-reactivity with those on nontoxigenic commensals.

II. CELLULAR ANTIGENS

A. Capsules

In a study of 12 strains of toxigenic and nontoxigenic strains of *F. necrophorum,* virulent organisms (toxigenic, AB biovar) grown on solid or in liquid medium produced mucoid capsules, whereas B biovars did not.[38] With prolonged cultivation *in vitro,* the degree of encapsulation decreased, as it did after 3 d growth in liquid medium. While it is possible that the capsule is polysaccharide in content, little is known of its antigenicity or its structure. It is also possible that anticapsular antibodies agglutinate the organism, but this remains to be confirmed. By analogy with another anaerobe, *Bacteroides fragilis,* where encapsulated organisms are more virulent *in vivo,* and anticapsular immunity is protective,[7] a similar situation could exist for *F. necrophorum.*

B. Outer-Membrane Proteins (OMP)

F. necrophorum possesses a cell wall typical of Gram-negative bacteria in which divalent cations stabilize the association between OMP and lipopolysaccharide. Extraction of large amounts of OMP from *F. necrophorum* has proven difficult; it appears unlikely that capsular material is responsible for the relative resistance of the cell wall to most extractive procedures, since nontoxigenic strains are also robust. In a trial of methods to extract OMP from *F. necrophorum,* only potassium isothiocyanate[8] (KSCN) and the detergents sodium dodecylsulfate, and Triton X100 with EDTA[9] were reasonably efficient, while lithium acetate, lithium di-iodosalicylate, lysozyme-EDTA, and detergents NP40, Brij 96, and deoxycholate[10] were less effective.[39]

The profile of OMP from several stains of *F. necrophorum* is shown in Figure 1. A large

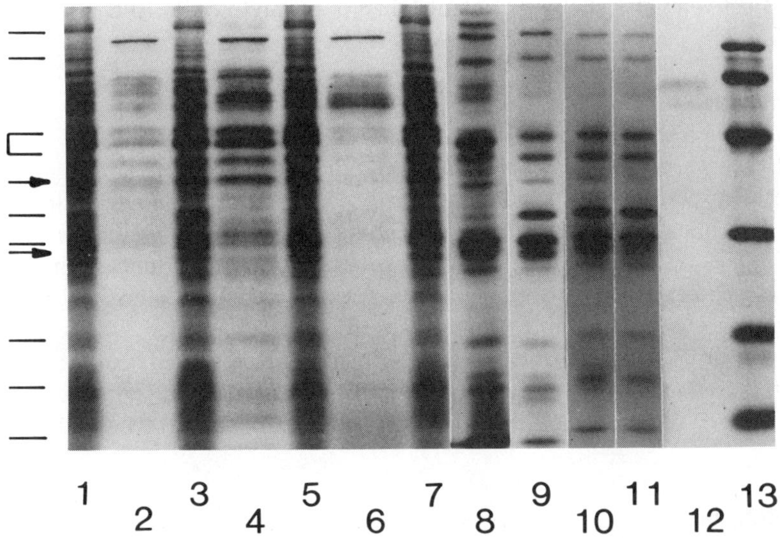

FIGURE 1. Extracts of *F. necrophorum* run under reducing conditions in 10 % sodium dodecyl sulphate polyacrylamide gels. Samples include membranes from sonicated whole cells of strains FnB-1, FnB-2, FnS-1, and FnS-2 (all AB biovars) in lanes 1, 3, 5, and 7, respectively, with their respective concentrated culture supernates in lanes 2, 4, and 6. A thiocyanate extract of membrane proteins (OMP) from FnB-1 is shown in lane 8, and autoradiographs of ^{125}iodinated (lactoperoxidase) OMP of FnB-1, FnS-1, and FnS-2 are run in lanes 9 to 11, respectively. Concentrated bacterial culture medium (Eugon Broth) is run in lane 12, and molecular weight standards of 97, 67, 43, 30, 20.1, and 14.4kDa are represented in lane 13. SDS-PAGE gels were stained with Coomassie blue. Arrows on the left denote molecular weights of proteins relevant to the text.

number of membrane proteins were extracted by KSCN,[8] and by comparison with ^{125}I-labeled cells. Seven or eight major groups of protein are membrane associated, and accessible to lactoperoxidase.[11] Proteins of 98, 75, 40 (2), 35, 30, 20, 16, and less than 14 kDa are common to all strains of *F. necrophorum* (Figure 1) and these bands were also identified in immunoblots of outer-membrane complex (OMC) proteins reacted with anti-*F. necrophorum* serum.[39] Other OMC proteins of around 38 and 28 kDa are less commonly shared between different strains of *F. necrophorum* (Figure 1).

C. Lipopolysaccharide (LPS)

F. necrophorum possess an endotoxin with biological potency equivalent to that of the *Enterobacteria.*[12] The LPS constitutes around 4% of the cell wall, and after isolation by phenol extraction,[13] consists of aggregates of trilaminar rods.[12] The LPS consists of Lipid A linked by 2-keto-3-deoxyoctonate (KDO) to heptose and the core sugars from which branch the polysaccharide side chains.[14] The content of KDO is low (0.8%) in fusobacterial LPS, but the endotoxin was lethal for mice with an LD^{50} of 0.002 μg (16.8mg/1kg[15]) and produced both localized and generalized Schwartzman reactions and biphasic pyrogenic responses in rabbits.[12] Limited studies on the LPS from *F. necrophorum* have indicated that A and B biovars possess endotoxin of equivalent biological potency,[16] but a recent report[17] on single strains from each biovar suggested that LPS from the A biovar was more potent.

Fusobacterial LPS possess a variety of oligosaccharide side chains of variable length (as detected in immunoblots) but appear to possess a similar antigenicity in all strains after LPS has been reduced and dissociated prior to electroblotting.[39] However, in its native state, LPS apparently dictates some of the somatic O antigenic specificities,[14] and IgM natural antibodies are directed towards the polysaccharide side chains.[18] In fusobacteria, the cell wall (containing LPS) activates the alternative complement pathway[19] while LPS also activates Hagemann factor.[20]

D. Pili/(Fimbriae)

By electron microscopic examination of *F. necrophorum,* fimbriae have been detected inconsistently on *F. necrophorum* and then only on strains of A biovar.[21] Examination of the pathogenicity of fimbriate hemagglutinating (HA^+) and nonhemagglutinating (HA^-) mutants of the same A-biovar strain revealed that (HA^-) mutants were less pathogenic.[21] This result implies that whereas HA activity was associated with virulence, fimbriae are not responsible for either HA activity or associated with virulence, although changes in the antigenicity of the fimbriae with the mutation were not discounted.[21] Where fimbriae have been isolated from *F. necrophorum,* the ultrastructure appears similar to those from other bacteria, but the molecular weight of any monomer has not been determined. In an examination of at least 12 A biovars of *F. necrophorum,* we have not observed structures analogous to fimbriae.

E. Hemagglutinin

The hemagglutinin of *F. necrophorum* is apparently associated with the cell wall, is heat labile, and possesses a subunit molecular weight of 19 kDa.[22] Its morphology resembles fusobacterial LPS, while chemical properties resemble those of pili.[22] More research on this component is required before its characteristics and pathogenetic activity can be fully appreciated.

III. SOLUBLE PRODUCTS

A. Exotoxins

Toxigenic strains of *F. necrophorum* belong to the A biovars and elaborate an exotoxin which lyses cultivated cell lines,[23,24] erythrocytes,[24,25] and leucocytes from several species, and induces necrotic cellulitis after s.c. inoculation into rabbits.[26,27] Two forms of the molecule have been reported; a high-molecular-weight (>300 kDa) complex which is heat labile and sensitive to proteolysis and which is produced by A and AB biovars during logarithmic growth *in vitro,*[25,28] and a low-molecular-weight species (< 500 Da) which is heat stable and is produced by all biovars during a 7 d growth period *in vitro.*[23,24] Partial purification of high-molecular-weight exotoxin by ultrafiltration and gel filtration indicated that after immunoblotting and probing with neutralizing antisera, toxigenic strains possessed moieties of 13, 14, and 103 kDa, which appeared associated with the leucocidin; nontoxigenic B biovars possessed a 13.5 kDa fragment.[29] Whether the larger exotoxin, which may form aggregates with LPS in culture supernates[29] is degraded to a low-molecular-weight form has not been resolved. For practical purposes, while extracts containing leucocidin were prophylactic for mice, variable results have been achieved in ruminants.[4,30,31] In other instances, cytoplasmic preparations of *F. necrophorum* which contained leucocidin, could be used to immunize mice and ruminants against a challenge infection.[32,33] Additional studies on the capacity of sera from immunized animals to neutralize the leucocidin revealed antigenic heterogeneity in the leucocidins from several fusobacterial strains.[29]

The leucocidal, hemolytic, and cytotoxic activities of *F. necrophorum* may be manifestations of the activity of the same molecule.[29] The leucocidal and hemolytic titers of culture supernates are correlated, as is the virulence of the strain of *F. necrophorum*, for mice.[29,34,35] The hemolysin exhibits biologic properties consistent with the activity of a phospholipase A and lysophospholipase,[36] but the enzymic properties of the leucocidin have not been investigated.

IV. OTHER FACTORS

Although *F. necrophorum* elaborates a factor which aggregates platelets,[37] the nature of the molecule has not been determined. There is also suggestive evidence that supernates from some strains of *F. necrophorum* have limited proteolytic activity. In addition, all strains produce lipase (see Chapter 6).

V. CONCLUSION

F. necrophorum possesses a range of virulence determinants which are antigens associated with exotoxins, the cell well, and capsule. However, the serogrouping of the organism is not well developed, and presently, is not predictive of protective immunity. In essence, agglutination tests at present have little practical value, as agglutination titers are not predictive of immunity[5] and the lack of a suitable model for experimental assessment of protective immunity (if any exists) has hampered progress towards identification of antigens useful for protection, diagnosis, and for assaying serological responses. With further technical advances to purify and evaluate antigens for *F. necrophorum*, many of these problems should be resolved.

REFERENCES

1. **Simon, P. C. and Stovell, P. L.,** Diseases of animals associated with *Sphaerophorums necrophorus:* characteristics of the organism, *Vet. Bull.,* 39, 311, 1969.
2. **Langworth, B. F.,** *Fusobacterium necrophorum:* its characteristics and role as an animal pathogen, *Bacteriol. Rev.,* 41, 373, 1977.
3. **Egerton, J. R., Roberts, D. S., and Parsonson, I. M.,** The aetiology and pathogenisis of ovine footrot. A histological study of the bacterial invasion, *J. Comp. Pathol.,* 79, 207, 1969.
4. **Emery, D. L., Vaughan, J. A., Clark, B. L., and Stewart, D. J.,** Virulence determinants of *Fusobacterium necrophorum* and their prophylactic potential in animals, in *Footrot in Ruminants,* Proc. Workshop, Melbourne 1985, Stewart, D. J., Peterson, J. E., McKern, N. M., and Emery, D. L., Eds., CSIRO Division Animal Health/Australian Wool Corp., Glebe, N.S.W., 1986, 267.
5. **Fievez, L.,** *Comparative Study of Strains of Shaerophorus Necrophorus Isolated from Man and Animals,* European Academic Press, Brussels, 1963.
6. **Emery, D. L., Vaughan, J. A., Clark, B. L., Dufty, J. H., and Stewart, D. J.,** Cultural characteristics and virulence of strains of *Fusobacterium necrophorum* isolated from the feet of cattle and sheep, *Aust. Vet. J.,* 62, 43, 1985.
7. **Kasper, D. L.,** Bacterial capsules — old dogma and new tricks, *J. Infect. Dis.,* 153, 407, 1986.
8. **Mukker, T.K.S.,** Immunogenicity of a chaotropically extracted protective antigen(s) of *Pasteurella multocida* type A (bovine origin) against experimental pasteurellosis in mice, *J. Gen. Microbiol.,* 113, 37, 1979.
9. **Schnaitman, C. A.,** Effect of ethylene diamine tetraacetic acid, triton X-100 and lysozyme on the morphology and chemical composition of isolated cell walls of *Escherichia coli, J. Bacteriol.,* 108, 553, 1971.
10. **Helenius, A. and Simons, K.,** Solubilization of membranes by detergents, *Biochem. Biophys. Acta,* 415, 29, 1975.
11. **Emery, D. L., Clark, B. L., Stewart, D. J., O'Donnell, I. J., and Hewish, D. R.,** Analysis of the outer membrane proteins of *Bacteroides nodosus,* the causal organism of ovine footrot, *Vet. Microbiol.,* 9, 155, 1984.
12. **Garcia, M. M., Alexander, D. C., and McKay, K. A.,** Biological characterization of *Fusobacterium necrophorum* cell fractions in preparation for toxin and immunization studies, *Infect. Immun.,* 11, 609, 1975.
13. **Clark, K., Garay, G. W., and Reaveley, D. A.,** The cell walls of *Pseudomonaus aeruginosa, Biochem. J.,* 105, 749, 1967.
14. **Hofstad, T.,** The distribution of heptose and 2-keto-3-deoxy octonate in *Bacteroidaceae, J. Gen. Microbiol.,* 85, 314, 1974.

15. **Warner, J. F., Fales, W. H., Sutherland, R. C., and Teresa, G. W.,** Endotoxin from *Fusobacterium necrophorum* of bovine hepatic abscess origin, *Am. J. Vet. Res.*, 36, 1015, 1975.
16. **Berg, J. N. and Scanlan, C. M.,** Studies of *Fusobacterium necrophorum* from bovine hepatic abscesses: biotypes, quantitations, virulence and antibiotic susceptibility, *Am. J. Vet. Res.*, 43, 1580, 1982.
17. **Inoue, T., Kanoe, M., Goto, N., Matsumura, K., and Nakano, K.,** Chemical and biological properties of lipopolysaccharides from *Fusobacterium necrophorum* biovar A and biovar B strains, *Jpn. J. Vet. Sci.*, 47, 639, 1985.
18. **Dahlen, G., Nygren, H., and Hansson, H. A.,** Immunoelectron microscopic localization of lipopolysaccharides in the cell wall of *Bacteroides oralis,* and *Fusobacterium nucleatum, Infect. Immun.*, 19, 265, 1978.
19. **Bjornson, A. B.,** Role of complement in host resistance against members of the *Bacteroidaceae, Rev. Infect. Dis.*, 6(Suppl. 1), S34, 1984.
20. **Bjornson, H. S.,** Enzymes associated with the survival and virulence of gram-negative anaerobes, *Rev. Infect. Dis.*, 6(Suppl. 1), S21, 1984.
21. **Shinjo, T. and Kiyoyama, H.,** Pathogenicity of a non-haemagglutinating mutant strain of *Fusobacterium necrophorum* biovar A in mice, *Jpn. J. Vet. Sci.*, 48, 523, 1986.
22. **Nagai, S., Kanoe, M., and Toda, M.,** Purification and partial characterization of *Fusobacterium necrophorum* haemagglutinin, *Zentralbl. Bakteriol. Microbiol. Hyg.*, A 258, 232, 1984.
23. **Fales, W. H., Warner, J. F., and Teresa, G. W.,** Effects of *Fusobacterium necrophorum* leukotoxin on rabbit peritoneal macrophages *in vitro, Am. J. Vet. Res.*, 38, 491, 1977.
24. **Scanlan, C. M., Berg, J. N., and Campbell, F. F.,** Biochemical characterization of the leukotoxins of three bovine strains of *Fusobacterium necrophorum, Am. J. Vet. Res.*, 47, 1422, 1986.
25. **Emery, D. L., Dufty, J. H., and Clark, B. L.,** Biochemical and functional properties of a leucocidin produced by several strains of *Fusobacterium necrophorum, Aust. Vet. J.*, 61, 382, 1984.
26. **Roberts, D. S.,** Toxic, allergic and immunogenic factors of *Fusiformis necrophorus, J. Comp. Pathol.*, 80, 247, 1970.
27. **Beveridge, W. I. B.,** A study of twelve strains of *Bacillus necrophorus* with observations on the oxygen intolerance of the organism, *J. Pathol. Bacteriol.*, 38, 467, 1934.
28. **Coyle-Dennis, J. E. and Lauerman, L. H.,** Biological and biochemical characteristics of *Fusobacterium necrophorum* leukocidin, *Am. J. Vet. Res.*, 39, 1790, 1978.
29. **Emery, D. L., Edwards, R. D., and Rothel, J. S.,** Studies on the purification of the leucocidin of *Fusobacterium necrophorum* and its neutralization by specific antisera, *Vet. Microbiol.*, 11, 357, 1986.
30. **Emery, D. L. and Vaughan, J. A.,** Generation of immunity against *Fusobacterium necrophorum* in mice inoculated with extracts containing leucocidin, *Vet. Microbiol.*, 12, 255, 1986.
31. **Clark, B. L., Emery, D. L., Stewart, D. J., Dufty, J. H., and Anderson, D. A.,** Studies into immunisation of cattle against interdigital necrobacillosis, *Aust. Vet. J.*, 63, 107, 1986.
32. **Garcia, M. M., Dorward, W. J., Alexander, D. C., Magwood, S. E., and McKay, K. A.,** Results of a preliminary trial with *Sphaerophorus necrophorus* toxoids to control liver abscesses in feedlot cattle, *Can. J. Comp. Med.*, 38, 222, 1974.
33. **Garcia, M. M. and McKay, K. M.,** Intraperitoneal immunization against necrobacillosis in experimental animals, *Can. J. Comp. Med.*, 42, 121, 1978.
34. **Coyle-Dennis, J. E. and Lauerman, L. H.,** Correlations between leukocidin production and virulence of two isolates of *Fusobacterium necrophorum, Am. J. Vet. Res.*, 40, 274, 1979.
35. **Scanlan, C. M., Berg, J. N., and Fales, W. H.,** Comparative *in vitro* leukotoxin production of three bovine strains of *Fusobacterium necrophorum, Am. J. Vet. Res.*, 43, 1329, 1982.
36. **Abe, P. M., Kendall, C. J., Stauffer, C. R., and Holland, J. W.,** Hemolytic activity of *Fusobacterium necrophorum* culture supernates due to presence of phospholipase A and lysophospholipase, *Am. J. Vet. Res.*, 40, 92, 1979.
37. **Forrester, L. J., Campbell, B. J., Berg, J. N., and Barrett, J. T.,** Aggregation of platelets by *Fusobacterium necrophorum, J. Clin. Microbiol.*, 22, 245, 1986.
38. **Emery, D. L. and Vaughan, J. A.,** Unpublished data.
39. **Emery, D. L.,** Unpublished data.

Chapter 9

HOST RESPONSES TO FOOTROT AND FOOT ABSCESS

David L. Emery

TABLE OF CONTENTS

I. INTRODUCTION

The two most important infectious foot diseases affecting ruminants are footrot and foot abscess. Gram-negative, obligate, anaerobic bacteria are considered responsible for the production and maintenance of the final clinical lesion, although other organisms may participate in the initial pathogenesis. *Bacteroides nodosus* is readily isolated from lesions of footrot in sheep and cattle; the causal role of this bacterium is evidenced by the therapeutic or prophylactic effects of *B. nodosus* whole-cell vaccines for both conditions.[1,2] Foot abscess can be reproduced experimentally by the inoculation of toxigenic organisms of *Fusobacterium necrophorum* into the interdigital subcutis of ruminants[3,4] and the organism is readily isolated from clinical lesions, in combination with an inconsistent variety of other microbial flora. In ruminants, *F. necrophorum* is also associated with several other clinical conditions including calf diphtheria, and the liver abscess/rumenitis complex,[5,6] and immunization against these infections has been less successful than for footrot (reviewed in Chapter 4).

The immune response of the definitive host against components of the causal organisms of disease, especially the "virulence determinants" provides valuable insights into those components of the microorganism which might be essential to the pathogenesis of infection. In addition, it is often possible to identify those components which stimulate protective immunity and allow the causal agent to suppress or evade host defenses. An analysis of host responses is particularly relevant where susceptible or resistant individuals exist within flocks or breeds of ruminants which are subjected to natural-challenge infection. It is the purpose of this review to describe the cellular and humoral immune responses of cattle and sheep during natural and experimental infections with *B. nodosus* and *F. necrophorum* as an introduction to chapters on specific vaccination. As is the case with many bacterial vaccines, the evolution of footrot vaccines has been largely empirical, such that the apparent continuity between host immune responses and the development of antigens for stimulating protective immunity is often retrospective.

II. OVINE FOOTROT

A. Background

That ovine footrot occurs in the temperate regions of Australia in two seasons per year (spring and autumn, weather permitting) is tangible evidence that sheep fail to mount, following natural infections, a protective immune response of sufficient magnitude and longevity to prevent the seasonal recurrence of the infection. This, in turn, results from the location of the infection and the insidious nature of its progress, especially during the early pathogenesis of the disease. The pathogenesis and pathology of ovine footrot is adequately described in Chapter 1, and only those features of the infection relevant to host immunity will be reiterated. Footrot commences with an interdigital dermatitis, and inflammatory response resulting from bacterial colonization of the macerated epithelial surface. Subsequently, the infection with *B. nodosus* progressively underruns the horn of the heel, with the advancing infective focus restricted to the relatively avascular stratum spinosum of the epidermis. As the support for the outer cornified layers of the hoof are eroded, separation of the horny layers of the hoof from the germinal tissue ensues, with attendant inflammation, pain, and lameness. By this time, usually 3 to 8 weeks after initial infection, the fate of both host and infection is usually determined by prevailing weather conditions and flock management, but with treatment, resolution occurs rapidly.

To establish infection, *B. nodosus* possesses a range of virulence determinants, which have been identified by their capacity to duplicate pathogenetic effects *in vitro* (viz. adherence to cells, hemagglutination or destruction of ovine cells, or proteolysis of hoof material) or induce protective immunity when purified antigens are used as immunogens. For *B. nodosus*, the virulence determinants currently studied include pili (or fimbriae), outer-membrane proteins, endotoxin (lipopolysaccharide — LPS), and proteases. It is against these virulence determinants that the quantity and quality of the immune response have been assayed in situations of natural or acquired immunity, and following experimental- or natural-challenge infections.

B. Nonspecific Host Defenses

Phagocytic cells (including polymorphonuclear leuococytes [PMN] and macrophages) and serum opsonins (complement components, antibody, and β-lysins) are the major nonspecific effector mechanisms against Gram-negative, anaerobic bacteria in ruminants. *In vitro, B. nodosus* is readily phagocytosed by ovine and bovine PMN and by peripheral blood monocytes which are adherent to plastic.[7] Greater than 90% of *B. nodosus* are engulfed by PMN within 30 min of mixing. The rate of phagocytosis was enhanced when preliminary opsonization of *B. nodosus* was performed in purified immunoglobulins from sheep immunized with pili or piliated whole cells of *B. nodosus* prior to washing and addition to phagocytes. In particular, the rate of phagocytosis by ovine PMN was most significantly enhanced by antipilus IgG_2; the activities of antipilus IgG_1 and IgM were equivalent, and comparable to that of IgG_1, IgG_2, and IgM from normal ovine serum. For adherent macrophages, antipilus IgM and IgG_1 enhanced phagocytosis most effectively,[7] since ovine macrophages possess F_c' receptors for IgG_1[8], while ovine PMN possess receptors for IgG_2[9]. In contrast, ovine peripheral blood monocytes in suspension did not phagocytose *B. nodosus*.[7] Bovine PMN and adherent or activated macrophages also engulfed bovine and ovine strains of *B. nodosus* in the absence of serum and with an efficiency comparable to that of ovine cells (as measured by chemiluminescence).[52]

Several other salient features of the phagocytic response were noted

1. Virulent and benign strains of *B. nodosus* were phagocytosed with equal facility by PMN and macrophages from ruminants.
2. The rate of phagocytosis of *B. nodosus* was similar in PMN derived from breeds of sheep which were resistant and susceptible to ovine footrot (see Section II).
3. The rate of phagocytosis was not affected during a concurrent footrot infection.

It was concluded from the above results that professional phagocytes did not prevent the establishment and progress of footrot infection and did not constitute a major mechanism of resistance in the natural infection,[7] but were probably involved in resolution of the lesion. This suggestion is supported by the lack of an intense inflammatory response to the early underrunning lesion in the hoof, implying that this stage is relatively sequestered from the host defenses (nonspecific and specific). In resistant sheep, infection is usually restricted to an initial interdigital dermatitis and resistance manifests as a failure of the infection to underrun the horn of the hoof. It is during the incitement of the dermatitis that inflammatory cells first contact the invading microflora, and in resistant sheep, perhaps it is their activity against *B. nodosus* that prevents extension of the pathology. In addition, in sheep vaccinated with *B. nododus,* immunity manifests either as a failure of interdigital lesions to progress or develop, or resolution of the underrunning lesion at the point of the initial separation of the external horny layers of the hoof from the subcutis. Both these stages represent times when inflammatory cells and serum components have the most ready access to the causal organism(s). As presented below, the time at which physical separation of the hoof wall occurs in footrot provides the maximum stimulus to the host immune response.

The bactericidal activity of serum against Gram-negative aerobes has been widely recognized in ruminants.[10] The binding of IgM and IgG in ovine serum to the LPS of *Escherichia coli* followed by the lysis of the organism by the activation of the alternative pathway of complement is one bactericidal mechanism.[10] A second involves activation of the classical complement pathway by the binding of serum Ig to OMPs (reviewed in Chapter 11). For anaerobic bacteria in ruminants, few examples are available. In ovine footrot, the bactericidal activity of adult-ovine serum has been demonstrated against several strains of *B. nodosus* which have been propagated *in vitro*,[12] and the titers of the antibodies involved have been boosted by infection. However, although a range of laboratory-cultivated virulent and benign strains of *B. nodosus* readily succumb to bacteriolysis, the growth *in vitro* of virulent strains recently isolated from clinical lesions has not been impeded by the addition of up to 40% (v/v) newborn-lamb serum or serum from infected or immunized sheep.[13] To understand completely the mechanism of disease resistance, this feature of ovine immunity requires further study, as it does not preclude a role for bactericidal mechanisms *in vivo*, and may simply reflect technical shortcomings in the *in vitro* model. However, the relatively weak bactericidal activity of ovine serum against *B. nodosus* may reflect the biological impotency and reduced toxicity of *B. nodosus* LPS,[14] and the relatively weak cross-protective immunity engendered by immunization with cell-wall preparations of *B. nodosus*,[15] which has been proposed to operate through the stimulation of bactericidal mechanisms.[16]

Bactericidal mechanisms which operate by the deposition of complement components on the bacterial cell wall rely on the subsequent destruction of bacteria by phagocytic cells. Such cells, especially PMN, are capable of killing anaerobes under both aerobic and anaerobic conditions,[17] and the membranes of PMN possess receptors for C3b (CR1 receptor) and iC3b (CR3 receptor) which facilitate adherence or phagocytosis depending on whether the receptors are resting or activated, respectively.[18] Macrophages also phagocytose opsonized particles via β-glucan receptors to which bacteria opsonized via the alternate complement pathway can bind.[19] For *B. nodosus*, the structure facilitating such activity is probably LPS.

C. Innate Resistance

A comparison betwen the immune responses of individual breeds or strains of animals which differ markedly in susceptibility to infection with specific pathogens is a plausible method for the identification of protective mechanisms and the microbial antigens which elicit them. In sheep, coarse-wooled British breeds are more resistant to footrot than are fine-wooled Merinos under conditions of moderate transmission (i.e., where the virulence of the infecting strain is reduced, or when climatic conditions are less than optimal for the spread of infection and less than 50% of feet become infected).[20-22] Severe challenges with highly virulent organisms abrogated innate resistance.[22] Macroscopically, resistance is manifested as the failure of interdigital lesions to progress or to underrun the horn of the hoof. An analysis of nonspecific effector mechanisms between Romney and Merino sheep did not reveal differences in the rates of phagocytosis and bactericidal activity that could account for the phenomenon.[7,13] In addition, both breeds of sheep possessed comparable levels of hemolytic complement.[52] Two observations suggested that the structure of the interdigital skin was central to the resistance of Romney sheep. Firstly, mild scarification of the interdigital skin at the time of experimental challenge with *B. nodosus* rendered Romney sheep as susceptible to infection as their Merino counterparts.[22] Similar results were obtained by Parker et al.,[23] who scarified half the foot of each individual sheep and observed that the majority of severe lesions developed in the treated feet. Secondly, an analysis of the kinetics and magnitude of the antibody response emanating in lymph from the popliteal lymph node draining feet during the development of footrot revealed comparable responses in Merinos and Romney sheep.[22] Both these observations argue that resistance is not due to acquired immunity. It may be possible to define this phenomenon further, to the extent that some biochemical marker could be used to select resistant animals.

Within breeds of sheep, the progeny of individual sires exhibit differing degrees of resistance to ovine footrot. In seasons of variable challenge, ranging from moderate to severe (low or high transmission), the heritability of resistance averaged 0.30.[24] While there was no association between hemoglobin type and footrot resistance in other preliminary studies,[7] an association between resistance to footrot and the ovine Class I antigen SY 6 of the major histocompatibility complex has been elucidated.[53] The specificity was found in 2/3 of resistant progeny from susceptible sires in which the SY 9 specificity predominated.[54] If verified in larger flocks, these findings could be exploited for the genetic selection of sheep. It has been observed for many years that an interaction between innate and acquired immunity to footrot meant that lesser quantities of protective vaccinal antigen formulated in less potent adjuvants was required to immunize innately resistant sheep compared to susceptible animals. Thus, innately resistant sheep exhibit a similar degree of protection against footrot with lower levels of serum antibody against *B. nodosus* antigens than do susceptible Merinos.

D. The Specific Immune Response during Natural Infection

This section will discuss the development of immune responses following natural or experimental challenges of naive, unvaccinated sheep. To examine the evolution of acquired immunity a variety of assays have been utilized.

1. Antibody Responses

Agglutination and hemagglutination (HA) tests — These assays have utilized formaldehyde-fixed pilated whole cells (agglutination) in which antibody detected K or pilus/fimbrial antigens or somatic O antigens.[25] Both K- and O- type agglutination is readily accomplished by ovine IgM, IgG_1, and IgG_2 antibodies.[7] The K-agglutinating epitope on *B. nodosus* pili forms the basis of the current serogroup classification of bacterial isolates; since protective immunity is also largely serogroup specific when sheep are vaccinated with pilus antigen, the assay has a practical and predictive value.[26,27] HA assays have been used principally to detect *Bacteroides* LPS absorbed to erythrocytes,[14] but have largely been replaced by solid-phase immunoassays.

Enzyme-linked immunoassay (ELISA) and radioimmunoassays — As technical advances enabled greater purification of bacterial antigens, solid-phase immunoassays replaced the relatively insensitive agglutination and precipitation tests.[28] The development of monoclonal antibodies (MAb) to ovine-Ig isotypes has facilitated greater definition of the specificity and induction of protective immunity against ovine footrot.

Immunoblotting — The transfer of electrophoretically separated proteins and polysaccharides of *B. nodosus* to nitrocellulose paper[29,30] has been utilized to examine the reactivity of ovine sera with individual proteins of *B. nodosus* when these have not been available in purified form. Since many conformational epitopes on proteins are lost by reduction, dissociation, and denaturation of antigens during electrophoretic separation and transfer of proteins (e.g., the agglutinating epitope on pili), the procedure has proven useful to check the nature of epitopes recognized by MAb. The procedure has also been utilized to examine those antigens of *B. nodosus* which are shared among bacteria from heterologous serogroups.

Enzyme inhibition — This procedure utilizes serum to inhibit the activity of proteases from *B. nodosus.*[31] Since the protease is also inhibited by α_2 macroglobulin in normal ovine serum, pretreatment of test serum with rivanol (ethoxy-6,9-diaminoacridine lactate) has been employed to permit evaluation of Ig titers in the absence of the macroglobulin.[31] The antigenicity of the protease is common to all strains of *B. nodosus,* although the isoenzyme pattern and activity of molecules differ among strains and are associated with virulence.[32,33]

2. Cell-Mediated Immunity

Apart from phagocytic assays, and those which monitor the proliferative responses of helper T lymphocytes to specific antigen,[34] there has been no evidence to date that *B. nodosus* invokes delayed-type hypersensitivity reactions in sheep affected with footrot. The causal organism does not proliferate locally following s.c. inoculation into sheep.

3. Kinetics, Magnitude, and Specificity of the Immune Response

Three features of the immune response against ovine footrot argue that the major protective response is antibody mediated.

1. In sheep vaccinated with pilated *B. nodosus* cells or purified pili, serogroup-specific immunity is correlated with the acquisiton of agglutination titers of greater than 3000.[35] Chemical modification of the agglutinating epitope and other conformational epitopes abrogates the capacity of pili to elicit protective immunity.[36]
2. Effective induction of prolonged immunity requires the use of adjuvants to potentiate the immune response, and susceptibility returns as the serum titer declines.[37]
3. Immunity can be transferred passively with γ-globulins from immunized sheep to naive recipients.[25]

Sheep which have been reared in areas free of ovine footrot possess very low to undectable titers of serum antibody against *B. nodosus,* most of which probably arise from cross-reactions between membrane proteins of *B. nodosus* and other Gram-negative organisms in the gastrointestinal tract (as revealed by ELISA and immunoblotting). Following natural infection (field infection on irrigated pasture with natural transmission) or experimental infection (application of pure cultures of *B. nodosus* to the interdigital skin)[38] the immune response was monitored in serum and in lymph leaving the popliteal lymph node draining the infected foot.[22] The typical response is shown for reactions against membrane proteins and pili[22] and *B. nodosus* outer-membrane complex (OMC), pili, and protease (see also References 25 and 31) (Figure 1). During the first 3 weeks after experimental infection (up to the stage of underrunning lesions of the hoof), the relatively slight increases in antibody titers in lymph and serum probably reflect the relative sequestration of the lesion from the immune system. The blastogenic response of the popliteal lymph node is minimal by comparison with responses to conventional protein antigens, and is commensurate with the response to carbohydrate antigens or nonimmunogenic haptens.[22,55] Throughout this period, the specific antibody titers against *B. nodosus,* OMC, pili, or protease increase less than five-fold (Figure 1), and if the infection fails to progress, decline rapidly following resolution of the lesion.[22] With the development of underrunning lesions and the separation of the horn of the hoof from the underlying germinal epithelium (severe footrot — Chapter 1), the titers increased 10- to 20-fold over the levels prior to infection, and the titers and lymph and serum were comparable (Figure 1). The increased titers aginst *B. nodosus,* OMC, and pili were almost entirely of the IgG_1 isotype,[7] whereas an initial increase in antiprotease IgG_2 and a subsequent increase in specific IgG_1 was reported during experimental infection.[31] In natural field infections, the magnitude of the titers are similar, but the increases are more protracted as lesions take longer to develop.[22] It should be noted that the maximum antipilus-agglutinin titers attained during a natural infection are usually less than 2000, well below that deemed necessary for the provision of protection to Merino sheep, about 3000.[35] In addition, the titers decline rapidly following resolution of the lesion, a situation which contrasts markedly to that in which sheep vaccinated with pili in oil adjuvants maintain serum titers in excess of 3000 for up to a year.[39,40]

III. OVINE FOOT ABSCESS

This disease can be readily reproduced by the s.c. inoculation of toxigenic strains (A and AB biovars) of *F. necrophorum* into the interdigital skin; nontoxigenic, B biovars elicit only a transient swelling in the feet of cattle and sheep.[41,42] Since the microflora of the gastrointestinal and urogenital tracts also contain strains of *F. necrophorum* of predominantly B biovars,[6] it is not surprising that the proportion of ruminants with agglutinin titers against

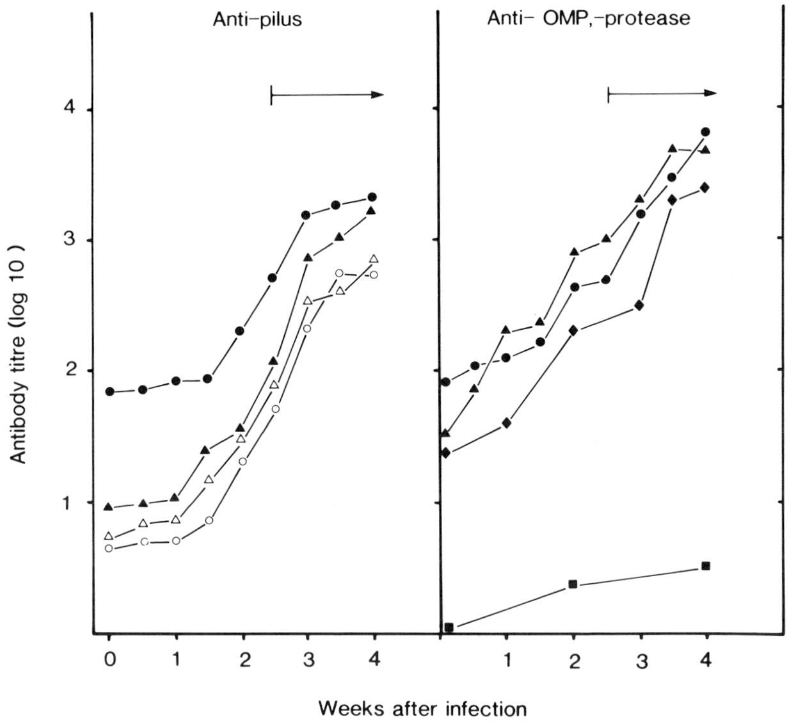

FIGURE 1. Kinetics of antibody responses to pili OMC, LPS, and protease of *B. nodosus* following the application of pure cultures of *B. nodosus* to the feet of four Merino sheep. Left: agglutination titers in lymph (△) from the popliteal lymph node and serum (○) and ELISA titers in lymph and (●) serum. Right: ELISA titers against OMC proteins in lymph (△) and serum (●) and serum ELISA titers against *B. nodosus* LPS (♦) and protease (■).→ onset of severe footrot lesions.

F. necrophorum increases with age.[43,44] In Australia, the periods of maximum incidence of foot abscess are spring and autumn, when up to 20% of sheep can be affected, principally ewes in lamb. There appears to be little durable immunity to the disease and experimental vaccines give only limited protection.[4]

Since the protective antigen(s) of *F. necrophorum* have not been defined, assays to monitor the development of "host immunity" are rather fictitious. Empirically, virulence determinants of *F. necrophorum* have been suggested as capsules, membrane proteins, LPS, and exotoxins (see Chapter 8) and it is against these antigens or their biological effects, that serological responses of ruminants have been measured. The typical results of ovine responses are presented in Table 1 where sera taken from adult sheep which resisted or succumbed to an experimental infection were examined piror to infection 1 month later. From this data, two main points can be observed:

1. Resistance to a challenge infection could not be predicted from any of the serological assays which measured agglutination (anticapsule or anti-LPS activity), ELISA (anti-OMP), or toxin neutralization. Within each experimental group, the variance of the titers against *F. necrophorum* rendered the group differences insignificant.
2. A similar proportional increase in serum titers for each assay occurred in all sheep, regardless of whether they became infected.

A number of factors complicate serology against *F. necrophorum*. Firstly, and most importantly, all adult sheep possess substantial antibody titers against antigens from *F.*

Table 1
ANTIBODY TITERS OF SHEEP FOLLOWING EXPERIMENTAL CHALLENGE WITH TOXIGENIC *FUSOBACTERIUM NECROPHORUM*

		Group mean titer ± SE					
		At challenge			30 d post challenge		
Experiment (Number of d)	Assay	Total	Infected	Resistant	Total	Infected	Resistant
1 (25)	Agglutination	16,266 ± 2,347	21,564 ± 3,550	6,850 ± 960	34,330 ± 7,250	41,420 ± 10,790	14,833 ± 1,860
	ELISA	6,488 ± 1,922	7,000 ± 2,176	4,025 ± 640	20,940 ±8,280	26,017 ±12,260	13,325 ± 4,035
	Toxin Neutralization[a]	15,625 ±6,073	18,170 ±5,094	8,630 ±1,990	35,800 ±8,675	42,960 ±12,890	16,110 ±3,560
2 (45)	Agglutination	102,107 ±28,960	91,078 ±22,540	112,380 ±30,630	—	—	—
	ELISA	58,780 ±10,370	53,300 ±7,935	63,900 ±13,335	—	—	—

Note: For assays, sera were grouped for the entire flock (Total) or for those sheep which resisted (Resistant) or succumbed to infection (Infected) after receiving s.c. inoculum of 2.5×10^8 toxigenic *F. necrophorum* (Strain FnB-1) in the interdigital skin.

[a] For comparison, newborn-lamb serum from six lambs had a neutralization titer of less than 500.

necrophorum. While most of these are probably mounted against shared or cross-reacting determinants on commensal strains, it frustrates the interpretation of serology. These antibodies could be considered not protective against infection, since the magnitude of titers are not predictive of immunity, and adult sheep are most at risk to natural infections. However, it is not known whether an Ig isotype against a restricted number of epitopes in the mixture of cellular antigens may be host protective. Such a consideration requires the use of purified antigens, and in this respect, although neutralizing IgG_2 antibodies are consistently elevated after immunization of sheep with leucocidin,[45] these also do not presage immunity against a challenge infection. It is also possible that if putative protective antigens on virulent *F. necrophorum* shared some epitopes with gut commensals, then antibodies to the shared (nonprotective) epitopes may be preferentially stimulated after clinical infections. (This would be analogous to the suppression of antihapten responses by a previous immunization with the carrier molecule.[46])

Secondly, there are several reports that immunization with complete Freund's adjuvant, nonspecifically enhances resistance to experimental infection with *F. necrophorum* in mice.[47,48] Since this immunity is manifested when the routes of immunization and challenge are similar, it may explain in part, encouraging results in these models. Comparing several reports, it appears that it is easier to limit the ability of *F. necrophorum* to cause liver abscesses, than to prevent foot abscess. In some cases of foot abscess, immunization with antigens from *F. necrophorum* exacerbated the experimental infection with the organism.[55]

Thirdly, the lack of a reliable model for reproducible experimental infections with *F. necrophorum* has hampered the evaluation of vaccine antigens and the predictive value of immune responses against the organism.

IV. BOVINE FOOTROT (*B. NODOSUS* INFECTION)

By inference, one causal organism of bovine footrot is *Bacteroides nodosus*, since autogenous vaccines significantly reduced the severity of lesions[2] and *B. nodosus* is routinely isolated from clinical lesions.[50] The incidence of interdigital dermatitis can approach 100%[50] and up to 70% of calves aged 3 to 12 weeks can be affected when kept on pasture with their dams.[44] Following infections in calves, the recurrence of lesions usually affects less than 20% of animals. Antibody titers against *B. nodosus* organisms from serogroups B and G[51] which generally cause bovine footrot in our CSIRO herd exhibit a transient increase of five to six-fold in agglutination titers and a similar ephemeral increase in ELISA titers against *B. nodosus* pili and OMC. Titers against *F. necrophorum* are also increased by the seasonal occurrence of bovine footrot (see below).

V. BOVINE FOOT ABSCESS

S.c. inoculation of pure cultures of toxigenic biovars of *F. necrophorum* into the interdigital skin induces a dose-dependent purulent cellulitis which is identical to that of clinical foot abscess in cattle.[41] Toxigenic A and AB biovars of *F. necrophorum* are readily isolated from clinical lesions.[42] In the temperate regions of Australia, the maximum incidence of foot abscess overlaps with that of interdigital dermatitis in the periods of high rainfall, humidity, and lush pasture growth.[41] This situation contrasts with the capacity of *F. necrophorum* to cause liver abscess in cattle throughout the year. Since the frequency of liver abscesses in the CSIRO herd was less than 1%, a longitudinal serological study was undertaken to examine the basis for differences in the incidence of foot-abscess infections in progeny from four inbred sire lines at the Institute. Serological responses in ELISA and agglutination tests are presented in Figure 2 for calves born in October to November (spring) and sampled at monthly intervals for the next 6 months. Three features of the serology presented in Figure 2 are relevant.

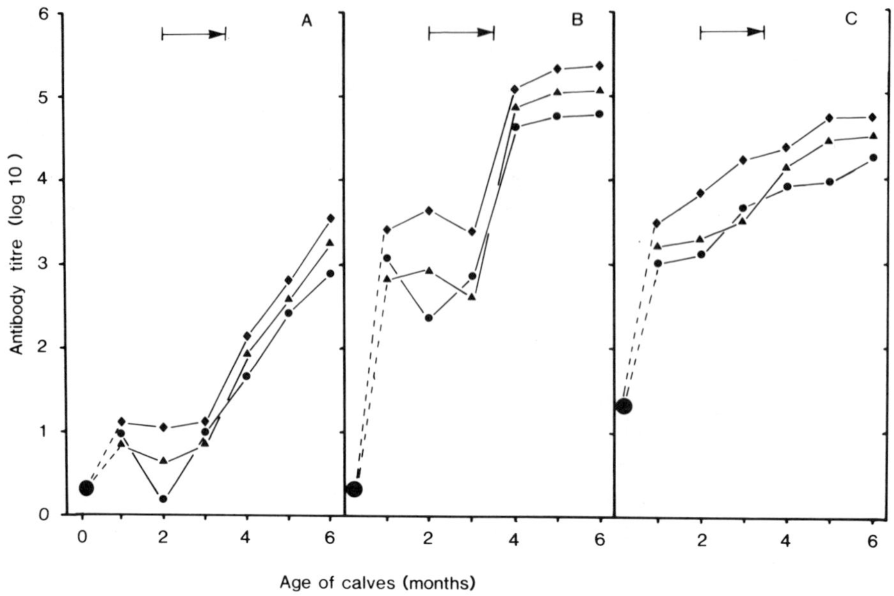

FIGURE 2. Serological responses of calves from three inbred sire lines at CSIRO during the first 6 months after birth. The titers of fetal-bovine serum are given at time zero and responses were monitored in (A) agglutination assays, (B) ELISA, and (C) leucocidin-neutralization tests. The period covering the appearance of clinical foot abscess is denoted (→). Sire lines include those producing high (◆), intermediate (▲), or low (●) antibody titers after immunization with ovalbumin and several unrelated antigens.[56]

1. As was the situation for sheep with foot abscess, during the period of maximum morbidity the titers of antibody against *F. necrophorum* increase whether or not cattle develop clinical disease.
2. All calves receive anti-*F. necrophorum* antibodies from their dams via colostrum and the maternally derived titers decline over the ensuing month. With October calvings, the specific antibody titers are low at the time of maximum incidence of both footrot and foot abscess in November to December.
3. Titers of antibodies which neutralize *F. necrophorum* leucocidins increase rapidly after birth and persist at high levels after the first transmission period.
4. In all cases, calves from "high responder" sires elaborate higher titers of antibody in all three assays than do calves from the other two sire lines. However, whether this attribute is directly responsible for the reduced incidence of clinical foot abscess in calves aged >6 months in this group is not known, since it is also possible that these calves may suffer more severe lesions as young calves than animals from the other two groups.[56]

At present, the necessity for a vaccine against foot abscess is less urgent than for the prevention of liver abscess. This reflects the limited morbidity of the infection, its relatively rapid development and resolution (unless the joint becomes infected), and the therapeutic effect of antibiotics. Any change in the livestock industry's perception of the importance of controlling foot abscess could lead to a reappraisal of the virulence determinants and serology of *F. necrophorum*.

VI. CONCLUSIONS

Footrot and foot abscess in ruminants are infections caused by anaerobic bacteria to which the host's immune response is insufficient to prevent the seasonal recurrence of the disease. In the case of footrot, the industrial impetus for a vaccine has enabled the conversion of a subliminal immunity to a durable protective response through systemic immunization with pilus or protease antigens. In this case, protection is predicted by serological classification of the organism. For foot abscess, it appears that a proportion of protective immunity is nonspecific and cell mediated, but augmentation of this mechanism by specific immunization has realized limited success. For *F. necrophorum,* serological assays are not yet predictive of protective immunity, and are complicated by the ubiquity of nonpathogenic strains of *F. necrophorum* in the gut and environment. The possibility that several other organisms contribute to the pathogenesis of clinical foot abscess also renders the establishment of experimental models more complex, and necessitates further research on both the organism and the disease so that the present problems can be resolved.

REFERENCES

1. **Egerton, J. R. and Burrell, D. H.,** Prophylatic and therapeutic vaccination against ovine footrot, *Aust. Vet. J.,* 46, 517, 1970.
2. **Clark, B. L., Stewart, D. J., Emery, D. L., Dufty, J. H., and Jarrett, R. G.,** Immunisation of cattle against interdigital dermatitis (footrot) with an autogenous *Bacteroides nodosus* vaccine, *Aust. Vet. J.,* 63, 61, 1986.
3. **Clark, B. L., Emery, D. L., Stewart, D. J., Dufty, J. H., and Anderson, D. A.,** Studies into immunisation of cattle against interdigital necrobacillosis, *Aust. Vet. J.,* 63, 107, 1986.
4. **Emery, D. L., Vaughan, J. A., Clark, B. L., and Stewart, D. J.,** Virulence determinants of *Fusobacterium necrophorum* and their prophylatic potential in animals, in *Footrot in Ruminants,* Proc. Workshop, Melbourne 1985, Stewart, D. J., Peterson, J. E., McKern, N. M., and Emery, D. L., Eds., CSIRO Division Animal Health/Australian Wool Corp., Glebe, N.S.W., 1986, 267.
5. **Langworth, B. F.,** *Fusobacterium necrophorum:* its characteristics and role as an animal pathogen, *Bateriol. Rev.,* 41, 373, 1977.
6. **Scanlan, C. M. and Hathcock, T. L.,** Bovine rumentitis — liver abscess complex: a bacteriological review, *Cornell Vet.,* 73, 288, 1983.
7. **Emery, D. L. and Stewart, D. J.,** Phagocytosis of *Bacteroides nodosus* by ovine peripheral blood leucocytes, *Vet. Microbiol.,* 9, 169, 1984.
8. **Yasmeen, D.,** Antigen-specific cytophilic activity of sheep IgG_1 and IgG_2 antibodies, *Aust. J. Exp. Biol. Med. Sci.,* 59, 297, 1981.
9. **Watson, D. L.,** The effect of cytophilic IgG_2 on phagocytosis by ovine polymorphonuclear leucocytes, *Immunology,* 31, 159, 1976.
10. **Mittal, K. R. and Ingram, D. G.,** Bactericidal and opsonic activities of normal sheep serum against gram-negative bacteria, *Am. J. Vet. Res.,* 36, 1189, 1975.
11. **Taylor, P. W.,** Bactericidal and bacteriolytic activity of serum against gram-negative bacteria, *Microbiol. Rev.,* 47, 46, 1983.
12. **Egerton, J. R. and Merritt, G. C.,** The occurrence of bactercidal antibodies against *Fusiformis nodosus* in sheep serum, *J. Comp. Pathol.,* 80, 369, 1970.
13. **Emery, D. L.,** Reactivity of sera from sheep immunised with individual outer membrane proteins of *Bacteroides nodosus* against heterologous bacterial strains, *Vet. Microbiol.,* 9, 45, 1984.
14. **Stewart, D. J.,** Biochemical and biological studies on the lipopolysaccharide of *Bacteroides nodosus, Res. Vet. Sci.,* 23, 319, 1977.
15. **Stewart, D. J., Clark, B. L., Emery, D. L., Peterson, J. E., and Fahey, K. J.,** A *Bacteroides nodosus* immunogen, distinct from the pilus, which induces cross-protective immunity in sheep vaccinated against footrot, *Aust. Vet. J.,* 60, 83, 1983.
16. **Emery, D. L., Stewart, D. J., Clark, B. L., Elleman, T. C., and McKern, N.,** Identification of protective antigens of *Bacteroides nodosus* by the immune response against footrot in sheep, *Annu. Proc. Sheep Vet. Soc.,* 9, 23, 1985.
17. **Ingham, H. R., Sisson, P. R., Middleton, R. L., Narang, H. K., Codd, A. A., and Selkon, J. B.,** Phagocytosis and killing of bacteria in aerobic and anaerobic conditions, *J. Med. Microbiol.,* 14, 391, 1981.

18. **Wilson, M. E., Genco, R. J., and Snyderman, R.,** The phagocytic cell, *Rev. Infect. Dis.*, 7, 387, 1985.
19. **Czop, J. K.,** Phagocytosis of particulate activators of the alternative complement pathway: effects of fibronectin, *Adv. Immunol.*, 38, 36, 1986.
20. **Egerton, J. R. and Morgan, I. R.,** Treatment and prevention of footrot in sheep with *Fusiformis nodosus* vaccine, *Vet. Rec.*, 91, 453, 1972.
21. **Skerman, T. M., Erasmuson, S. K., and Every, D.,** Differentiation of *Bacteroides nodosus* biotypes and colony variants in relation to their virulence and immunoprotective properties, *Infect. Immun.*, 32, 788, 1981.
22. **Emery, D. L., Stewart, D. J., and Clark, B. L.,** The comparative susceptibility of five breeds of sheep to footrot, *Aust. Vet. J.*, 61, 85, 1984.
23. **Parker, C. F., Cross, R. F., and Hamilton, K. L.,** Genetic resistance to footrot in sheep, *Annu. Proc. Sheep Vet. Soc.*, 9, 16, 1985.
24. **Skerman, T. M.,** Genetic variation and inheritance of susceptibility to footrot in sheep, in *Footrot in Ruminants,* Proc. Workshop, Melbourne, 1985, Stewart, D. J., Peterson, J. E., McKern, N. M. and Emery, D. L., Eds., CSIRO Division Animal Health/Australian Wool Corp., Glebe, N.S.W., 1986, 73.
25. **Egerton, J. R. and Merritt, G. C.,** Serology of footrot: antibodies against *Fusiformis nodosus* in normal, affected, vaccinated and passively immunised sheep, *Aust. Vet. J.*, 49, 139, 1973.
26. **Stewart, D. J.,** The role of various antigenic fractions of *Bacteroides nodosus* in eliciting protection against footrot in vaccinated sheep, *Res. Vet. Sci.*, 24, 14, 1978.
27. **Stewart, D. J., Clark, B. L., Peterson, J. E., Griffiths, D. A., and Smith, E. F.,** Importance of pilus-associated antigen in *Bacteroides nodosus* vaccines, *Res. Vet. Sci.*, 32, 140, 1982.
28. **Fahey, K. J., McWaters, P. G., Stewart, D. J., Peterson, J. E., and Clark, B. L.,** Quantitation by ELISA of pili and sheep antibodies to the pili of *Bacteroides nodosus, Aust. Vet. J.*, 60, 111, 1983.
29. **Renart, J., Reiser, J., and Stark, G. R.,** Transfer of proteins from gels to diazobenzoxymethyl paper and detection with antisera: a method for studying antibody specificity and antigen structure, *Proc. Natl. Acad. Sci. (U.S.A.),* 76, 3116, 1979.
30. **O'Donnell, I. J., Stewart, D. J., and Clark, B. L.,** Serological identification of pilus antigen and other protein antigens of *Bacteroides nodosus* using electro-blot radio-immunoassay after electrophoretic fractionation of the proteins on sodium dodecyl sulphate polyacrylamide gels, *Aust. J. Biol. Sci.*, 36, 15, 1983.
31. **Merrit, G. C. and Egerton, J. R.,** IgG_1 and IgG_2 immunoglobulins to *Bacteroides (Fusiformis) nodosus* protease in infected and immunized sheep, *Infect. Immun.*, 22, 1, 1978.
32. **Depiazzi, L. J. and Richards, R. B.,** A degrading protease test to distinguish benign and virulent ovine isolates of *Bacteroides nodosus, Aust. Vet. J.*, 55, 25, 1979.
33. **Every, D.,** Proteinase isoenzyme patterns of *Bacteroides nodosus:* distinction between ovine virulent isolates, ovine benign isolates and bovine isolates, *J. Gen. Microbiol.*, 128, 809, 1982.
34. **Emery, D. L., MacHugh, N. D., and Ellis, J. A.,** Properties and functional activity of macrophages from bovine afferent lymph, *Immunology,* 62, 177, 1987.
35. **Thorley, C. M. and Egerton, J. R.,** Comparison of alum-absorbed and non-alum absorbed oil emulsion vaccines containing either pilate or non-pilate *Bacteroides nodosus* cells in inducing and maintaining resistance of sheep to experimental footrot, *Res. Vet. Sci.*, 30, 32, 1981.
36. **Emery, D. L., Stewart, D. J., and Clark, B. L.,** The structural integrity of pili from *Bacteroides nodosus* is required to elicit protective immunity against footrot in sheep, *Aust. Vet. J.,* 61, 237, 1984.
37. **Reed, G. A.,** The role of footrot vaccines in Australia, in *Footrot in Ruminants,* Proc. Workshop, Melbourne, 1985, Stewart, D. J., Peterson, J. E., McKern, N. M., and Emery, D. L., Eds., CSIRO Division Animal Health/Australian Wool Corp., Glebe, N.S.W., 1986, 173.
38. **Egerton, J. R., Roberts, D. S., and Parsonson, I. M.,** The aetiology and pathogenesis of ovine footrot. A histological study of the bacterial invasion, *J. Comp. Pathol.,* 79, 207, 1969.
39. **Chetwin, D. H., Liardet, D. M., Kingsley, D. F., and Hindmarsh, F. H.,** The serological responses of sheep to immunisation at varying intervals with a multistrain footrot vaccine, in *Footrot in Ruminants,* Proc. Workshop, Melbourne, 1985, Stewart, D. J., Peterson, J. E., McKern, N. M., and Emery, D. C., Eds., CSIRO Division Animal Health/Australian Wool Corp., Glebe, N.S.W., 1986, 165.
40. **Stewart, D. J., Clark, B. L., Emery, D. L., Peterson, J. E., and Kortt, A. A.,** The phenomenon of cross protection against footrot induced by vaccination of sheep with *Bacteroides nodosus* vaccine, in *Footrot in Ruminants,* Proc. Workshop, Melbourne, 1985, Stewart, D. J., Peterson, J. E., McKern, J. E., and Emery, D. L., Eds., CSIRO Division Animal Health/Australian Wool Corp., Glebe, N.S.W., 1986, 185.
41. **Clark, B. L., Stewart, D. J., and Emery, D. L.,** The role of *Fusobacterium necrophorum* and *Bacteroides melaninogenicus* in the aetiology of interdigital necrobacillosis in cattle, *Aust. Vet. J.*, 62, 47, 1985.
42. **Emery, D. L., Vaughan, J. A., Clark, B. L., Dufty, J. H., and Stewart, D. J.,** Cultural characteristics and virulence of strains of *Fusobacterium necrophorum* isolated from the feet of cattle and sheep, *Aust. Vet. J.*, 62, 43, 1985.

43. **Fievez, L.,** *Comparative Study of Strains of Shaerophorus Necrophorus Isolated from Man and Animals,* European Academic Press, Brussels, 1963.
44. **Emery, D. L.,** Immunity against anaerobic bacterial infections, *Vet. Immunol. Immunopathol.,* 15, 1, 1987.
45. **Emery, D. L., Edwards, R. D., and Rothel, J. S.,** Studies on the purification of the leucocidin of *Fusobacterium necrophorum* and its neutralization by specific antisera, *Vet. Microbiol.,* 11, 357, 1986.
46. **Herzenbeng, L. A. and Tokuhisa, T.,** Epitope-specific regulation. I. Carrier-specific induction of suppression for IgG anti-hapten antibody responses, *J. Exp. Med.,* 155, 1730, 1982.
47. **Conlon, P. J., Hepper, K. P., and Teresa, G. W.,** Evaluation of experimentally induced *Fusobacterium necrophorum* infections in mice, *Infect. Immun.,* 15, 510, 1977.
48. **Emery, D. L. and Vaughan, J. A.,** Generation of immunity against *Fusobacterium necrophorum* in mice inoculated with extracts containing leucocidin, *Vet. Microbiol.,* 12, 255, 1986.
49. **Quintavolla, F., Zannetti, G., Licata, E., Tagliavini, M., and Pasini, A.,** Therapeutic use of a *Bacteroides nodosus* vaccine in various diseases of the bovine foot: preliminary note, *Alt. Soc. Ital. Buitria,* 17, 237, 1985.
50. **Buczek, J. and Cygan, Z.,** Outbreak of interdigital dermatitis in cattle. I. Prevalence and course, *Med. Weter.* 42, 208, 1986.
51. **Claxton, P. D., Ribeiro, L. A., and Egerton, J. R.,** Classification of *Bacteroides nodosus* by agglutination tests, *Aust. Vet. J.,* 60, 331, 1983.
52. **Emery, D. L. and Rothel, J. S.,** Unpublished data.
53. **Outteridge, P. M. and Stewart, D. J.,** Unpublished data.
54. **Outteridge, P. M. and Skerman, T. M.,** Unpublished data.
55. **Emery, D. L.,** Unpublished data.
56. **Dufty, J. and Emery, D.,** Unpublished data.

Chapter 10

ANTIGENIC CLASSIFICATION OF *BACTEROIDES NODOSUS*

P. D. Claxton

TABLE OF CONTENTS

I. INTRODUCTION

Antigenic variation among strains of *Bacteroides nodosus* was first reported by Beveridge[1] in 1941. Using agglutination tests, he observed serological heterogenicity among isolates of *B. nodosus* from Australia and the U.S. Egerton[2] subsequently described the occurrence of surface (K) and somatic (O) antigens and a system for typing *B. nodosus* isolates based on K-agglutination tests. He was able to classify 33 of 46 isolates into one of three types designated A, B, and C and suggested that more than one K antigen may be present on some isolates. A third antigen, a proteolytic enzyme produced by the organism, was demonstrated by Egerton and Merritt,[3] who found it capable of stimulating the production of antiprotease antibodies in sheep.

The K-agglutination reaction has been shown to result from the reaction between the pilus antigens of *B. nodosus* and specific antibody.[4,5] The pili, long proteinaceous appendages, constitute both the type-specific serotyping antigen and the major protective immunogen.[6,7] This knowledge has stimulated interest in the serotyping of *B. nodosus*, since it has also been demonstrated that properly formulated vaccines can provide effective protection against homologous challenge[8] but less effective or no protection against heterologous challenge.[2,9,10]

The K-agglutination test has now formed the basis for a number of reports on the serological classification of *B. nodosus*.[2,11-18] Differences among the published classifications may be attributable in part to the methods employed for preparation of antisera and antigens and to the test method itself. Geographical differences in distribution of serotypes are also likely to be a contributing factor.[15] These studies all, however, confirmed that the pilus antigens of *B. nodosus* are quite diverse, with up to 9 serogroups or 18 serotypes.[17]

II. SELECTION OF ISOLATES

Isolates devoid of pili cannot be serotyped nor are they suitable for the preparation of antisera. Selection can be based on electron microscopic examination for the presence of pili or more simply on colony characteristics on solid media. Organisms from fimbriate colonies on media containing 4% agar are piliated, whereas those from nonfimbriate colonies are either poorly piliated or are devoid of pili.[19]

III. PREPARATION OF ANTIGENS

Antigens for K-agglutination tests have been produced from organisms grown on solid medium[2,9,13,14,16,17] or in liquid medium.[9,11,13,17,18] In order to retain the pili, antigens have been prepared from unwashed cultures killed with formalin, although a centrifugation step[9,13] or isoelectric precipitation at pH 4.5 has been employed, when necessary, to concentrate broth cultures.[18] Claxton[9] and Claxton et al.[13] standardized all antigen suspensions to an optical-density standard rather than by cell count.

IV. PRODUCTION OF ANTISERA

A. Using Whole-Cell Antigens

Antisera for use in the K-agglutination test have been prepared by several different methods, usually from unwashed whole cells. Egerton[2] employed antigens prepared from formalin-killed cultures grown on solid medium and raised antisera in sheep and rabbits. The bacterial suspensions were preserved with 0.25% formalin after they had been adjusted to contain about 10^{10} cells per ml. Adult rabbits were injected i.v. twice weekly with 0.25 ml of suspension for 3 weeks. Sera were collected 14 d after the last injection. Formalinized suspensions of *B. nodosus* emulsified in oil adjuvant were used to raise antisera in sheep.

Two injections were given i.m. at an interval of 1 week, each dose containing approximately 2.5×10^9 cells. The sheep were bled 14 d after the second dose.

Thorley[11] used formalized overnight cultures in liquid medium absorbed onto 2.5% potassium aluminum phosphate or emulsified in Freund's complete adjuvant. He employed two doses of 2 ml of the alum-adjuvated vaccine in rabbits at an interval of 6 weeks, with sera being collected 14 d later. Alternatively, he injected 2 ml of the emulsion vaccine followed 6 weeks later by an i.v. injection of 2 ml of aqueous antigen, with serum collection 14 d later.

Schmitz and Gradin[12] used a method similar to that of Egerton,[2] except that they employed antigens prepared from broth cultures.

Claxton[9,17] and Claxton et al.[13] used antigens prepared from cultures on either solid or in liquid medium inactivated with 1.25% formalin and emulsified in Freund's complete adjuvant. Rabbits were injected with 0.5 ml i.m. in each hind leg and bled 21 to 28 d later for antiserum.

Day et al.[18] and Thorley and Day[15] raised antisera in rabbits using formalin-killed piliated cultures of *B. nodosus* concentrated by isoelectric precipitation at pH 4.5. These concentrates were incorporated into alum-oil-adjuvanted vaccines as described by Thorley and Egerton,[8] except that Alhydrogel® (Superfos a/s, Denmark) was used in place of potash alum in some instances. The rabbits were given two 0.2 ml s.c. injections of vaccine at an interval of about 8 weeks and were bled 3 weeks after the second vaccination.

Kingsley et al.[16] raised antisera in sheep by immunizing with semi-purified pili, the sheep receiving two 1 ml s.c. injections at an interval of 6 weeks. Blood collections were obtained 58 to 63 d after the initial injection, at which time peak titers were reached.

B. Antipilus Antisera

Day et al.[18] purified pili from *B. nodosus* broth cultures by multiple isoelectric precipitations and magnesium chloride precipitations at 4°C.[20] Pili so obtained were formulated into alhydrogel-oil vaccines and antisera were raised using a similar regimen to that employing whole-cell antigens.[18]

C. Absorption of Antisera

Most workers[2,9,11-14,16,17] have employed unabsorbed antisera for serotyping *B. nodosus.* However, in order to eliminate most of the low-titer cross-serotype reactions observed in tube-agglutination tests in earlier work, Thorley and Day[15] routinely absorbed each serum with formalin-killed cells of two heterologous serotypes, followed by a slide-agglutination test to confirm specificity. If cross-reactions were still observed, a further absorption step with the cross-reacting organisms was carried out.

V. TYPING BY SLIDE AGGLUTINATION

In order to provide a convenient method of screening large numbers of isolates of *B. nodosus,* a rapid-agglutination test has been developed[9,13] that can be carried out on clean, glass microscope slides. The appropriate undiluted serum is dispensed, using a full 2.0-mm nichrome wire loop. The antigen, at a concentration ten times that used in the tube-agglutination test, is dispensed with a full 3.0-mm loop and thoroughly mixed with the serum on the slide. The slide is rocked back and forth several times and examined. A positive reaction is recorded when heavy floccular agglutination occurs within 10 s of mixing the serum and antigen. Using this method, Claxton et al.[13] examined 1267 isolates of *B. nodosus* from 292 sheep in 58 Australian flocks. Typing was commenced using antisera prepared against Egerton's[2] three reference strains, viz:

Serotype	Designation	Collection No.
A	McMaster 198	VCS1001
B	McMaster 199	VCS1007
C	McMaster 217	VCS1008

However, strain VCS1007 did not produce uniformly fimbriate colonies so a closely related isolate VCS1006 (McMaster 215) was used as the serotype B reference strain for all further work. As isolates that did not react with any of these three sera were detected, antisera were prepared against them in rabbits. Grouping of isolates was then made on the basis of slide-agglutination patterns. Using this procedure, it was possible to group 1260 of the 1267 isolates into eight serogroups that were designated A to H.

Most antigens, when correctly standardized, reacted with only one group antiserum. However, a number of isolates reacted with both group B and group F antisera and a small number reacted with group A and group F antisera. Claxton[17] subsequently proposed a further serogroup, group I, and also reported strains reacting with group C and E antisera, groups B and I antisera, or group D and E antisera.

The slide-agglutination test has also been employed by other workers. Hindmarsh and Fraser,[14] using antisera prepared in sheep, against semi-purified pili, were able to classify all of their 449 British isolates into the original eight serogroups of Claxton[9] and Claxton et al.[13] Kingsley et al.[16] reported that they had successfully typed 2126 isolates of *B. nodosus* from six countries by using the slide-agglutination test and pili antisera. These isolates could all be successfully allocated to one of the eight serogroups (A to H) defined by Claxton et al.[13]

VI. TYPING BY TUBE AGGLUTINATION

Thorley[11] found that agglutination reactions between unheated formalin-treated cell suspensions of *B. nodosus* and antisera fell into two broad groups, relatively high titer and relatively low titer. The high-titer agglutination, consisting of large, loose cell aggregates was characteristic of the reaction between a strain and its homologous antiserum. This reaction, shown to be pilus mediated,[19] was considered to be essentially the same as that described by Egerton[2] for the interaction of unwashed agar-grown cell suspensions with homologous antiserum and designated by him as K agglutination. Short et al.[19] also suggested that low-titer heterologous aggutination, which usually appeared as finely granular aggregates, was due to other surface antigens. By employing cross-tube-agglutination tests, Thorley[11] was able to divide 13 isolates from Great Britain into at least 7 different K-agglutination types distinct from Egerton's[2] A and C, 6 of these being isolated from sheep in one flock.

Schmitz and Gradin[12] examined 97 isolates of *B. nodosus* from sheep in 20 flocks in Oregon, U.S. By employing cross-tube-agglutination tests, they were able to identify 14 serotypes, including isolates that were serologically similar to Australian serotypes A, B, and C. These 14 serotypes were designated I to XIV. Serotypes X and XIV were closely related to type A; serotypes IX, XII, and possibly XIII to type B; and serotype VI to type C.

Australian isolates examined by the slide-agglutination test were also subjected to tube-agglutination tests.[9,13] Cross-tube-agglutination tests carried out on 44 isolates and their antisera supported the existence of at least the 8 serogroups (A to H), but also provided evidence that most serogroups could be divided into subgroups or serotypes.[13] Subsequent work indicated the existence of a further serogroup (I) and subgroups within group C, bringing the number of serogroups in this classification to 9 and the number of subgroups to 18[17] (Table 1).

Table 1
SEROGROUPS OF *BACTEROIDES NODOSUS* DETERMINED BY CROSS-TUBE AGGLUTINATION REACTIONS[9,13,17]

Prototype Strain	Serogroup	VCS 1001	VCS 1251	VCS 1006	VCS 1208	VCS 1190	VCS 1125	VCS 1008	VCS 1202	VCS 1172	VCS 1137	VCS 1114	VCS 1017	VCS 1244	VCS 1220	VCS 1004	VCS 1215	VCS 1057	VCS 548
VCS 1001	A1	10240	2560	*	*	*	*	*	*	*	*	*	*	*	*	*	*	*	*
VCS 1251	A2	2560	10240	*	*	*	*	*	*	*	*	*	*	*	*	*	*	*	*
VCS 1006	B1	*	*	10240	10240	1280	1280	*	*	*	*	*	*	*	*	*	*	*	*
VCS 1208	B2	*	*	5120	10240	2560	320	*	*	*	*	*	*	*	*	*	*	*	NT
VCS 1190	B3	*	*	5120	10240	20480	5120	*	*	*	*	*	*	*	*	*	*	*	NT
VCS 11250	B4	*	*	5120	1280	5120	20480	*	*	*	*	*	*	*	*	*	*	*	NT
VCS 1008	C1	*	*	*	*	*	*	10240	1280	*	*	*	*	*	*	*	*	*	*
VCS 1202	C2	*	*	*	*	*	*	2560	20480	*	160	*	*	*	*	*	*	*	*
VCS 1172	D	*	*	*	*	*	*	*	*	10240	*	*	*	*	*	*	*	*	*
VCS 1137	E1	*	*	*	*	*	*	*	*	*	10240	2560	*	*	*	*	*	*	*
VCS 1114	E2	*	*	*	*	*	*	*	*	*	1280	20480	*	*	*	*	*	*	*
VCS 1017	F1	*	*	*	*	*	*	*	*	*	*	*	20480	320	*	*	*	*	*
VCS 1244	F2	*	*	*	*	*	*	*	*	*	*	*	320	5120	*	*	*	*	*
VCS 1220	G1	*	*	*	*	*	*	*	*	*	*	*	*	*	5120	1280	*	*	*
VCS 1004	G2	*	*	*	*	*	*	*	*	*	*	*	*	*	640	10240	*	*	*
VCS 1215	H1	*	*	*	*	*	*	*	*	*	*	*	*	*	*	*	5120	640	*
VCS 1057	H2	*	*	*	*	*	*	*	*	*	*	*	*	*	*	*	320	5120	*
VRS 548	I	*	*	*	320	320	320	*	*	*	*	*	*	*	*	*	*	*	10240

Note: *Titers ≤80; NT, not tested.

A recent survey of *B. nodosus* strains in Great Britain and Western Europe has identified the presence of the eight serogroups as defined by Claxton et al.[13] and an additional nine serotypes which were considered to be distinct from each other and from serogroups A to H.[15,18]

Although no complete comparisons have been undertaken, the data presented by each of the above workers can in fact be presented in such a way as to indicate the existence of between 8 and 11 serogroups with two or more subgroups or serotypes within some or most serogroups. The tube-agglutination-test results presented by each group of workers suggest the presence of shared as well as type-specific surface or pilus antigens among the members of some serogroups. Supporting evidence has also been presented from studies employing absorbed antisera.

Claxton[9] examined 30 isolates of *B. nodosus* classified by slide agglutination into serogroup A. Using the tube-agglutination test he could divide these into two distinct subgroups or serotypes, which he designated A1 and A2. Eleven isolates were classified as A1 and 19 as A2 (Table 2). In cross-agglutination tests using absorbed and unabsorbed antisera, he was able to demonstrate that absorption of antiserum with the homologous antigen removed all K-agglutination activity against both subtype antigens, whereas absorption of antiserum with the heterologous antigen removed all K-agglutination activity against that antigen but only partially removed K-agglutination activity against the homologous antigen (Table 3).

The picture was not as clear, however, with serogroup B, which could be divided into four subgroups.[9,13] Studies employing a representative of each subgroup (B1 to B4) and absorbed and unabsorbed antisera suggested the presence of at least one "major" group-specific surface antigen, with each of the four isolates apparently carrying one or more "minor" surface antigens, which may or may not be shared by one or more of the other isolates.[9] Whether these "minor" antigens were pili or pilus associated was not determined, although the agglutination reactions concerned were of the typical pilus type.

Day et al.[18] reported the results of some more exhaustive studies on strains classified by Claxton[9] as belonging to serogroup B subtypes B1, B2, B3 and B4 together with strains belonging to their serotypes K, L, and N. On the basis of slide-agglutination reactions with specific antisera, they classified their serotypes B, K, L, and N as B-group-analogous strains. In microtiter agglutination assays, they demonstrated that such strains exhibit high titer reactions with homologous antisera but also demonstrate fairly high titers, in some instances as high as the homologous titer, against sera of other serotypes within the B group. However, although strains classified by Claxton as belonging to serogroup B subtypes B1, B2, B3, and B4 exhibited many strong cross-reactions with strains from their serotypes K, L, and N, cross-agglutination and absorption studies failed to reveal any particular homology between strains from the two serogrouping systems.

Beesley et al.[21] suggested that more than one species of antigen may exist on an individual pilus. Day et al.[18] presented evidence from immunogold labeling studies with unabsorbed and homologously or heterologously absorbed antisera, that three distinct antigens existed on the pili of each of a serotype B, a serotype K, and a serotype N strain. Each of these strains possessed one unique antigen. One of the other two antigens on each strain was shared with a second strain and the remaining antigen was shared with the third strain only. Taken in conjunction with the results of cross-agglutination studies on absorbed and unabsorbed antisera, they have also deduced the presence of a further antigen common to all three strains. Their data therefore suggest at least seven different pilus antigens are shared variously among the three strains.

The existence of such antigenic diversity is certainly consistent with the results of attempts at serological classification by either slide- or tube-agglutination tests. The low-titer cross-reactions sometimes recorded between members of different serogroups may also be explained by the existence of more than one antigen on the pilus.

Table 2
SERUM TUBE-AGGLUTINATION REACTIONS OF 30 SEROGROUP-A ANTIGENS WITH ANTISERA AGAINST STRAINS 1001 (A1) AND 1251 (A2) K-AGGLUTININ TITERS[9]

		Antiserum	
Subgroup	Antigen	1001	1251
A1	1001	10240	1280
	1016	5120	1280
	1056	10240	1280
	1068	10240	1280
	1247	10240	1280
	1248	10240	640
	1254	10240	1280
	1293	10240	1280
	1298	10240	1280
	1299	10240	1280
	1305	10240	640
A2	1014	1280	10240
	1015	2560	10240
	1133	1280	5120
	1191	2560	5120
	1194	1280	10240
	1199	2560	5120
	1203	2560	10240
	1205	2560	10240
	1213	2560	10240
	1233	2560	10240
	1250	2560	5120
	1251	1280	5120
	1261	1280	5120
	1264	2560	5120
	1265	2560	5120
	1269	2560	5120
	1303	2560	5120
	1335	2560	10240
	1336	2560	5120

In the face of such antigenic complexity within a serogroup, the relative titers of typing sera against the various component antigens of a strain may well be important in deciding to which subgroup or serotype that strain should be assigned. These relative titers may depend on the relative frequency of the different antigens on the pili of the strain being serotyped and the strain against which antiserum is raised. The immunogenicity of each component antigen would also be important.

Such pilus-mediated cross-reactivity would make accurate serotyping of *B. nodosus* a very difficult procedure. Since the pilus is the major immunogen of *B. nodosus,* affording serotype-specific protection,[6] this cross-reactivity among strains of different serotypes also has implications in vaccine production.[18] In fact, the prime aim of the serogrouping studies of Claxton[9] and Claxton et al.[13] was to develop a system to permit the selection of a limited number of isolates for the production of vaccines capable of providing protection across a wide range of serotypes. Studies were, in fact, carried out to determine whether cross

Table 3
RESULTS OF CROSS-AGGLUTINATION TESTS USING ABSORBED AND UNABSORBED ANTISERA AGAINST *B. NODOSUS* STRAINS 1001(A1) AND 1251(A2) K-AGGLUTININ TITERS[9]

Anti-serum	Absorbed with	K-Agglutinin titer against antigen	
		1001	1251
1001	—	10240	1280
1001	1001	<80	<80
1001	1251	1280	<80
1251	—	1280	5120
1251	1001	<80	640
1251	1251	<80	<80

protection could be demonstrated within a serogroup between members of different subgroups. Claxton[9] demonstrated under experimental challenge conditions that a vaccine produced from a subgroup A1 strain was capable of producing good protection against challenge with both A1 and A2 virulent strains.

Studies with B-group isolates, however, were again less clear cut. A B1 vaccine provided good protection against challenge with B1 and B2 strains but was less effective against a B3 or B4 challenge.

The inclusion of a large number of distinct serotypes (>10) of *B. nodosus* in a vaccine would be a commercial impossibility and antigenic competition might well lead to inadequate response to individual antigens. It may well be possible, however, to exploit the existence of cross-reacting pilus antigens in the selection of strains for vaccine production if it could be demonstrated that effective protection against a wider range of antigenic types was produced in the field.

VII. DISTRIBUTION OF ANTIGENIC TYPES.

A. Within Flocks and Within Sheep

It is quite apparent from the published reports that multiple serotype infections are common within flocks of sheep and within individual feet or lesions. Schmitz and Gradin,[12] in identifying 14 serotypes of *B. nodosus* among 97 isolates derived from 20 Oregon flocks, reported multiple serotype infections in 13 of the 15 flocks from which more than one isolate was examined. They found up to seven different serotypes in a single flock and two different serotypes in each of three sheep. In Australia, Claxton et al.[13] found multiple serogroup infections in 19 of 32 flocks from which two or more isolates were collected. Up to six serogroups were isolated from a flock and up to four from an individual infected foot.

Similar results were obtained in the U.K., where Hindmarsh and Fraser[14] examined isolates from 34 flocks throughout Scotland and England. They reported that 13.8% of the 297 sheep examined had multiple serogroup infections and over 61% of the flocks were infected with more than one serogroup. Thorley and Day[15] examined isolates from 58 farms in the U.K. and 24 in Western Europe (France, Holland, Belgium, Italy, Sicily, and Sardinia). They also reported up to three serotypes from individual feet and the frequent isolation of multiple serotypes from a single farm. Some 37 of 65 New Zealand farms (57%) sampled also yielded more than one serotype.[16]

Table 4
GEOGRAPHIC DISTRIBUTION OF SEROGROUPS OF *B. NODOSUS*

Country	No. of isolates	Percentage of isolates in serogroup									Ref.
		A	B	C	D	E	F	G	H	NA[a]	
Australia	1267	10.8	28.2	6.1	13.6	5.5	11.2	12.6	11.5	0.6	13
New Zealand	1547	5.3	56.2	8.5	10.9	8.0	3.2	0.5	7.5	—	16
United Kingdom	449	2.0	40.8	2.0	10.9	2.0	8.7	7.6	26.5	—	16
Australia	113	16.8	32.7	2.7	12.4	7.1	9.7	10.6	8.0	—	16

[a] Not agglutinated.

B. Geographic

In the small number of studies carried out employing the serogrouping system described by Claxton et al.[13] it has become apparent that the eight serogroups are represented in each country sampled.[13-16] Each country appears to have an individual profile of isolation frequencies for the serogroups, although Group B is by far the most common in each country (Table 4).

C. Among Species of Animals

Bacteroides nodosus is the etiological agent of benign and virulent footrot of sheep and also of footrot in goats and cattle. Although there is a relationship between proteolytic activity and virulence in sheep, there is no relationship between serotype and virulence. Claxton[9] found that six of nine isolates from benign ovine footrot could be grouped using the slide-agglutination test. He was also able to serogroup 19 of 23 bovine isolates and all 5 available goat isolates. Five of the eight serogroups were represented among the benign sheep isolates (A, B, E, G, H); six among the cattle isolates (A, B, C, F, G, H) and one among the goat isolates (F). The three untyped ovine and four untyped bovine isolates were subsequently found to belong to serogroup I.

In a subsequent study, Claxton and O'Grady[22] examined 148 isolates of *B. nodosus* from seven flocks of goats in New South Wales, Australia. Using the slide-agglutination test, they were able to classify 139 of these isolates into serogroups A, B, C, F, G, H, or I. The nine untypable isolates were all isolated from interdigital lesions in goats from a single flock.

VIII. CONCLUSIONS

The system of antigenic classification based on the K-agglutination reaction is one of convenience. By the use of the slide-agglutination test, large numbers of isolates can be screened and classified into a limited number of serogroups. The tube-agglutination test can be used to further explore the antigenic relationships among isolates of *B. nodosus*.

The data presented here demonstrate clearly the diversity of the pilus antigens of *B. nodosus*. The finding that more than one species of antigen may exist on an individual pilus,[18,21] is supported by Mattick et al.[23] and Anderson et al.[24] They reported that isolated pili contain two prominent antigens, the structural subunit of the pilus strand (mol wt approximately 17 kDa) and a basal protein (mol wt approximately 80 kDa). Employing matrix western transfer serological analysis, they demonstrated that the pilus subunit was recognized by the homologous antiserum but not generally by heterologous antisera, whereas the basal antigen reacted randomly across serological boundaries.[24,25] This basal antigen(s) may be responsible for the cross-reactions seen between some members of some serogroups. However, they too demonstrated some interesting cross-reactions associated with the pilus subunit. Their data indicated that some serogroups have a closer relationship than others.

They also suggested that the various serogroups and subtypes of *B. nodosus* comprise a series of overlapping sets of antigenically related strains with the prototype isolates representing convenient and/or historical serological reference points.

The value, then, of the antigenic classification based on the K-agglutination reaction will depend on its usefulness as an epidemiological tool and on its usefulness in selecting strains of *B. nodosus* for incorporation in multivalent vaccines to provide broad-spectrum protection against field infection.

It is interesting to note that the information available demonstrates that most isolates of *B. nodosus* from widely separated geographic areas can be grouped, using the slide-agglutination test proposed by Claxton et al.[13] Multivalent vaccines based on the serogroups proposed by these authors have been developed in Australia and New Zealand and, in general, provide high levels of protection for sheep against field infection. Suitable vaccines have, however, not yet been developed for use in cattle, and the efficacy of the sheep vaccines for goats is still not well defined.

REFERENCES

1. **Beveridge, W. I. B.,** Foot-rot in sheep: a transmissible disease due to infection with *Fusiformis nodosus* (N. sp.), *Bull. Aust. Counc. Sci. Ind. Res. (Melbourne),* No. 140, 1941.
2. **Egerton, J. R.,** Surface and somatic antigens of *Fusiformis nodosus, J. Comp. Pathol.,* 83, 151, 1973.
3. **Egerton, J. R. and Merritt, G. D.,** Serology of footrot: antibodies against *Fusiformis nodosus* in normal, affected, vaccinated and passively immunised sheep, *Aust. Vet. J.,* 49, 139, 1973.
4. **Stewart, D. J.,** An electron micropscopic study of *Fusiformis nodosus, Res. Vet. Sci.,* 14, 132, 1973.
5. **Walker, P. D., Short, J. A., Thomson, R. O., and Roberts, D. S.,** The fine structure of *Fusiformis nodosus* with special reference to the location of antigens associated with immunogenicity, *J. Gen. Microbiol.,* 77, 351, 1973.
6. **Stewart, D. J.,** The role of antigenic fractions of *Bacteroides nodosus* in eliciting protection against foot-rot in vaccinated sheep, *Res. Vet. Sci.,* 24, 14, 1978.
7. **Stewart, D. J.,** Studies on the antigenic structure of *Bacteroides nodosus, Res. Vet. Sci., 24, 293, 1978.*
8. **Thorley, C. M. and Egerton, J. R.,** Comparison of alum-absorbed or non-alum-absorbed oil emulsion vaccines containing either pilate or non pilate *Bacteroides nodosus* cells in inducing and maintaining resistance of sheep to experimental foot-rot, *Res. Vet. Sci.,* 30, 32, 1981.
9. **Claxton, P. D.,** Studies on *Bacteroides nodosus* vaccines with particular reference to factors associated with their efficacy in protecting against ovine virulent foot-rot, Ph.D. thesis, University of Sydney, Sydney, 1981.
10. **Stewart, D. J., Clark, B. L., Emery, D. L., Peterson, J. E., and Fahey, K. J.,** A *Bacteroides nodosus* immunogen, distinct from the pilus, which induces cross-protective immunity in sheep vaccinated against footrot, *Aust. Vet. J.,* 60, 83, 1983.
11. **Thorley, C. M.,** A simplified method for the isolation of *Bacteroides nodosus* from ovine foot-rot and studies on its colony morphology and serology, *J. Appl. Bacteriol.,* 40, 301, 1976.
12. **Schmitz, J. A. and Gradin, J. L.,** Serotypic and biochemical characterisation of *Bacteroides nodosus* isolates from Oregon, *Can. J. Comp. Med.,* 44, 440, 1980.
13. **Claxton, P. D., Ribeiro, L. A., and Egerton, J. R.,** Classification of *Bacteroides nodosus* by agglutination tests, *Aust. Vet. J.,* 60, 331, 1983.
14. **Hindmarsh, F. and Fraser, J.,** Serogroups of *Bacteroides nodosus* isolated from ovine footrot in Britain, *Vet. Rec.,* 116, 187, 1985.
15. **Thorley, C. M. and Day, S. E. J.,** Serotyping survey of 1296 strains of *Bacteroides nodosus* isolated from sheep and cattle in Great Britain and Western Europe in *Footrot in Ruminants,* Proc. Workshop Melbourne 1985, Stewart, D. J., Peterson, J. E., McKern, N. M., and Emery, D. L., Eds., CSIRO Division Animal Health/Australian Wool Corp., Glebe, N.S.W., 1986, 135.
16. **Kingsley, D. F., Hindmarsh, F. H., Liardet, D. M., and Chetwin, D. H.,** Distribution of serogroups of *Bacteroides nodosus* with particular reference to New Zealand and the United Kingdom, in *Footrot in Ruminants,* Proc. Workshop, Melbourne 1985, Stewart, D. J., Peterson, J. E., McKern, N. M., and Emery, D. L., Eds., CSIRO Division Animal Health/Australian Wool Corp., Glebe, N.S.W., 1986, 134.

17. **Claxton, P. D.,** Serogrouping of *Bacteroides nodosus* isolates, in *Footrot in Ruminants,* Proc. Workshop, Melbourne, 1985, Stewart, D. J., Peterson, J. E., McKern, N. M., and Emery, D. L., Eds., CSIRO Division Animal Health/Australian Wool Corp., Glebe, N.S.W., 1986, 131.
18. **Day, S. E. J., Thorley, C. M., and Beesley, J. E.,** Serotyping of *Bacteroides nodosus:* proposal for 9 further serotypes (J-R) and a study of antigenic complexity of *B. nodosus* pili, in *Footrot in Ruminants,* Proc. Workshop, Melbourne 1985, Stewart, D. J., Peterson, J. E., McKern, N. M., and Emery, D. L., Eds., CSIRO Division Animal Health/Australian Wool Corp., Glebe, N.S.W., 1986, 147.
19. **Short, J. A., Thorley, C. M., and Walker, P. D.,** An electron microscope study of *Bacteroides nodosus:* ultrastructure of organisms from primary isolates and different colony types, *J. Appl. Bacteriol.,* 40, 311, 1976.
20. **Every, D.,** Purification of pili from *Bacteroides nodosus* and an examination of their chemical, physical and serological properties, *J. Gen. Microbiol.,* 115, 309, 1979.
21. **Beesley, J. E., Day, S. E. J., Betts, M. P., and Thorley, C. M.,** Immunocytochemical labelling of *Bacteroides nodosus* pili using an immunogold technique, *J. Gen. Microbiol.,* 130, 1481, 1984.
22. **Claxton, P. D. and O'Grady, K. C.,** Footrot in goats and characterisation of caprine isolates of *Bacteroides nodosus,* in *Footrot in Ruminants,* Proc. Workshop, Melbourne 1985, Stewart, D. J., Peterson, J. E., McKern, N. M., and Emery, D. L., Eds., CSIRO Division Animal Health/Australian Wool Corp., Glebe, N.S.W., 1986, 125.
23. **Mattick, J. S., Anderson, B. J., Mott, M. R., and Egerton, J. R.,** Isolation and characterisation of *Bacteroides nodosus* fimbriae: structural subunit and basal protein antigens, *J. Bacteriol.,* 160, 740, 1984.
24. **Anderson, B. J., Egerton, J. R., and Mattick, J. S.,** Electrophoretic and antigenic variation in the structural subunit and basal protein of *Bacteroides nodosus* fimbriae, in *Footrot in Ruminants,* Proc. Workshop, Melbourne 1985, Stewart, D. J., Peterson, J. E., McKern, N. M., and Emery, D. L., Eds., CSIRO Division Animal Health/Australian Wool Corp., Glebe, N.S.W., 1986, 219.
25. **Anderson, B. J., Mattick, J. S., Cox, P. T., Kristo, C. L., and Egerton, J. R.,** Western blot (immunoblot) analysis of the fimbrial antigens of *Bacteroides nodosus, J. Bacteriol.,* 169, 4018, 1987.

Chapter 11

VACCINATION AGAINST FOOTROT AND FOOT ABSCESS

D. J. Stewart

TABLE OF CONTENTS

I. FOOTROT

A. Introduction

Beveridge[1,2] concluded, after some preliminary immunization experiments, that it was unlikely that an effective vaccine could be developed against footrot, a contagious disease of the epidermal tissues of the hooves of sheep caused by *Bacteroides nodosus*. The reasons for this lack of success were attributed to natural attack not conferring immunity and that the lesions were too superficial to involve mechanisms of cellular or humoral immunity.

Further research into seeking an immunological solution to footrot was only resumed in Australia in the late 1960s by CSIRO when it was found that eradication was impracticable under certain conditions, so that more cost-effective and less labor-intensive methods of eradicating the disease were required. The impetus for this work was the finding that, although *B. nodosus* was confined to the avascular epidermis, an inflammatory exudate was present within the *stratum granulosum*, suggesting that substances might be able to diffuse from the vascular system to the infected area of the hoof.[3] This was first confirmed with successful parenteral antibiotic therapy.[4] Secondly, the intriguing possibility that antibodies may also be able to diffuse from capillaries to reach the site of infection was vindicated when Parsonson[5] cured sheep of footrot by the single i.v. inoculation of suspensions of *B. nodosus*.

B. Empirical Vaccines

Egerton and colleagues[6,7] found that sheep, immunized i.m. with monovalent *B. nodosus* incorporated in an oil-emulsion adjuvant, were resistant to homologous artificial challenge in pen experiments and also to homologous natural field infection. Vaccination was also effective in slowly curing recently infected sheep in a high proportion of cases over a 6-week period.[6,7] These important findings were obtained with organisms cultured on solid or biphasic medium containing ground-hoof particles. This approach was unacceptable to commercial vaccine manufacturers who preferred a less labor-intensive, nonparticulate, liquid culture system for bulk growth of *B. nodosus*. In England[8] and New Zealand,[9,10] this led to the development of simplified techniques for growing the organism in liquid medium containing no hoof particles. In New Zealand, monovalent vaccine prepared from each of four strains grown in Skerman's culture medium were therapeutically and prophylactically effective in field trials involving seven flocks of Romney, Corriedale, or Merino crossbred sheep.[11] Vaccines derived from bacterial cells and cell-free components of the broth cultures contained protective immunogenic activity. In most of the experiments, single injections of oil-adjuvant vaccine were as effective as double doses given 14 to 28 d apart. However, under more severe challenge conditions, when over 60% of controls were affected, double vaccination with an interval of 3 weeks protected more animals than single doses.[11] Even when the interval was as short as 1 week, Egerton and Burrell[7] found that one dose of vaccine was slightly less effective than two doses. Immunity was of short duration, since Romney sheep were found not to be resistant to footrot 5 months after vaccination.[11] Roberts et al.,[12] in England, also prepared vaccines from formalinized liquid cultures of *B. nodosus*, but in this case potash alum was used as adjuvant. In field trials, an 80% reduction in prevalence of footrot was recorded 2 to 4 weeks following the second dose of a commercial footrot vaccine (Clovax®, Burroughs Wellcome & Co., now Coopers Animal Health), and good protection was achieved with this vaccine in young crossbred sheep heavily predisposed to footrot and housed in partly covered yards with straw litter.[13] Following each of three doses of the alum-precipitated vaccine, there was an incremental increase in agglutinating titers, although there was a considerable range in values for individual sheep. Encouraging results were also obtained with this alum-precipitated vaccine in West Germany,[14] Holland,[15] and Yugoslavia,[16] although Eikelenboom[15] recommended that supportive measures such as paring and footbathing also be used. However, since 1968, research in Australia had focused on using oil-adjuvant vaccines, since satisfactory protection was not obtained in Merinos with alum-precipitated vaccines in small field trials.[17]

It was first recognized that antigenic disparity may be important in the subsequent production of effective footrot vaccines when strains antigenically distinguishable from those in the vaccine were isolated from affected sheep on two of ten properties where an experimental oil-adjuvanted bivalent vaccine was ineffective. The field trial was conducted in Victoria, Australia.[18,19] Pen challenge experiments with one of these strains confirmed that this was the case.[20] Consequently, at least one vaccine manufacturer incorporated this strain (217) in their vaccine to supplement the prototype McMaster strain 198. On most of the remaining eight properties in the Victorian field trial, results were not as impressive as originally obtained with the homologous challenge strain.

By 1974 in Australia, there was growing dissatisfaction from state authorities and farmers about the performance of commercially manufactured vaccines not living up to the promise of early, experimental vaccines. Commercial vaccines had become available to the sheep industry in New south Wales and Victoria in 1971 and 1972 on a prescription or permit basis and the restricted sale was used to help assessment of the vaccine.[21,22] Clearly, the development of a commercial footrot vaccine, effective for all circumstances, was a complex one and was proving far more difficult and challenging than originally anticipated.[22]

Because of criticism about the efficacy of the footrot vaccine, two field evaluation trials were conducted by the N.S.W. Department of Agriculture in 1975, with a view to possible registration for unrestricted release if satisfactory results were obtained. Although homologous challenge by natural infection was used in both trials, disappointing results were obtained with the commercial vaccines.[21] In the second of these trials, an improved experimental CSIRO vaccine gave a high level of protection for the duration of the trial and reconfirmed that adequate protection could be achieved. These trials also demonstrated that there was considerable scope for further improvement of commercial vaccines. Consequently, the sale of footrot vaccines was prohibited in Australia due to their questionable value and the realization that their full potential had not been attained. As a result, various government, university, and vaccine-company research organizations in Australia and overseas felt obliged and encouraged to conduct further research for the development of better footrot vaccines.

C. Improved Footrot Vaccines

A number of factors have contributed to a better understanding of the interaction of ovine footrot vaccines with the animal host.

1. Identification of the Major Host-Protective Immunogen

The important discovery that sheep could be vaccinated against footrot[6] stimulated considerable interest in determining the nature of the immunogen inducing the host-protective response. Egerton[23] described the occurrence of a heat-labile, serotype-specific surface or K antigen exhibiting a characteristic agglutination reaction which was easily removed by washing and distinguishable from a heat-stable, indirect hemagglutinating (IHA) antigen common to different serotypes. In conjunction with this work, ultrastructural studies were undertaken in several laboratories to examine the morphology of the organism and locate the antigen responsible for stimulating immunity to footrot. *B. nodosus* was found to possess long, extracellular, filamentous appendages termed pili or fimbriae,[24-26] which constituted the major serotyping antigen. The pili, which are more abundant on freshly isolated strains and on strains cultured on solid than in liquid medium, reacted with specific antibody to give a typical K-agglutination reaction.[24-27] The pili were shown to consist of aggregates of a single protein subunit using SDS-PAGE,[26,28] which was later confirmed by other investigators.[29,30]

To determine the importance of various antigens for inducing resistance against footrot, vaccine experiments were initially performed with crude extracts.[26,31] Sheep vaccinated with organisms from which the pili had been removed by blending and heating at 60°C were susceptible to footrot, whereas sheep vaccinated with the washings fraction were resistant. It was also apparent that resistance could be induced with a high-molecular-weight antigen liberated into the culture supernatant and purified by isoelectric precipitation, but not with the culture-supernatant fraction from which this antigen had been removed by ultrafiltration. The protective antigen was heat-labile, since sheep immunized with boiled cells were susceptible to footrot.

Additional vaccination studies were performed to elucidate more fully the location and nature of the protective antigen(s).[26,31] Various gentle-extraction procedures were employed for the isolation of the somatic O antigen in its native state. The pilus antigen was purified by procedures used to purify pili from other Gram-negative bacteria. It was found that sheep vaccinated with the pilus fraction were resistant to footrot, whereas sheep immunized with the somatic-O antigen were still susceptible. Sheep vaccinated with pili developed very high agglutinin titers to this antigen but had low titers to the somatic-O antigen and protease component. On the other hand, somatic-O-antigen extracts induced a small increase in O-agglutinin titers and a substantial rise in IHA-antibody levels. Agar-gel diffusion studies revealed that the resistant sheep, immunized with pili, developed precipitins to the pili and no other antigen of *B. nodosus*. Thus, the protective antigen appeared to be a heat-labile, high-molecular-weight antigen associated with or identical to the pili.[26,31]

Serological studies with strains from three different serotypes confirmed that the protective immunogen is identical to or intimately associated with the pili and not the lipopolysaccharide located in the cell wall or extracellular protease.[26,28] Although *B. nodosus* possesses amorphous, fuzzy, capsular structures which were seen in thin sections[26,32] and which are probably analogous to the diffuse polar material,[33] the purified-pilus preparation was considered to be free of this material and, consequently, responsible for inducing resistance against footrot.

Further studies by Every,[34] Every and Skerman,[35] Stewart et al.,[36-39] and Lee et al.[30] have confirmed that the pilus is the major antigen responsible for protecting sheep against challenge with the homologous strain of *B. nodosus*. Every and Skerman[35] used pili purified to greater than 99% homogeneity, and sheep vaccinated with this material developed significant resistance to artificial footrot infection with the homologous strain when challenged 3 weeks after vaccination, as well as high serum-agglutinating antibody titers. Stewart et al.[37] showed in a field experiment that three groups of sheep vaccinated with two purified-pilus oil-adjuvant vaccines and a well-piliated whole-cell oil-adjuvant vaccine had significantly higher antipilus-antibody responses and were subsequently more resistant to footrot naturally transmitted by homologous *B. nodosus* than sheep vaccinated with sparsely piliated cells. The two pili vaccines induced significantly better antibody levels and protection than the well-piliated-cell vaccine. The pili vaccines induced adequate protective immunity and high agglutinin titers for at least 28 weeks following the second dose. A marked increase in mean agglutinating-antibody titers occurred after the second inoculation, being highest between 4 to 8 weeks following vaccination.

Unequivocal evidence that the pilus antigen is the major homologous host-protective antigen has been provided by the application of recombinant-DNA technology. Expression of pili on the surface of *Pseudomonas aeruginosa* has supplied a source of pili, free of other endogenous *B. nodosus* antigens,[40,41] and in vaccination experiments, recombinant-DNA-produced pili were just as effective as native pili or whole cells of *B. nodosus* in eliciting immunity against homologous challenge.[40,42,43] The use, in vaccination experiments, of highly purified native pili from *B. nodosus* did not exclude the trace contamination (<2%) with a 75- to 80-kDa protein antigen[39,44] located in the outer-membrance complex and termed protein 1[45] or the basal protein.[46] However, it would seem improbable that this contaminating antigen was responsible for the high level of homologous protection achieved with purified native pili. Firstly, washed whole cells had a much lower efficacy than pili purified from their supernatant,[37] and secondly, purified pili were effective at very low doses.[38,39,47,48]

As with other Gram-negative bacteria, such as *Neisseria gonorrhoeae*[49] and *Moraxella bovis*,[50] preliminary studies suggested that there was an association between virulence, colony type, and degree of piliation for three stable culture variants derived from a virulent *B. nodosus* isolate. One of these, the spreading, corroding-colony type, induced footrot and possessed more pili than two other avirulent-colony types.[26] This work was extended by Thorley and colleagues, who demonstrated that colonies from primary isolates possess a spreading granular or fimbriate edge[51] and contain highly piliated cells.[27] Following indiscriminate serial subculture, smooth nonfimbriate colonies containing poorly piliated cells become evident.[27,51] Vaccines prepared from highly piliated *B. nodosus* were superior to those containing few pili.[52]

Skerman and co-workers, using different culture medium, confirmed the relationship between virulence, colony type, and immunoprotection. Papillate or beaded (B)-type colonies were highly virulent, whereas mucoid (M)-type colonies were less pathogenic, and circular (C)-type colonies were avirulent. *In vitro* changes from prototype B-type colonies to M- and C- variants were nonreversible and were accelerated in liquid cultures. Strains differed in their stability to these changes.[53] B-type colonies contained highly piliated bacteria, those from M-type colonies usually bore low numbers of pili, whereas organisms from avirulent C-type colonies were devoid of pili.[34,54] Vaccines prepared from B-type colonies induced higher K-antibody responses and a greater degree of protective immunity than did organisms from the other clonal variants.[53]

There is a reasonable correlation between serum-antipilus-agglutinin titers and resistance of vaccinated sheep to homologous infection.[26,30,31,36,37,52,54] Thorley and Egerton[52] considered that groups of sheep with mean titers above 5000, 12 weeks after vaccination were usually resistant, whereas, susceptible sheep had titers below 3000. The results of Stewart et al.[36,37] were in close agreement in that during the 28 weeks following the second vaccination, 91% of all sheep with titers of 3,200 or less at 12 weeks postvaccination developed severe footrot of greater than 3 weeks duration, compared to 25% of all sheep with titers of 12,800 or higher. Although there was a close relationship between vaccine protection and group mean titers, this was not always the case on a group or an individual-sheep basis. Thus, the agglutinin response to vaccination was not universally associated with resistance or susceptibility to severe footrot[36,37,52] and this finding has been confirmed with *B. nodosus* pili produced by recombinant DNA in *P. aeruginosa*.[43]

Although pilus vaccines containing 250 μg of pilus protein per dose induced significantly better protection than a well-piliated whole-cell vaccine containing 5×10^9 cells per dose,[37] whole-cell vaccines containing the same quantity of pili as purified-pilus vaccines gave comparable protection.[35,38,47] However, despite the similar content of pili, the pilus vaccines induced higher, more persistent mean titers than whole-cell vaccines.[35,47]

Dose-response experiments[39] using three dose levels in either complete Freund's adjuvant (CFA) or incomplete Freund's adjuvant (IFA) showed that a vaccine containing 38 μg of purified pili in IFA was nearly optimal for homologous protection against footrot in sheep vaccinated on two occasions with a 4-week interval between injections. A fourfold increase (154 μg) did not improve protection but 38 μg was significantly better than 10 μg. There was no difference between the adjuvants or an adjuvant-dose effect. Furthermore, when used at a dose rate of 50 μg, *B. nodosus* pili produced by recombinant DNA in *P. aeruginosa* were just as effective as native pili in vaccinating sheep against footrot.[42]

The capacity of animals to respond differently to viral and bacterial antigens in terms of the level of antibody produced is a well-documented phenomenon and this spectrum of responsiveness also occurs with the pilus antigen of *B. nodosus*. Stewart et al.[37] reported that there was considerably more variation in homologous pilus-agglutinating-antibody titers in Merino sheep of the Peppin strain vaccinated with whole cells in oil adjuvant than in Merinos of the same genetic background vaccinated with pili in oil. In the former group, there was also a much greater proportion of poor responders. Later it was shown that differences in antibody responses also exist between strains of the same breed as well as between different breeds. Thus, agglutinating-antibody titers after two doses of well-piliated *B. nodosus* whole cells were significantly lower in the Peppin Merino than in the Saxon Merino, Romney Marsh, Dorset Horn, or Border Leicester breeds for the first 6 weeks, while, after vaccination with cells virtually devoid of pili, agglutinating-antibody titers remained elevated longer in the Border Leicester and Saxon Merino that the other breeds. These differences may be a characteristic of the strain or breed of sheep or due to the use of a particular sire or line of breeding ewes.[55]

Two independent studies have confirmed that within breeds there is a genetic influence on the degree of antibody responsiveness to *B. nodosus* vaccines. Lines of sheep selected for resistance to footrot from the Targhee[56] and Corriedale[57] breeds had higher agglutinating-antibody responses after vaccination than unselected susceptible lines within the same breeds.[56,58] In the study with the Targhees, it was not reported whether the sheep received one or two doses of vaccine,[56] but in the Corriedales the difference in response was only significant after the first vaccination.[58] Nevertheless, an association has now been established between high antibody responses and class 1 ovine lymphocyte antigen-type of the major histocompatability complex.[58] This may provide a useful marker for the selection and breeding of sheep more responsive to the pilus antigen in *B. nodosus* vaccines.

2. *Requirement for Potent Adjuvants*

Effective protective immunity against footrot depends on high levels of circulating antibodies because of the necessity for their effective diffusion into the epidermis.[6] Immunity to footrot can be passively acquired by unvaccinated animals following the i.v. injection of large amounts of γ-globulins from vaccinated sheep,[59] as well as by colostral transfer to lambs immediately after birth.[60] In the latter study, lambs with high levels of maternal agglutinating antibody, derived from Border Leicester × Merino dams vaccinated with a *B. nodosus* alum-precipated oil-adjuvanted vaccine, were resistant to homologous challenge for the first 8 weeks. However, the lambs were susceptible to footrot at 12 weeks of age when their antibody titers had fallen to relatively low levels. At parturition, the mean antibody titer of the vaccinated dams had fallen to 40% of the level of that of vaccinated nonpregnant ewes, due to selective transfer of IgG_1 antibody into the colostrum.[60] Initially, it was reported that bactericidal IgG antibody was important in host protection,[6,61] but it is now thought that the pilus-agglutinating antibody of the IgG_1 isotype[26,31] plays the major role in immunity.[45,62]

From the outset of vaccine research in the late 1960s, oil adjuvants have been preferred in Australia to alum-precipitated vaccines because of their better efficacy and duration of protection resulting from higher, more persistent antibody titers.[17] However, the disadvantage of water-in-oil-emulsion vaccines is that they tend to cause chronic, discharging granulomata at the injection site, so that vaccine manufacturers initially favored alum-based adjuvants, although producing an immunity of shorter duration.[12] A number of investigators have confirmed the superior protective activity of oil-adjuvanted vaccines. It was reported that oil-adjuvanted vaccines were more effective than an alum vaccine when a strategic biannual vaccination was combined with a footbathing program in a Corriedale flock.[63] Also, higher cure rates were obtained in Merino sheep with oil-adjuvanted than alum-precipitated vaccines[64] and more durable homologous protection was achieved with the former.[65] Sheep given two doses of oil-adjuvanted vaccine were resistant at 4 weeks and partially resistant after 8 weeks, although they became susceptible after 12 weeks. Two doses of alum-precipitated vaccine or one dose of oil-emulsion vaccine followed by a dose of alum-precipitated vaccine induced protection at 4 but not at 8 weeks. However, at this time, footrot was less severe than in the controls. The agglutinating-antibody titers were significantly lower than for the oil-emulsion vaccine. There was little difference in the antibody response of sheep primed with oil emulsion and boosted with either oil-emulsion or an alum-precipitated vaccine. Antibody levels fell sharply, so that 8 weeks after the second dose, antibody levels were similar in all vaccinated groups.[65] In sheep challenged 8 weeks after vaccination, Lee et al.[30] induced protection with purified pili in oil adjuvant but not with pili in alum, and this difference was reflected in their antibody titers.

Supplementation with Quil A, the active component of the adjuvant saponin, induced higher agglutinin-antibody titers and improved the curative effect of alum-precipitated vaccines in Merinos.[66] However, Claxton[60] concluded that although the addition of Quil A to alum-precipitated vaccine enhanced resistance of Merino sheep to pen challenge with the homologous strain, protection was still shortlived, being negligible 8 weeks after revaccination. Despite this, Thorley et al.[67] showed in a field trial that five to six times more nonvaccinated British-breed sheep became affected than those injected with a saponin-adjuvanted, aluminum-hydroxide-precipitated multistrain footrot vaccine. Vaccination conferred a distinct benefit for up to 15 weeks after the second inoculation.[67] There was no difference in the performance of the saponin derivative obtained from different commercial sources. There was only a relatively minor serological response after the primary inoculation, but an appreciable anamnestic response occurred on subsequent inoculations. Thereafter, agglutinin titers declined at about the same rate as oil-adjuvanted vaccines, with a half-life of 6 weeks, except that for the latter, the antibody peak remained higher for longer.[67,68]

A further increase in antibody levels and duration of protection over that with whole cells emulsified in oil adjuvant was achieved by using the dual-adjuvant system incorporating alum-absorbed, highly piliated cells in the oil phase.[52] The piliated-alum-oil vaccine (PAO) induced better protection against homologous challenge at 12 weeks after the second dose than piliated cells in oil. After a single dose, the PAO vaccine protected sheep for 14 weeks, and for 22 weeks after three doses. A beneficial effect was still evident after 40 weeks.[52] Not only does aluminum hydroxide in oil enhance the homologous antibody response to the pilus antigen, but with purified pili the dose can be reduced to 5 μg and the antibody response remains comparable to that from the higher dose of pili (50 μg) in oil vaccine.[48] Using a crude whole-cell antigen in an enzyme-linked immunosorbent assay (ELISA) to assess serological responses of two commercially manufactured products, the double-adjuvanted (aluminum hydroxide-oil) vaccine (Pro-vac®, formerly Wellcome Australia Ltd., now Coopers Animal Health) promoted higher antibody levels than the single oil-adjuvanted vaccine (Footvax®, formerly ICI Australia Ltd., now Coopers Animal Health). When the crossed immunoelectrophoretic patterns of serums from vaccinated sheep were compared, Pro-vac® generated antibodies to 8 to 12 antigens, whereas Footvax® produced antibodies against 2 to 4 antigens.[69] However, these results must be viewed with some caution as the quality and quantity of antigens in the two types of vaccines may have contributed to the differences in antibody responses.[69]

The degree of tissue reactivity at the injection site depends on a number of factors.

a. Type of Adjuvant

Skerman and Cairney[11] noted that oil-based vaccines led to the formation of a firm lump 1 to 3 cm in diameter at the injection site and in 20% of cases the lesion discharged its contents at the skin surface, with the loss of a small amount of wool. The lesion ultimately healed, leaving a small, fibrous scar. An alum-adjuvant vaccine produced a slightly smaller lesion that healed more rapidly,[11] although Roberts et al.[12] considered this type of vaccine to be free from unacceptable reactions. Ross and Titterington[70] confirmed that an oil-based commercial vaccine produced significantly larger lesions than a commercial alum-precipitated vaccine at injection sites in the neck and brisket. The lesions were caused by the irritant action of sterile vaccine which was supported by negative bacteriological results. Skin disinfection had no effect on lesion frequency or size. There was variation in the size of the reaction depending on the site selected for inoculation. Lesions at the neck site were smaller than those at the brisket, in ewes, but there was no difference in rams, possibly due to the greater difficulty in rams of ensuring that the injection was not deposited i.m. Firm nodules or plaques with indurated edges and a maximum dimension of 1 cm or more were still common 5 to 6 weeks after injection. A variable proportion exhibited a liquefied central area of necrosis and later discharged necrotic material. Ross and Titterington[70] concluded that vaccines should not be administered to sheep slaughtered within the next 2 to 3 months, since carcasses blemished by a vaccination reaction are downgraded in value and cost more to process.

Vaccines containing IFA caused less reaction at the site of the injection than pili incorporated in CFA.[39] The CFA vaccine produced more persistent granulomas at the first site of inoculation, as well as more discharging lesions at the second site. After the first injection, the majority of CFA lesions persisted for at least 6 months, with a smaller proportion of IFA lesions still being evident. Reactions to the second tended either to be smaller, or in the case of CFA lesions, to decline in size more rapidly due to rupture and discharge of contents. Resolution of lesions at the second inoculation site was more common in the groups vaccinated with IFA containing low (10 μg) and medium (38 μg) concentrations of pilus antigens. In both IFA and CFA vaccine groups, the majority of lesions contained caseous material and most of these were at the first inoculation site. Adjuvant was still present in some of the lesions 6 months after the first inoculation.[39]

Adverse tissue reactions were not observed in Australian Merino flocks following s.c. injection of Quil A-alum precipitated *B. nodosus* vaccines.[66] However, Thorley et al.[67] observed that mild, transient, local reactions were produced in some British flocks with this vaccine, the most noticeable feature being tenderness and pain at the injection site between 24 and 48 h, resulting in some stiffness of gait or disinclination to walk. There was also some swelling at the injection site, which developed more slowly, and, rarely, abscess formation at about 2 weeks.

There is some difference of opinion as to whether dual-adjuvant vaccines containing aluminum salts incorporated in oil are more reactive than vaccines containing oil without the former component. This is mainly because there is only limited experimental data available where a direct comparison has been made. Thorley et al.[71] considered that reactions to alum in oil whole-cell vaccines resolved in 3 to 8 weeks and that repeated doses do not affect the severity or duration of the reaction. Stewart et al.,[38,47] in Merinos, found no difference in the size, number, and persistence of reactions at the first and second inoculation sites induced by whole cells in oil or whole cells absorbed with aluminum hydroxide and emulsified in oil, although with the latter more swellings burst and discharged their contents. In field trials conducted by Lambell,[72] vaccination reactions produced in Merino × Border Leicester ewes by a commercial dual-adjuvant vaccine (Pro-vac®), were compared with those in Romney Marsh crossbred ewes caused by another commercial vaccine in oil adjuvant alone (Footvax®). Pro-vac® produced larger reactions than Footvax®, with a greater number discharging their contents and a far higher proportion of lesions persisted for 6 months. These severe reactions were not associated with any management problems such as interfering with shearing or predisposing to flystrike.[73] However, this comparison is not entirely valid because the trials were conducted in sheep with different genetic backgrounds on separate properties.

Thorley et al.[71] have suggested that there is some correlation between the severity of the reaction and protection afforded by Pro-vac®. This observation has been extended by the findings of Lambell,[73] who reported that a sheep factor exists which is not only responsible for the correlation between the occurrence and size of the reaction at both inoculation sites on the same sheep, but that the development of reactions was negatively correlated with the footrot status of the sheep. Thus, the proportion of sheep with no inoculation lesions was significantly higher among those affected with footrot than among those not affected. It was not determined whether this phenomenon was also associated with the magnitude of the agglutinating-antibody response of the vaccinated sheep.

The tissue irritant and immunostimulatory properties of 13 adjuvants have been compared in vaccines containing 200 μg per dose of partially purified pili. These adjuvants included a number of white mineral oils, single component straight-chain or branched-chain hydrocarbon compounds, peanut oil, aluminum hydroxide (Alhydrogel®, Superfos a/s, Denmark), Alhydrogel® in mineral oil, or Quil A in Alhydrogel®. There was a relationship between the magnitude of the homologous agglutinating-antibody response and the mean size of the reactions at the vaccine-injection site.[48] For the different adjuvants, the ranking of mean antibody titers and size of the lesion was highly significant.[74] The best antibody responses were obtained with Alhydrogel®-oil, but this adjuvant was the most tissue irritant,[48] Furthermore, higher antibody responses and greater tissue irritation were produced by IFA containing the mineral oil, Bayol F®, than with IFA containing Drakeol 6VR® (Penreco). The latter is a highly refined, special pharmaceutical grade mineral oil for use in human vaccines. There was no advantage, in terms of tissue reactivity or the pilus-antibody response, in using either a straight-chain (n-hexadecane $C_{16}H_{34}$) or two-branched-chain hydrocarbons (7-n-hexyloctadecane $C_{24}H_{50}$ and squalane $C_{30}H_{62}$) as adjuvants. Alhydrogel®, Alhydrogel®-Quil A, and peanut oil were less reactive adjuvants than mineral oil, but were inferior in terms of antibody-promoting activity. Peanut oil and Alhydrogel®-Quil A produced slightly higher, persistent titers than Alhydrogel®.[48]

b. Nature and Dose of Antigen

The severity of the reaction is proportional to the amount of bacterial antigen present in the adjuvant vaccine.[11] This observation was extended by Stewart,[26,31] who showed that a purified-pilus oil-adjuvant vaccine was far less irritant than various somatic-O-antigen vaccines, suggesting that the endotoxin of *B. nodosus* may facilitate the tissue-necrotizing properties of oil adjuvants.

The size of the reaction is also dependent on the quantity of protein present in the vaccine since 10 or 38 μg of purified pili produced less reaction than those containing 154 μg.[39] The difference in relation to protein dose was less marked with CFA than IFA.[39]

Pili (30 μg per dose) incorporated in IFA are also much less irritant than whole cells in the same adjuvant, with a substantial difference in the number, size, and persistence of vaccination lesions.[38,47] Purified pili in oil produced fewer, smaller, less persistent and discharging lesions than whole cells in the same adjuvant. Also, with pili incorporated in Alhydrogel® and emulsified with oil, fewer swellings at the first inoculation site ruptured than those produced by cells in oil or cells in Alhydrogel®-oil, although the reactions were of similar size. At the second inoculation site, a larger number of lesions discharged in the cells in Alhydrogel®-oil group than in the cells in oil and pili in Alhydrogel®-oil groups. Differences between the two successive vaccinations were small, although there was a tendency in groups with large reactions for the second one to be slightly smaller and resolve more quickly.[38,47] Contrary to these findings, Every and Skerman[35] found no difference between pilus and whole-cell preparations in vaccination-site reactions, although only small numbers of vaccinated sheep were compared and the amount of protein per dose was much higher (100 to 400 μg protein per dose). Tissue reactions in sheep inoculated with antigen-free emulsions were significantly less than those inoculated with antigen in oil.[35] Furthermore, oil alone is considered to be still less reactive than an oil emulsion containing only saline.[68]

An adverse effect of vaccination was reflected in bodyweight changes when sheep were injected in the maxillary area of the face. The decrease in mean bodyweight was significantly greater with whole-cell vaccines than purified pili in either IFA or Alhydrogel®-IFA, indicating that both nonpilus antigens, antigenic mass, or both interact with adjuvant to cause the weight loss. The difference in mean bodyweight was still evident 117 d after the second vaccine dose. However, systemic toxicity may have also caused a decrease in appetite, since the group vaccinated with pili in Alhydrogel®-IFA had vaccination reactions of similar size to those of the whole-cell vaccine groups but gained more weight.[38,47] Furthermore, in another experiment, vaccination of young sheep 12 to 14 months of age in the upper neck away from the facial region also reduced bodyweight gain, confirming that interference with mastication is not the only factor involved.[55]

c. Breed of Sheep

Skerman and Cairney[11] noted that vaccination reactions were more marked in Romney flocks than in Corriedale and Merino breeds and this breed effect was confirmed by Stewart et al.,[55] who found that lesions at inoculation sites with oil-adjuvant vaccines were generally larger in British breeds than Merinos. The reactions were largest, most numerous, and most frequently discharged their contents in the Dorset Horn than in Border Leicester and Romney Marsh. There did not appear to be any relationship between the pilus-agglutinating-antibody titers of individual sheep and the size of their reactions at the vaccination sites. It was suggested that the greater reactivity of the British breeds, particularly the Dorset Horn, was in some way related to the tighter attachment of the skin to the underlying s.c. tissues or to differences in the inflammatory reaction to oil adjuvant.[55] It seems likely, but not proven, that reactions to aggressive dual adjuvants such as Alhydrogel®-oil would be even more pronounced in British-breed than Merino sheep.

3. Recognition of Antigenic Variation

Antigenic heterogeneity among *B. nodosus* isolates was first recognized by Beveridge.[1] However, Egerton[23] was the first to attempt serological classification of strains. He categorized 33 of 46 isolates into three major types A, B, and C. Since then the complexity of typing *B. nodosus* has become increasingly apparent. Thorley,[51] using cross-tube-agglutination tests, recognized seven different K-agglutinin types in Great Britain which were distinct from those of Egerton.[23] On the other hand, Schmitz and Gradin[75] in Oregon, with 97 isolates from 20 flocks, have demonstrated the occurrence of at least 14 serotypes designated I to XIV, including those serologically similar to Egerton's. However, some serotypes shared major antigens and, consequently, were more closely related. This cross-reactivity with Australian strains was recognized by Claxton[60,76] and Claxton et al.[77] and it was incorporated into a simplified classification scheme consisting of eight major serogroups, A to H, most of which could be subdivided into two subgroups on the basis of cross-absorption-agglutination tests. Serogroup B was the most complex and contained four subgroups. Thus, Australian isolates could be divided into 8 serogroups and 16 subgroups but considerable cross-reactions occurred among subgroups within a serogroup.

Claxton's serological studies were performed in Egerton's laboratory and covered 762 isolates of *B. nodosus* from 221 sheep on 47 properties in New South Wales,[76] On the basis of slide agglutination, 755 could be grouped into eight serogroups. Seven isolates could not be typed. Ribeiro,[78] also working in Egerton's laboratory, tested 1048 isolates from 144 sheep on nine properties in New South Wales and typed all into the same serogroups of Claxton. In New Zealand, Skerman[79] classified 86% of 172 strains into 8 distinct groups with 24 strains being nontypable. Later, Claxton et al.[77] identified serogroup B on 39.7% of farms out of 1267 isolates, with A and F (25.9% of farms, each) being the next most frequent. A relatively small number of isolates could not be classified into the eight serogroups. Most of these were subsequently found to belong to a new serogroup (I)[80] and many were benign isolates.[80-82] In another New Zealand survey, Kingsley et al.[83] obtained 1547 isolates from 487 sheep on 65 properties and all were typable by Claxton's scheme, with serogroup B predominating (56.2%). They found that 113 Australian isolates represented each serogroup, with 37% belonging to serogroup B.[83] In the U.K., Hindmarsh and Fraser[84] found that out of 449 isolates from 166 sheep, 40.4% were serogroup B, 26.5% H, 10.8% D, 8.7% F, and 7.6% G. A serotyping survey of 1296 isolates in Great Britain and western Europe was conducted by Thorley and Day.[85] They typed 60% of their 849 isolates from Great Britain as either D or H, with type H in 68%, type B analogs 46%, and type D in 61% of those counties tested in Great Britain. In western Europe, types D and H also predominated.[85] Thorley and Day[85] and Day et al.[86] have extended the classification A to H of Claxton to include a further 9 distinct serotypes J to R, giving a total of 17 serotypes. This was because a large number of untypable isolates were found when serotype-specific rabbit antisera, absorbed with heterologous serotypes, were used in a slide-agglutination test. Unabsorbed rabbit antisera was used by Claxton et al.[77] for typing strains in a slide-agglutination test. Serogroup Q was of minor significance, since it has been found on only one farm, being isolated from cattle in Sweden.[85,86] In Claxton's typing scheme, some cross-reactions among subgroups within a serogroup were of fairly low titer compared to the homologous titer. Strains belonging to these subgroups may have been assigned by Thorley and Day[85] and Day et al.[86] to new serotypes outside A to H, illustrating the need for a unified *B. nodosus* typing scheme.[86]

The first reports of antigenic differences between the vaccine strains and challenge strains influencing vaccine performance in the field were those of Egerton et al.[18] and Egerton and Morgan.[19] These were later confirmed with one of the isolates in a pen experiment using artificial challenge with pure cultures,[20] and extended by other investigators.[36,71,76,79,87] Sheep vaccinated with serotypes distinct from that of the challenge strain were not resistant.[79] In further pen experiments, Claxton[76] immunized sheep with a polyvalent Alhydrogel®-oil vaccine containing strains belonging to serogroups A, B, C, D, and F and then challenged with lesion material containing serogroup B as well as G and H not represented in this vaccine. The vaccine afforded substantial protection against serogroup B but little if any protection against challenge with serogroups G and H. In a separate field experiment with the same vaccine, protection was achieved against serogroup B but not against serogroup G in a flock infected with these two serogroups.[76] Field-evaluation trials of experimental batches of commercially manufactured vaccines also indicated a requirement for the incorporation of a wide range of serotypes for effective footrot prophylaxis. Adequate protection for purposes of meeting the minimum requirements for registration was not obtained against strains not represented in the vaccine.[71] Another commercial vaccine containing strains from five different serogroups produced satisfactory protection and cure rates on six properties, whereas a bivalent vaccine used previously had given unsatisfactory results.[87]

Confirmation that the pilus is responsible for serogroup-specific protection was obtained in pen and field experiments. In pen experiments with purified and partially purified pilus preparations, the results revealed that a single-serogroup pilus vaccine did not provide adequate cross-protective immunity to seven distinct serogroups, although good homologous protection was obtained.[36,88] In a field experiment using the same strains compared by Egerton,[20] good homologous protection was obtained following vaccination with purified pili but resistance to heterologous challenge was substantially less.[38,47] In the second field experiment, protection acheived with homologous purified pili was significantly better than that achieved with pili from three different heterologous serogroups.[48] In the third experiment, a purified *B. nodosus* pilus preparation produced by recombinant DNA in *P. aeruginosa* induced good homologous protection and cure comparable to that induced by native pili. By comparison, pili purified from a heterologous serogroup did not.[42]

4. Exploitation of Cross-Protective Immunity

a. Protection within Serogroups

Currently, commercially available vaccines contain eight to ten strains representative of the most common subgroups (subtypes, serotypes) in each of the serogroups to provide broad-spectrum cover against the majority of field strains. Field evaluation of the efficacy of these commercially manufactured multivalent whole-cell vaccines is discussed in Chapter 1, Section VI.

Because there are a relatively large number of subgroups or serotypes,[75,76,77,86] inclusion of all these (14 to 17) would be a manufacturing impossibility. Consequently, the selection of strains with appropriate cross-reacting pilus antigens is advantageous.[85,86] The rationale for the inclusion of just one strain from each serogroup is that related subgroups with shared-pilus antigens will induce adequate cross-protective immunity against the other subgroup from the same serogroup. This has in fact been established for several serogroups. Sheep vaccinated with serogroup A, subgroup 1, were protected against challenge with the homologous subgroup and against serogroup A, subgroup 2. Mean titers to the heterologous A2 strain, although elevated, were about four fold lower than to the homologous A1 subgroup.[76] This has been confirmed in a field experiment using purified pili from a benign strain, belonging to subgroup A2, to vaccinate sheep, followed by challenge with virulent subgroup A1. Again, titers to the homologous strain were about five fold higher than to the heterologous strain.[42] Cross protection among subgroups within some other serogroups has also been achieved,[89] but it is not known whether this is universally applicable to all serogroups. Skerman[79] did not obtain protection in a pen experiment where vaccine and challenge strains showed partial similarity in their pilus antigens. However, the degree of serological relatedness and the serogroup/subgroup identity of the strains was not stated. The minimum amount of shared pilus required to induce protection between subgroups needs to be determined before an effective serogrouping system can be established.

An alternative approach is to choose vaccine strains based on their prevalence in serological surveys of different regions or countries, omitting those which seemed less frequently encountered than others. Such an approach was used by Thorley et al.[67] to select the 9 most commonly encountered serotypes out of 17 to incorporate in the less-reactive Quil A-Alhydrogel® adjuvant in order to meet the potential requirements of British and western European markets. On this basis, it was presumed that 70% of farms in Great Britain and western Europe would be completely covered, 23% of farms would receive some specific cover, and 7% no cover.[67]

b. Protection Between Serogroups

The concept for the development of a cross-protective vaccine is that *B. nodosus* possess a critical common antigen or epitope, shared among all the serogroups and that the immune response induced by this antigen is sufficient to provide sheep with protection against footrot. Because commercial vaccines against ovine footrot are relatively complex, containing strains from at least eight different serogroups, knowledge of the nature of cross-protective antigens in *B. nodosus* may facilitate the more rational formulation of vaccines containing fewer strains, and the production of these antigens by recombinant-DNA technology with a broad spectrum of activity against a wide range of serogroups. The impetus for this approach was the suggestion from a number of research groups that an antigen, possibly distinct from the pilus, was responsible for inducing heterologous cure and protection. A considerable curative effect has been demonstrated with footrot vaccines, despite very low agglutinin levels to heterologous infecting strains of *B. nodosus*.[64,66,90] Furthermore, in pen experiments, sheep vaccinated with three doses of piliated strain 198 cells in alum oil and challenged with the heterologous strain 217, were generally resistant to footrot 6 weeks after the third dose of vaccine,[52] and some homologous protection was obtained in sheep vaccinated with cells from a nonpiliated-colony type.[53] Results from field experiments confirmed these suggestions, in that whole cells were more effective than purified pili against a heterologous strain.[38,47,48]

c. Factors Influencing Cross Protection

Factors contributing to the degree of cross protection obtained with nonpilus antigens included the type of adjuvant, severity of challenge, virulence of the challenge strain, breed of sheep, and the nature of the immunogen.

Type of adjuvant — Purified-pilus vaccines in oil adjuvant do not provide cross protection of any consequence against highly virulent strains of *B. nodosus*. However, when pili are incorporated in the dual-adjuvant system, Alhydrogel®-oil, some cross protection is evident against these strains, although substantially less than obtained with whole cells.[38,47] This enhanced protection is associated with higher outer-membrane complex (OMC), ELISA titers.[47] Western blot analysis has implicated a role in cross protection for outer-membrane proteins, in particular a 78 kDa protein.[47] This antigen probably represents the major outer-membrane protein[44,45] or basal protein for anchoring pili to the cell wall.[46] Thus, the results of serological and Western blot analyses indicate that the powerful dual adjuvant, Alhydrogel®-oil, potentiates the antibody response to the basal protein and possibly other membrane proteins, when present in small quantities in preparations of purified pili. The benefit of using Alhydrogel®-oil as an adjuvant, in contrast to oil alone, was abrogated when whole cells containing larger quantities of the cross-protective immunogens were incorporated in either adjuvant, since both vaccines induced cross protection of similar magnitude, and better than that achieved with purified pili in Alhydrogel®-oil.[38,47]

Severity of challenge — The severity of challenge can also influence the degree of protection and cure obtained with nonpilus antigens. A high curative rate against heterologous strains was only achieved when climatic conditions did not favor transmission of footrot.[19,71] In the face of heavy challenge, the extent of cure was low. It has also been reported that protection in vaccinated sheep against footrot, caused by unidentified field strains, could be overwhelmed when vaccinated sheep were exposed to very severe challenge on wet, swampy ground and that removal of sheep to drier pastures or revaccination led to a more rapid resolution of footrot lesions in vaccinated sheep compared to controls.[63] Supportive evidence has been obtained in a field experiment, under light to moderate challenge conditions, in which Merino sheep vaccinated with cells virtually devoid of pili were as resistant to footrot as those immunized with a whole-cell vaccine containing pili, despite the significantly higher antipilus-agglutinin titers in the latter group.[55] However, under severe challenge conditions, well-piliated cells induced a high level of protection and poorly piliated cells a low level.[37] In the latter experiment, 80% of feet in the unvaccinated Merinos became affected with severe footrot compared with about 40% in the former. Thus, cells virtually devoid of pili can provide protection against light to moderate homologous footrot challenge. However, under severe challenge, well-piliated cells containing 30 to 40 μg of pili, or a similar quantity of purified pili[37-39,47] are required for adequate protection.

Virulence of the challenge strain — The degree of cross protection obtained with whole cells appears to be associated with the virulence of the challenge strain. In a pen experiment, poorly piliated cells provided better protection against heterologous strain 243 (serogroup F) than against the highly virulent homologous strain (serogroup A).[88] The latter appeared to be more virulent since it caused in the control unvaccinated sheep, significantly more underrunning lesions 2 weeks after challenge, and a higher number of severely affected feet 4 weeks after challenge. Thus, the greater the virulence of the strain, the lower the protection induced by cross-protective immunogens under the severe challenge conditions usually experienced by penned sheep and less often by sheep under field conditions. Moreover, in a field experiment, the cross protection obtained against heterologous strain 217 (serogroup

C) with purified pili in Alhydrogel®-oil may have been due to its decreased virulence. The lower virulence of this strain in comparison to the homologous one was shown in the control sheep by the fewer feet affected with underrunning lesions of the abaxial wall. This decreased severity was also reflected in their lower pilus-agglutinating and OMC ELISA antibody titers.[47]

Breed of sheep — In field trials conducted by Egerton et al.[18] and Egerton and Morgan,[19] with a bivalent oil-adjuvant vaccine, the incidence, severity, and duration of footrot tended to be less in vaccinated and unvaccinated crossbred Merinos than in pure-bred Merinos against challenge with both presumed and demonstrated heterologous strains. In pen experiments, Skerman et al.[91] demonstrated that there was a marked difference between vaccinated Merinos and Romneys in their duration of resistance. Following vaccination with a strain that was only moderately piliated and incorporated in oil adjuvant, immunity in Merinos lasted only 4 to 5 weeks. On the other hand, Romneys were relatively resistant to homologous challenge up to 16 but not 20 weeks. Agglutinating-antibody titers were generally low and bore little relationship with resistance.[91] Although vaccines prepared from either piliated or depiliated cells provided protection in all breeds of sheep, there were significantly more lesions of footrot in Merinos than in Romney Marsh, Dorset Horn, and Border Leicester sheep,[55] reflecting the greater susceptibility of the Merino to footrot. Definitive experiments have not been performed but these results suggest that the protection induced by vaccination would be superior in British-breed sheep than in Merinos against heterologous strains of *B. nodosus*.

d. Potential Cross-Protective Antigens

Cell-associated antigen — Although a high level of homologous protection was achieved with either purified or partially purified pili vaccines, the protection against challenge with heterologous strains from six of seven different serogroups, with these types of vaccines, was inadequate. However, against strain 216 (serogroup E), a significant level of cross protection was obtained with a partially purified strain 198 (serogroup A) vaccine which was probably due to contaminating OMC-protein antigens.[88] This observation was extended by the demonstration that a strain 198 whole-cell vaccine induced a significant level of cross protection against three heterologous strains, 216 (E), 217 (C), and 243 (F) but not against 234 (B), 240 (D), and 265 (H). Immunoblotting revealed a strong reactivity for a 78 kDa protein among strains 198, 216, 234, and 243 and which had a similar molecular weight to OMC protein 1,[44,45] or the basal protein.[46] Another protein (30 kDa) was also heavily labeled on strains 198, 216, and 243. However, the patterns of antigenicity did not exactly coincide with the spectrum of cross protection.[62,88]

Evidence for some differences in the degree of cross protection achieved with three heterologous whole-cell vaccines was obtained in another experiment using natural field challenge with a known strain. These differences may reflect antigenic variation in nonpilus immunogens which in turn affect the level of cross protection obtained. A vaccine prepared from heterologous strain 217 (C) produced better protection 7 weeks after challenge than that obtained with the other heterologous strains, 234 (B) and 265 (H), but this difference was not so apparent at 15 weeks. In all of the heterologous whole-cell and pilus-vaccine groups, there were significantly fewer feet affected with severe footrot than in the control, unvaccinated group. Furthermore, protection obtained with all the heterologous whole-cell vaccines was better than that obtained with purified pili from the same strains, the difference between 234 whole cells and pili being highly significant.

There appear to be several cross-protective antigens in *B. nodosus*. At least one of these appears to be cell associated and is probably located in the OMC, since washed cells devoid of pili can provide some protection against challenge with a virulent strain from the homologous serogroup[55] and heterologous serogroups.[48] Furthermore, the cross protection obtained with a pili in Alhydrogel®-oil vaccine could be related to antibody titers against OMC proteins and not anti-lipopolysaccharide antibody levels.[47] In particular, OMC protein 1 (75 kDa),[44,45] also known as the basal protein,[46] was implicated by immunoblotting.[47] Some protection of vaccinated sheep against heterologous challenge has been obtained with the OMC fraction.[48]

Extracellular proteases — The extracellular proteases of *B. nodosus* are the most promising candidates as cross-protective antigens. The proteases were purified by gel filtration and ion-exchange chromatography[92] and emulsified in oil or Alhydrogel®-oil adjuvant. These have been shown to provide good protection against the homologous virulent strain and also against a number of different heterologous strains. In fact, the protection with the protease vaccines was not significantly different from that obtained with either purified pili or whole cells against field challenge with a highly virulent homologous strain. A vaccine containing a mixture of proteases produced slightly better protection than a vaccine with only a single protease isoenzyme. This superior performance was reflected in the higher protease ELISA antibody titers with the most-resistant sheep generally having the highest titers. The purified proteases did not induce pilus-agglutinating-antibody titers in vaccinated sheep. Agar-gel diffusion studies showed that the sheep immunized with the mixture of proteases developed precipitating antibodies to an additional protease antigen and these antibodies were not present in sheep vaccinated with the single isoenzyme. These results were confirmed by the immunoblotting procedure. ELISA, immunoblotting, and immunodiffusion also showed that identical protease antigens were shared by strains in different serogroups.[48] Previously, neutralizing antibodies with activity against crude, unpurified protease in culture supernatants have been demonstrated.[93] A complete patent specification has been filed covering their use for vaccination against footrot of proteases produced either by *B. nodosus* or through recombinant-DNA technology.[94]

D. Genetically Engineered Vaccines

The evolution of footrot vaccines is likely to continue. Recent technological developments in genetic engineering, molecular genetics, and industrial biotechnology are generating alternative scientific and commercial approaches to vaccine manufacture. The production of pili by recombinant DNA in a host that has less fastidious growth requirements and is able to grow to a much higher cell density than *B. nodosus* offers the possibility of an improved vaccine against footrot. The pilin gene of *B. nodosus* has been cloned, characterized, and expressed at a high level in *Escherichia coli*.[95-98] However, the pilin subunit became embedded in the cell membrane and was not assembled into pilus filaments. The pilin, when extracted from within the cell and used to vaccinate sheep, gave only limited protection, initially retarding progression of infection with the homologous strain[98] and produced a small increase in agglutinating-antibody titers, which may have accounted for the initial decreased severity of the disease. The pilin gene was then cloned into *P. aeruginosa* with a wide host-range plasmid containing the pilin gene under the transcriptional control of a powerful promoter. Expression in *P. aeruginosa* has resulted in the polymerization of *B. nodosus* pilin into mature pili on the exterior surface of this host organism.[40,41] In vaccination experiments, agglutinating-antibody titers achieved by vaccination with recombinant-pilus vaccines were comparable to those in sheep receiving native pili and similar levels of homologous protection[40,42,43] and cure[42,43] against footrot were obtained. Although homologous whole cells have been shown to cure footrot,[6,7] the curative effect of pilus vaccines had not previously been reported. Thus, pilus vaccines have both immunotherapeutic as well as immunoprophylactic value against footrot. Since the curative effect against both early and chronic lesions is evident after only one vaccination, this has important implications for the future application of footrot vaccines.[42] Provided vaccines contain sufficient pili and the challenge is homologous, the first dose of vaccine can be administered during an outbreak after transmission has commenced. By contrast, in the majority of sheep, one dose of poorly piliated cells did not prevent infection, and interdigital skin lesions progressed to underrunning of the hoof, although two doses induced a cure in most cases.[42] Thus, in this respect, recombinant-pilus vaccines offer a great deal of promise.

An alternative but complementary approach is the utilization of proteases, produced either by *B. nodosus* itself or through recombinant-DNA technology, as cross-protective antigens for providing broad-spectrum cover against the complete range of serogroups. This strategy, using proteases as the sole protective antigen, or in combination with selected recombinant-pilus antigens may overcome some of the difficulties associated with the production of multivalent vaccines.

E. Conclusions

The prospects are highly promising for more refined vaccines based on the results of recent research with the recombinant-DNA-produced pilus vaccines as well as the possible exploitation of proteases as cross-protective antigens. These third generation vaccines should greatly assist the development of more effective and less labor-intensive strategies for controlling and eradicating footrot.

II. FOOT ABSCESS

A. Introduction

After footrot, foot abscess is the second most important bacterial disease of the ovine foot, causing lameness and loss of productivity in adult ewes, particularly in late pregnancy,[99] and rams.[100] In cattle, foot abscess is the most prevalent cause of lameness and is most common in dairy cows in early lactation and in feedlot cattle.[101] Foot abscess is a mixed bacterial infection characterized by necrosis and suppuration of the dermal and subdermal tissues of the ruminant foot. The disease is caused principally by *Fusobacterium necrophorum* in both sheep[99,102-105] and cattle,[104,105] although other organisms such as *Corynebacterium pyogenes*,[99] *Bacteroides melaninogenicus*,[106] and *B. nodosus*[107-112] may assist the development of the lesion.

B. *F. necrophorum* Vaccines

Although commercial vaccines are available for foot abscess, their value has not been conclusively established.[113,114] In fact, there is only scanty information in the scientific literature on evaluation of vaccines for *F. necrophorum* infections in target species. In Australia, a killed *F. necrophorum* adjuvanted vaccine (FRA vaccine®, CSL) has been registered for use in cattle for many years but no claims are made for its efficacy in sheep. Jensen et al.,[115] using a series of i.p. inoculations of sterile, polyvalent-culture filtrates and a final inoculation of formalinized, whole-fluid culture of *F. necrophorum*, failed to protect sheep against lung and liver abscesses resulting from intraportal challenge. Although Gilder[116] reported that the use of vaccine is followed by a marked reduction in the incidence of foot abscess in cattle herds, data on control unvaccinated animals was not included as a comparison.

C. Potential Protective Antigens

Recent studies have shown that pure cultures of toxigenic strains of *F. necrophorum* caused the disease when injected through the interdigital skin of cattle[105] and sheep.[103] Virulent strains (biovars A or AB) are readily isolated from lesions of foot abscess and are highly toxigenic, whereas nontoxigenic strains (biovar B) are avirulent and commonly present in the environment and on normal feet as a commensal organism derived from the ruminant intestinal tract.[117] When injected through the interdigital skin, young (18 h) cultures produce more severe lesions than older cultures and this is related to their exotoxin content.[105] Apart from exotoxin production, the A- and AB-type biovars can be distinguished from the nontoxigenic B biovars by their colony morphology, type of growth in liquid medium, and hemolytic activity.[117] The exotoxin causes an inflammatory response in rabbit and guinea-pig skin and is destructive for leucocytes, erythrocytes, and cultured cells.[118-120] There is also an association between toxin production and the capacity to produce abdominal abscesses and deaths in rodents.[103,121,122] The cytolytic effects are probably mediated by the same molecule, since there is a relationship between the titers of leucocidin, hemolysin, and cytotoxicity to cultured cells and their respective neutralization titers of antisera.[120,123] Serum from immunized calves vary in their capacity to neutralize the leucocidins from heterologous strains of *F. necrophorum*, which may indicate antigenic variation.[123]

Since the leucocidal toxin is a major virulence factor of *F. necrophorum*, causing destruction of leucocytes and necrosis of tissues, induction of an immune response to this candidate host-protective antigen may be important in protecting cattle and sheep against foot abscess. Accordingly, a substantial effort has been directed towards purification of the molecule for vaccination studies in sheep and cattle. Garcia et al.[124] found that feedlot cattle could be partially protected against liver abscess with the use of alum-precipitated, toxoided, cytoplasmic protein but not with toxoided, sonicated, unfractionated cells, the degree of protection being related to the amount of protein injected. Previous studies had indicated that alum adjuvant elicited an antibody response to the cytoplasmic toxoid equivalent to or better than Freund's complete adjuvant. More severe local reactions were observed with whole cells in each of the adjuvants than sonicated cells or the cytoplasmic fraction.[125] The cytoplasmic fraction contained a cell-bound protein toxin with hemolytic activity and was strongly antigenic. The capacity of cytoplasmic toxoid, after a series of i.p. injections, to induce protective immunity against abdominal abscesses following challenge with *F. necrophorum*, was confirmed in mice and sheep.[126] The culture filtrate of *F. necrophorum* did not contain any detectable toxic activity.[127] By contrast, 18-h concentrated-culture supernatants containing leucocidin and incorporated in oil adjuvant were more effective in reducing the prevalence of experimentally induced foot abscess in cattle than vaccines containing washed whole cells or cytoplasmic fraction.[128] All of Clark's preparations[128] contained lipopólysaccharide (LPS), outer-membrane proteins (OMP), capsular antigens, and leucocidin, the latter being present in small amounts in the cytoplasmic fraction. Semipurified leucocidin from the same strain but not preparations containing LPS and OMP protected mice against homologous challenge with a toxigenic bovine strain, although LPS and OMP induced immunity to a nontoxigenic strain.[129]

The results obtained by immunization of cattle and mice were sufficiently encouraging to explore the capacity of extracts and whole cells of a toxigenic sheep strain of *F. necrophorum* to protect sheep, given two doses of oil-adjuvant vaccine, against homologous challenge. A similar number of abscesses developed in all the vaccinated groups and saline-adjuvant control group, although in the washed-whole-cell-vaccine group there was a 60% decrease in the severity of lesions as measured by a reduction in the aggregate score for the size of the swellings in the feet. For those groups vaccinated with either homologous concentrated supernatant, LPS, or semipurified leucocidin, there was only a small decrease in the severity of the lesions. However, in a second experiment, the reduction in severity of lesions in the whole-cell vaccine group was only slightly better than the controls.

D. Conclusions

There are a number of conclusions and recommendations emanating from research on foot abscess in sheep and cattle.

1. Emery et al.[103] have suggested that immunity against *F. necrophorum* may operate more effectively in the peritoneal cavity than at the extremities. Ruminants can be

more easily protected against liver disease[124,126] than foot abscess. Experimentally induced foot abscess is a relatively abrupt disease of rapid onset which resolves quickly once the discharge of abscess content begins. Sequestration of the lesion occurs before immune-effector mechanisms become fully operative. The mechanism of immunity remains unclear, but besides antibody,[113] the nonspecific activation of macrophages may be involved.[129-131]

2. There is a need to develop a more natural, less severe method of challenge for the induction of foot abscess.
3. There is a need to purify and evaluate the role of other antigens such as capsular polycaccharide[103] in initiating abscess formation and host protection. In other anaerobic abscess-forming bacteria, polysaccharide capsules can cause abscesses when purified material is inoculated.[132,133]
4. Although *F. necrophorum* alone will produce the condition, the contribution of other bacteria such as *C. pyogenes*,[134] *B. melaninogenicus*,[106] microaerophilic cocci,[135] and *B. nodosus* need to be further investigated to determine whether protection might be augmented if these were included in the vaccine. In this regard, promising results have been obtained by Polish investigators, in a dairy herd, using a commercial alum-precipated *B. nodosus* vaccine (Clovax®). Over a 2.5 year period, there was a threefold reduction in the incidence of lameness, in comparison to unvaccinated controls, in cattle given two doses of vaccine followed by a booster every 6 months.[111] An Australian study has also shown a significant reduction in the prevalence of interdigital dermatitis and a lower prevalence of foot abscess in Hereford cattle vaccinated with an autogenous oil-adjuvanted vaccine formulated with two benign *B. nodosus* strains.[112] However, larger field trials in both sheep and cattle are required to substantiate the benefit of using *B. nodosus* vaccines in the prevention of foot abscess.

REFERENCES

1. **Beveridge, W. I. B.,** Foot-rot in sheep: a transmissible disease due to infection with *Fusiformis nodosus* (n. sp.). Studies on its cause, epidemiology and control, *J. Counc. Sci. Ind. Res., (Melbourne),* 140, 14, 1941.
2. **Beveridge, W. I. B.,** Foot-rot of sheep: its epidemiology and control, *Bull. Off. Int. Epizoot.,* 59, 1537, 1963.
3. **Parsonson, I. M.,** Unpublished data, 1966.
4. **Egerton, J. R. and Parsonson, I. M.,** Parenteral antibiotic treatment of ovine footrot, *Aust. Vet. J.,* 42, 97, 1966.
5. **Parsonson, I. M.,** Unpublished data, 1966.
6. **Egerton, J. R. and Roberts, D. S.,** Vaccination against ovine foot-rot, *J. Comp. Pathol.,* 81, 179, 1971.
7. **Egerton, J. R. and Burrell, D. H.,** Prophylactic and therapeutic vaccination against ovine foot-rot, *Aust. Vet. J.,* 46, 517, 1970.
8. **Roberts, D. S.,** Vaccine preparation, British Patent 1375 544, 1974.
9. **Skerman, T. M.,** Improvements in or relating to methods of preparation and use of vaccine from *Bacteroides nodosus,* British Patent 1351343, 1974.
10. **Skerman, T. M.,** Determination of some *in vitro* growth requirements of *Bacteroides nodosus, J. Gen. Microbiol.,* 87, 107, 1975.
11. **Skerman, T. M. and Cairney, I. M.,** Experimental observations on prophylactic and therapeutic vaccination against foot-rot in sheep, *N.Z. Vet. J.,* 20, 205, 1972.
12. **Roberts, D. S., Foster, W. H., Kerry, J. B., and Calder, H. A. McC.,** An alum-treated vaccine for the control of foot-rot in sheep, *Vet. Rec.,* 91, 428, 1972.
13. **Kerry, J. B. and Craig, G. R.,** Effect of vaccination against footrot in young sheep wintered in straw yards, *Vet. Rec.,* 98, 446, 1976.
14. **Reuss, U., Conrad, P., and Traeder, W.,** Vakzinationsversuche gegen die moderhinke der schafe, *Tieraerztl. Umsch.,* 29, 504, 1974.

15. **Eikelenboom, G.,** Voorlopig resultaat van een onderzoek naar het effect van een vaccin tegen rotkreupel bij schapen, *Tijdschr. Diergeneeskd.* 101, 1207, 1976.
16. **Sterk, V., Beslin, R., and Vulic, I.,** Prilog Poznavanju Zastitne Vrednosti Vakcine protiv Zarazne Sepavosti Ovaca, *Vet. Glas.,* 12, 1009, 1976.
17. **Egerton, J. R.,** Ovine foot-rot vaccines, *N.S.W. Vet. Proc.,* 10, 36, 1974.
18. **Egerton, J. R., Morgan, I. R., and Burrell, D. H.,** Foot-rot in vaccinated and unvaccinated sheep, *Vet. Rec.,* 91, 447, 1972.
19. **Egerton, J. R. and Morgan, I. R.,** Treatment and prevention of foot-rot in sheep with *Fusiformis nodosus* vaccine, *Vet. Res.,* 91, 453, 1972.
20. **Egerton, J. R.,** Significance of *Fusiformis nodosus* serotypes in resistance of vaccinated sheep to experimental foot-rot, *Aust. Vet. J.,* 50, 59, 1974.
21. **Keogh, J.,** Footrot vaccine trials, *N.S.W. Vet. Proc.,* 12, 35, 1976.
22. **Stewart, D. J.,** Progress in the development of effective ovine foot-rot vaccines, *Vic. Vet. Proc.,* 1980, 10.
23. **Egerton, J. R.,** Surface and somatic antigens of *Fusiformis nodosus, J. Comp. Pathol.,* 83, 151, 1973.
24. **Stewart, D. J.,** An electron microscopic study of *Fusiformis nodosus, Res. Vet. Sci.,* 14, 132, 1973.
25. **Walker, P. D., Short, J., Thomson, R. O., and Roberts, D. S.,** The fine structure of *Fusiformis nodosus* with special reference to the location of antigens associated with immunogenicity, *J. Gen. Microbiol.,* 77, 351, 1973.
26. **Stewart, D. J.,** Studies on the morphology and antigenic structure of *Fusiformis nodosus,* Ph.D. thesis, University of Sydney, Sydney, 1975.
27. **Short, J. A., Thorley, C. M., and Walker, P. D.,** An electron microscope study of *Bacteroides nodosus:* ultrastructure of organisms from primary isolates and different colony types, *J. Appl. Bacteriol.,* 40, 311, 1976.
28. **Stewart, D. J.,** Studies on the antigenic structure of *Bacteroides nodosus, Res. Vet. Sci.,* 24, 293, 1978.
29. **Every, D.,** Purification of pili from *Bacteroides nodosus* and an examination of their chemical, physical and serological properties, *J. Gen. Microbiol.,* 115, 309, 1979.
30. **Lee, S. W., Alexander, B., and McGowan, B.,** Purification, characterization, and serologic characteristics of *Bacteroides nodosus* pili and use of a purified pili vaccine in sheep, *Am. J. Vet. Res.,* 44, 1676, 1983.
31. **Stewart, D. J.,** The role of various antigenic fractions of *Bacteroides nodosus* in eliciting protection against foot-rot in vaccinated sheep, *Res. Vet. Sci.,* 24, 14, 1978.
32. **Stewart, D. J. and Egerton, J. R.,** Studies on the ultrastructural morphology of *Bacteroides nodosus, Res. Vet. Sci.,* 26, 227, 1979.
33. **Every, D. and Skerman, T. M.,** Ultrastructure of the *Bacteroides nodosus* cell envelope layers and surface, *J. Bacteriol.,* 141, 845, 1980.
34. **Every, D.,** Surface structures and proteases of *Bacteroides nodosus* in relation to virulence and immunoprotective properties in sheep, in *Ovine Footrot,* Rep. Workshops University of Sydney, May 3 to 6, 1981, 8.
35. **Every, D. and Skerman, T. M.,** Protection of sheep against experimental footrot by vaccination with pili purified from *Bacteroides nodosus, N.Z. Vet. J.,* 30, 156, 1982.
36. **Stewart, D. J., Clark, B. L., Peterson, J. E., Griffiths, D. A., and Smith, E. F.,** Current CSIRO foot-rot vaccine research: *Bacteroides nodosus* pili vaccines, in *Ovine Footrot,* Rep. Workshop University of Sydney, May 3 to 6, 1981, 51.
37. **Stewart, D. J., Clark, B. L., Peterson, J. E., Griffiths, D. A., and Smith, E. F.,** Importance of pilus-associated antigen in *Bacteroides nodosus* vaccines, *Res. Vet. Sci.,* 32, 140, 1982.
38. **Stewart, D. J., Clark, B. L., Emery, D. L., Peterson, J. E., and Fahey, K. J.,** A *Bacteroides nodosus* immunogen, distinct from the pilus, which induces cross-protective immunity in sheep vaccinated against footrot, *Aust. Vet. J.,* 60, 83, 1983.
39. **Stewart, D. J., Clark, B. L., Peterson, J. E., Griffiths, D. A., Smith, E. F., and O'Donnell, I. J.,** Effect of pilus dose and type of Freund's adjuvant on the antibody and protective responses of vaccinated sheep to *Bacteroides nodosus, Res. Vet. Sci.,* 35, 130, 1983.
40. **Elleman, T. C., Hoyne, P. A., Stewart, D. J., McKern, N. M., and Peterson, J. E.,** Expression of pili from *Bacteroides nodosus* in *Pseudomonas aeruginosa, J. Bacteriol.,* 168, 574, 1986.
41. **Mattick, J. S., Bills, M. M., Anderson, B. J., Dalrymple, B., Mott, M. R., and Egerton, J. R.,** Morphogenetic expression of *Bacteroides nodosus* fimbriae in *Pseudomonas aeruginosa, J. Bacteriol.,* 169, 33, 1987.
42. **Stewart, D. J. and Elleman, T. C.,** A *Bacteroides nodosus* pili vaccine produced by recombinant DNA for the prevention and treatment of footrot in sheep, *Aust. Vet. J.,* 64, 79, 1987.
43. **Egerton, J. R., Cox, P. T., Anderson, B. J., Kristo, C., Norman, M., and Mattick, J. S.,** Protection of sheep against footrot with a recombinant DNA-based fimbrial vaccine, *Vet. Microbiol.,* 14, 393, 1987.
44. **O'Donnell, I. J., Stewart, D. J., and Clark, B. L.,** Serological identification of pilus antigen and other protein antigens of *Bacteroides nodosus* using electro-blot radioimmunoassay after electrophoretic fractionation of the proteins on sodium dodecyl sulfate polyacrylamide gels, *Aust. J. Biol. Sci.,* 36, 15, 1983.

45. **Emery, D. L., Clark, B. L., Stewart, D. J., O'Donnell, I. J., and Hewish, D. R.,** Analysis of the outer membrane proteins of *Bacteroides nodosus,* the causal organism of ovine footrot, *Vet. Microbiol.,* 9, 155, 1984.
46. **Mattick, J. S., Anderson, B. J., Mott, M. R., and Egerton, J. R.,** Isolation and characterization of *Bacteroides nodosus* fimbriae: structural subunit and basal protein antigens, *J. Bacteriol.,* 160, 740, 1984.
47. **Stewart, D. J., Clark, B. L., Emery, D. L., Peterson, J. E., Jarrett, R. G., and O'Donnell, I. J.,** Cross-protection from *Bacteroides nodosus* vaccines and the interaction of pili and adjuvants, *Aust. Vet. J.,* 63, 101, 1986.
48. **Stewart, D. J., Clark, B. L., Emery, D. L., Peterson, J. E., and Kortt, A. A.,** The phenomenon of cross protection against footrot induced by vaccination of sheep with *Bacteroides nodosus* vaccines, in *Footrot in Ruminants,* Proc. Workshop, Melbourne, 1985, Stewart, D. J., Peterson, J. E., McKern, N. M., and Emery, D. L., Eds., CSIRO Division Animal Health/Australian Wool Corp., Glebe, N.S.W., 1986, 185.
49. **Swanson, J., Kraus, S. J., and Gotschlich, E. C.,** Studies on gonococcus infection. I. Pili and zones of adhesion: their relation to gonococcal growth patterns, *J. Exp. Med.,* 134, 886, 1971.
50. **Pedersen, K. B., Froholm, L. O., and Bovre, K.,** Fimbriation and colony type of *Moraxella bovis* in relation to conjunctival colonization and development of keratoconjunctivitis in cattle, *Acta Pathol. Microbiol. Scand. Sect. B,* 89, 911, 1972.
51. **Thorley, C. M.,** A simplified method for the isolation of *Bacteroides nodosus* from ovine foot-rot and studies on its colony morphology and serology, *J. Appl. Bacteriol.,* 40, 301, 1976.
52. **Thorley, C. M. and Egerton, J. R.,** Comparison of alum-absorbed or non-alum-absorbed oil emulsion vaccines containing either pilate or non-pilate *Bacteroides nodosus* cells in inducing and maintaining resistance of sheep to experimental footrot, *Res. Vet. Sci.,* 30, 32, 1981.
53. **Skerman, T. M., Erasmuson, S. K., and Every, D.,** Differentiation of *Bacteroides nodosus* biotypes and colony variants in relation to their virulence and immunoprotective properties in sheep, *Infect. Immun.,* 32, 788, 1981.
54. **Every, D. and Skerman, T. M.,** Surface structure of *Bacteroides nodosus* in relation to virulence and immunoprotection in sheep, *J. Gen. Microbiol.,* 129, 225, 1983.
55. **Stewart, D. J., Emery, D. L., Clark, B. L., Peterson, J. E., Iyer, H., and Jarrett, R. G.,** Differences between breeds of sheep in their responses to *Bacteroides nodosus* vaccines, *Aust. Vet. J.,* 62, 116, 1985.
56. **Parker, C. F., Cross, R. F., and Hamilton, K. L.,** Genetic resistance to footrot in sheep, *Annu. Proc. Sheep Vet. Soc.,* 9, 16, 1985.
57. **Skerman, T. M. and Moorhouse, S. R.,** Broomfield corriedales: a strain of sheep selectively bred for resistance to footrot, *N.Z. Vet. J.,* 35, 101, 1987.
58. **Outteridge, P. M., Windon, R. G., Dineen, J. K., Stewart, D. J., and Skerman, T. M.,** Breeding for parasite and footrot resistance in sheep using ovine lymphocyte antigen (OLA) markers, in *Merino Improvement Programs in Australia,* Proc. Natl. Symp., Leura, N. S. W., McGuirk, B. J., Ed., CSIRO Division Animal Health/Australian Wool Corp., Glebe, N.S.W., 1987, 377.
59. **Egerton, J. R. and Merritt, G. C.,** Serology of foot-rot: antibodies against *Fusiformis nodosus* in normal, affected, vaccinated and passively immunised sheep, *Aust. Vet. J.,* 49, 139, 1973.
60. **Claxton, P. D.,** Studies on *Bacteroides nodosus* Vaccines with Particular Reference to Factors Associated with their Efficiency in Protecting Against Ovine Virulent Footrot, Ph.D. thesis, University of Sydney, Sydney, 1981.
61. **Egerton, J. R. and Merritt, G. C.,** The occurrence of bactericidal antibodies against *Fusiformis nodosus* in sheep serum, *J. Comp. Pathol.,* 80, 369, 1970.
62. **Emery, D. L.,** Reactivity of sera from sheep immunised with individual outer membrane proteins of *Bacteroides nodosus* against heterologous bacterial strains, *Vet. Microbiol.,* 9, 453, 1984.
63. **Skerman, T. M.,** Present status of immunization against footrot in sheep: an evaluation of vaccination techniques for the control of ovine footrot in New Zealand, in *Proc. 20th World Vet. Cong.,* World Veterinary Congress Organising Committee, Thessaloniki, Greece, 1975, 2, 1391.
64. **Egerton, J. R., Thompson, J. J., and Merritt, G. C.,** Comparison of oil adjuvant and alum precipitated *Bacteroides nodosus* vaccines in treatment of foot-rot, *Vet. Rec.,* 104, 98, 1979.
65. **Egerton, J. R. and Thorley, C. M.,** Effect of alum-precipitated or oil-adjuvant *Bacteroides nodosus* vaccines on the resistance of sheep to experimental foot-rot, *Res. Vet. Sci.,* 30, 28, 1981.
66. **Egerton, J. R., Laing, E. A., and Thorley, C. M.,** Effect of Quil A, a saponin derivative, on the response of sheep to alum precipitated *Bacteroides nodosus* vaccines, *Vet. Sci. Commun.,* 2, 247, 1978.
67. **Thorley, C. M., Kerry, J. B., Calder, H. A. McC., Ripley, P. H., and Day, S. E. J.,** A cadidate saponin adjuvanted, aluminium hydroxide precipitated, multistrain footrot vaccine for non-Merino (British) breed sheep, in *Footrot in Ruminants,* Proc. Workshop, Melbourne, 1985, Stewart, D. J., Peterson, J. E., McKern, N. M., and Emery, D. L., Eds., CSIRO Division Animal Health/Australian Wool Corp., Glebe, N.S.W., 1986, 193.

68. **Thorley, M.,** Discussion of session 5, on vaccination against ovine footrot, in *Footrot in Ruminants,* Proc. Workshop, Melbourne, 1985, Stewart, D. J., Peterson, J. E., McKern, N. M., and Emergy, D. L., Eds., CSIRO Division Animal Health/Australian Wool Corp., Glebe, N.S.W., 1986, 203.
69. **Ferrier, G. R., Smith, S. C., and Spencer, T. L.,** Serological responses to antigens of *Bacteroides nodosus* in footrot infected and vaccinated sheep, in *Footrot in Ruminants,* Proc. Workshop, Melbourne, 1985, Stewart, D. J., Peterson, J. E., McKern, N. M., and Emery, D. L., Eds., CSIRO Division Animal Health/Australian Wool Corp., Glebe, N.S.W., 1986, 161.
70. **Ross, A. D. and Titterington, D. M.,** Injection site lesions of footrot vaccines in sheep, *N.Z. Vet. J.,* 32, 6, 1984.
71. **Thorley, C. M., Kieran, P. J., Egerton, J. R., and Geldard, H.,** The use of *Bacteroides nodosus* vaccines in the attempted control of foot-rot of sheep in the field in New South Wales, in *Ovine Footrot,* Rep. Workshop University of Sydney, May 3 to 6, 1981, 88.
72. **Lambell, R. G.,** Field experience with footrot vaccines in Victoria, in *Footrot in Ruminants,* Proc. Workshop, Melbourne, 1985, Stewart, D. J., Peterson, J. E., McKern, N. M., and Emery, D. L., Eds., CSIRO Division Animal Health/Australian Wool Corp., Glebe, N.S.W., 1986, 199.
73. **Lambell, R. G.,** A field trial with a commercial vaccine against foot-rot in sheep, *Aust. Vet. J.,* 63, 415, 1986.
74. **Stewart, D. J., Clark, B. L., Jarrett, R. G., and Lewis, V. M.,** Unpublished data, 1984.
75. **Schmitz, J. A. and Gradin, J. L.,** Serotypic and biochemical characterization of *Bacteroides nodosus* isolates from Oregon, *Can. J. Comp. Med.,* 44, 440, 1980.
76. **Claxton, P. D.,** Serogroupings of *Bacteroides nodosus* in relation to protection by vaccination, in *Ovine Footrot,* Rep. Workshop, University of Sydney, May 3 to 6, 1981, 216.
77. **Claxton, P. D., Ribeiro, L. A., and Egerton, J. R.,** Classification of *Bacteroides nodosus* by agglutination tests, *Aust. Vet. J.,* 60, 331, 1983.
78. **Ribeiro, L. A. O.,** Distribution of antigenic types of *Bacteroides nodosus* in Australian sheep, in *Ovine footrot,* Rep. Workshop, University of Sydney, May 3 to 6, 1981, 35.
79. **Skerman, T. M.,** A comparison of *Bacteroides nodosus* biotypes with particular reference to properties of importance in footrot immunoprophylaxis, in *Ovine Footrot,* Rep. Workshop, University of Sydney, May 3 to 6, 1981, 3.
80. **Claxton, P. D.,** Serogrouping of *Bacteroides nodosus* isolates, in *Footrot in Ruminants,* Proc. Workshop, Melbourne, 1985, Stewart, D. J., Peterson, J. E., McKern, N. M., and Emery, D. L., Eds., CSIRO Division Animal Health/Australian Wool Corp., Glebe, N.S.W., 1986, 131.
81. **Stewart, D. J., Peterson, J. E., Vaughan, J. A., Clark, B. L., Emery, D. L., Caldwell, J. B., and Kortt, A. A.,** Clinical and laboratory diagnosis of benign, intermediate and virulent strains of *Bacteroides nodosus,* in *Footrot in Ruminants,* Proc. Workshop, Melbourne, 1985, Stewart, D. J., Peterson, J. E., McKern, N. M., and Emery, D. L., Eds., CSIRO Division Animal Health/Australian Wool Corp., Glebe, N.S.W., 1986, 81.
82. **Stewart, D. J., Peterson, J. E., Vaughan, J. A., Clark, B. L., Emery, D. L., Caldwell, J. B., and Kortt, A. A.,** The pathogenicity and cultural characteristics of virulent, intermediate and benign strains of *Bacteroides nodosus* causing ovine foot-rot, *Aust. Vet. J.,* 63, 317, 1986.
83. **Kingsley, D. F., Hindmarsh, F. H., Liardet, D. M., and Chetwin, D. H.,** Distribution of serogroups of *Bacteroides nodosus* with particular reference to New Zealand and the United Kingdom, in *Footrot in Ruminants,* Proc. Workshop, Melbourne, 1985, Stewart, D. J., Peterson, J. E., McKern, N. M., and Emery, D. L., Eds., CSIRO Division Animal Health/Australian Wool Corp., Glebe, N.S.W., 1986, 143.
84. **Hindmarsh, F. and Fraser, J.,** Serogroups of *Bacteroides nodosus* isolated from ovine foot-rot in Britain, *Vet. Rec.,* 116, 187, 1985.
85. **Thorley, C. M. and Day, S. E. J.,** Serotyping survey of 1296 strains of *Bacteroides nodosus* isolated from sheep and cattle in Great Britain and western Europe, in *Footrot in Ruminants,* Proc. Workshop, Melbourne, 1985, Stewart, D. J., Peterson, J. E., McKern, N. M., and Emery, D. L., Eds., CSIRO Division Animal Health/Australian Wool Corp., Glebe, N.S.W., 1986, 135.
86. **Day, S. E. J., Thorley, C. M., and Beesley, J. E.,** Serotyping of *Bacteroides nodosus:* proposal for 9 further serotypes (J-R) and a study of the antigenic complexity of *B. nodosus* pili, in *Footrot in Ruminants,* Proc. Workshop, Melbourne, 1985, Stewart, D. J., Peterson, J. E., McKern, N. M., and Emery, D. L., Eds., CSIRO Division Animal Health/Australian Wool Corp., Glebe, N. S. W., 1986, 147.
87. **Reed, G. A., Shepherd, G. E., Astill, K. J., and Liardet, D. M.,** The field evaluation of an experimental *Bacteriodes nodosus* vaccine for sheep, in *Ovine Footrot,* Rep. Workshop University of Sydney, May 3 to 6, 1981, 113.
88. **Stewart, D. J., Clark, B. L., Peterson, J. E., Emery, D. L., Smith, E. F., Griffiths, D. A., and O'Donnell, I. J.,** The protection given by pilus and whole cell vaccines of *Bacteroides nodosus* strain 198 against ovine foot-rot induced by strains of different serogroups, *Aust. Vet. J.,* 62, 153, 1985.
89. **Egerton, J. R.,** Personal communication, 1987.

90. **Egerton, J. R.,** Treatment of ovine foot-rot by vaccination with the specific aetiological agent *Bacteroides nodosus, Comp. Immunol. Microbiol. Infect. Dis.,* 2, 61, 1979.
91. **Skerman, T. M., Erasmuson, S. K., and Morrison, L. M.,** Duration of resistance to experimental footrot infection in Romney and Merino sheep vaccinated with *Bacteroides nodosus* oil adjuvant vaccine, *N.Z. Vet. J.,* 30, 27, 1982.
92. **Kortt, A. A., Caldwell, J. B., Burns, J. E., Guthrie, R. E., and Stewart, D. J.,** Characterisation and substrate specificity of the extracellular proteases of *Bacteroides nodosus* from virulent and benign ovine isolates, in *Footrot in Ruminants,* Proc. Workshop, Melbourne, 1985, Stewart, D. J., Peterson, J. E., McKern, N. M., and Emery, D. L., Eds., CSIRO Division Animal Health/Australian Wool Corp., Glebe, N. S. W., 1986, 237.
93. **Merritt, G. C. and Egerton, J. R.,** IgG_1 and IgG_2 immunoglobulins to *Bacteroides (Fusiformis) nodosus* protease in infected and immunized sheep, *Infect. Immun.,* 22, 1, 1978.
94. **Stewart, D. J. and Kortt, A. A.,** *Bacteroides nodosus* vaccine, Australian Patent 51387/85.
95. **Anderson, B. J., Bills, M. M., Egerton, J. R., and Mattick, J. S.,** Cloning and expression in *Escherichia coli* of the gene encoding the structural subunit of *Bacteroides nodosus* fimbriae *J. Bacteriol.,* 160, 748, 1984.
96. **Elleman, T. C., Hoyne, P. A., Emery, D. L., Stewart, D. J., and Clark, B. L.,** Isolation of the gene encoding pilin of *Bacteroides nodosus* (strain 198), the causal organism of ovine footrot, *FEBS Lett.,* 173, 103, 1984.
97. **Elleman, T. C. and Hoyne, P. A.,** Nucleotide sequence of the gene encoding pilin of *Bacteroides nodosus,* the causal organism of ovine footrot, *J. Bacteriol.,* 160, 1184, 1984.
98. **Elleman, T. C., Hoyne, P. A., Emery, D. L., Stewart, D. J., and Clark, B. L.,** Expression of the pilin gene from *Bacteroides nodosus* in *Escherichia coli, Infect. Immun.,* 51, 187, 1986.
99. **Roberts, D. S., Graham, N. P. H., and Egerton, J. R.,** Infective bulbar necrosis (heel-abscess) of sheep, a mixed infection with *Fusiformis necrophorus* and *Corynebacterium pyogenes, J. Comp. Pathol.,* 78, 1, 1968.
100. **West, D.,** A study of naturally occurring cases of ovine foot abscess in New Zealand, *N.Z. Vet. J.,* 31, 152, 1983.
101. **Clark, B. L.,** this volume, chap. 4.
102. **Gregory, T. S.,** Foot-rot in sheep, *Aust. Vet. J.,* 15, 160, 1939.
103. **Emery, D. L., Vaughan, J. A., Clark, B. L., and Stewart, D. J.,** Virulence determinants of *Fusobacterium necrophorum* and their prophylactic potential in animals, in *Footrot in Ruminants,* Proc. Workshop, Melbourne, 1985, Stewart, D. J., Peterson, J. E., McKern, N. M., and Emery, D. L., Eds., CSIRO Division Animal Health/Australian Wool Corp., Glebe, N.S.W., 1986, 267.
104. **Flint, J. C. and Jensen, R.,** Pathology of necrobacillosis of the bovine foot, *Am. J. Vet. Res.,* 12, 5, 1951.
105. **Clark, B. L., Stewart, D. J., and Emery, D. L.,** The role of *Fusobacterium necrophorum* and *Bacteroides melaninogenicus* in the aetiology of interdigital necrobacillosis in cattle, *Aust. Vet. J.,* 62, 47, 1985.
106. **Berg, J. N. and Loan, R. W.,** *Fusobacterium necrophorum* and *Bacteroides melaninogenicus* as etiologic agents of foot-rot in cattle, *Am. J. Vet. Res.,* 36, 1115, 1975.
107. **Gupta, R. B., Fincher, M. G., and Bruner, D. W.,** A study of the etiology of foot-rot in cattle, *Cornell Vet.,* 54, 66, 1964.
108. **Egerton, J. R. and Parsonson, I. M.,** Isolation of *Fusiformis nodosus* from cattle, *Aust. Vet. J.,* 42, 425, 1966.
109. **Laing, E. A. and Egerton, J. R.,** The occurrence, prevalence and transmission of *Bacteroides nodosus* infection in cattle, *Res. Vet. Sci.,* 24, 300, 1978.
110. **Egerton, J. R.,** Diseases of cattle causing lameness, in *Refresher Course for Veterinarians, Proceedings No. 42: The J. D. Stewart Memorial Refresher Course in Cattle Diseases,* Universtiy of Sydney, Post-Graduate Committee in Veterinary Science, Sydney, 1979, 203.
111. **Cakala, S., Pilaszek, J., Lubiarz, J., and Gogolewski, L.,** Clinical and serological investigations on the application of Clovax vaccine in cattle, *Bull. Vet. Inst. Pulawy,* 27, 12, 1984.
112. **Clark, B. L., Stewart, D. J., Emery, D. L., Dufty, J. H., and Jarrett, R. G.,** Immunisation of cattle against interdigital dermatitis (foot-rot) with an autogenous *Bacteroides nodosus* vaccine, *Aust. Vet. J.,* 63, 61, 1986.
113. **Roberts, D. S.,** Toxic, allergenic and immunogenic factors of *Fusiformis necrophorus, J. Comp. Pathol.,* 80, 247, 1970.
114. **Beveridge, W. I. B.,** *Bacterial Diseases of Cattle, Sheep and Goats,* Animal Health in Australia, Vol. 4, Australian Government Publishing Service, Canberra, 1983, 108.
115. **Jensen, R., Flint, J. C., and Griner, L. A.,** Experimental hepatic necrobacillosis in beef cattle, *Am. J. Vet. Res.,* 15, 5, 1954.
116. **Gilder, R. P.,** Foot diseases of cattle, *Aust. Vet. J.,* 36, 151, 1960.

117. **Emery, D. L., Vaughan, J. A., Clark, B. L., Dufty, J. H., and Stewart, D. J.,** Cultural characteristics and virulence of strains of *Fusobacterium necrophorum* isolated from the feet of cattle and sheep, *Aust. Vet. J.*, 62, 43, 1985.
118. **Roberts, D. S.,** The pathogenic synergy of *Fusiformis necrophorus* and *Corynebacterium pyogenes*. I. Influence of the leucocidal exotoxin of *F. necrophorus, Br. J. Exp. Pathol.*, 48, 665, 1967.
119. **Fales, W. H., Warner, J. F., and Teresa, G. W.,** Effects of *Fusobacterium necrophorum* leukotoxin on rabbit peritoneal macrophages *in vitro, Am. J. Vet. Res.*, 38, 491, 1977.
120. **Emery, D. L., Dufty, J. H., and Clark, B. L.,** Biochemical and functional properties of aleucocidin produced by several strains of *Fusobacterium necrophorum, Aust. Vet. J.*, 61, 382, 1984.
121. **Coyle-Dennis, J. E. and Lauerman, L. H.,** Correlations between leukocidin production and virulence of two isolates of *Fusobacterium necrophorum, Am. J. Vet. Res.*, 40, 274, 1979.
122. **Berg, J. N. and Scanlan, C. M.,** Studies of *Fusobacterium necrophorum* from bovine hepatic abscesses: biotypes, quantitation, virulence, and antibiotic susceptibility, *Am. J. Vet. Res.*, 43, 1580, 1982.
123. **Emery, D. L., Edwards, R. D., and Rothel, J. S.,** Studies on the purification of the leucocidin of *Fusobacterium necrophorum* and its neutralization by specific antisera, *Vet. Microbiol.*, 11, 357, 1986.
124. **Garcia, M. M., Dorward, W. J., Alexander, D. C., Magwood, S. E., and McKay, K. A.,** Results of a preliminary trial with *Sphaerophorus necrophorus* toxoids to control liver abscesses in feedlot cattle, *Can. J. Comp. Med.*, 38, 222, 1974.
125. **Alexander, D. C., Garcia, M. M., and McKay, K. A.,** Assessment of various adjuvants in *Sphaerophorus necrophorus* toxoids, *Can, Vet. J.*, 14, 247, 1973.
126. **Garcia, M. M. and McKay, K. A.,** Intraperitoneal immunization against necrobacillosis in experimental animals, *Can. J. Comp. Med.*, 42, 121, 1978.
127. **Garcia, M. M., Alexander, D. C., and McKay, K. A.,** Biological characterization of *Fusobacterium necrophorum* cell fractions in preparation for toxin and immunization studies, *Infect. Immun.*, 11, 609, 1975.
128. **Clark, B. L., Emery, D. L., Stewart, D. J., Dufty, J. H., and Anderson, D. A.,** Studies into immunization of cattle against interdigital necrobacillosis, *Aust. Vet. J.*, 63, 107, 1986.
129. **Emery, D. L. and Vaughan, J. A.,** Generation of immunity against *Fusobacterium necrophorum* in mice inoculated with extracts containing leucocidin, *Vet. Microbiol.*, 12, 255, 1986.
130. **Abe, P. M., Holland, J. W., and Stauffer, L. R.,** Immunization of mice against *Fusobacterium necrophorum* infection by parenteral or oral administration of vaccine, *Am. J. Vet. Res.*, 39, 115, 1978.
131. **Conlon, P. J., Hepper, K. P., and Teresa, G. W.,** Evaluation of experimentally induced *Fusobacterium necrophorum* infections in mice, *Infect. Immun.*, 15, 510, 1977.
132. **Onderdonk, A. B., Kasper, D. L., Cisneros, R. L., and Bartlett, J. G.,** The capsular polysaccharide of *Bacteroides fragilis* as a virulence factor: comparison of the pathogenic potential of encapsulated and unencapsulated strains, *J. Infect. Dis.*, 136, 82, 1977.
133. **Kasper, D. L.,** Bacterial capsule — old dogmas and new tricks, *J. Infect. Dis.*, 153, 407, 1986.
134. **Roberts, D. S.,** The pathogenic synergy of *Fusiformis necrophorus* and *Corynebacterium pyogenes*. II. The response of *F. necrophorus* to a filterable product of *C. pyogenes, Br. J. Exp. Pathol.*, 48, 674, 1967.
135. **Slee, K. J.,** A microaerophilic coccus in pyogenic infections of ruminants, *Aust. Vet. J.*, 62, 57, 1985.

Section IV: Applied Biotechnology

Section IV

APPLIED BIOTECHNOLOGY

W. K. Yong, G. G. Riffkin, and J. R. Egerton

I. INTRODUCTION

The last 5 years have witnessed the birth and development of a new, highly lucrative biotechnology industry. The basic understanding of biotechnology has already reached the point in the human medical field where marketable products are on the shelves or on trial. The backbone of this industry stems from basic studies in two apparently disparate but essentially compatible areas of biological research.

II. HYBRIDOMA TECHNOLOGY

The first is in the area of immunology, where the studies of antibody diversity led to the development of a somatic cell-fusion technique for the production of monoclonal antibodies. Conventionally, antibodies have been produced in limited quantities by injecting an antigen into an experimental animal and later collecting its serum. Since most antigens carry several antigenic determinants, the serum will contain a mixture of antibodies, each one being produced by a different clone of B lymphocytes. In 1975, Kohler and Milstein* successfully fused the nuclei of cultured mouse myeloma cells with spleen cells from a mouse that had been immunized with sheep red blood cells. The resultant hybrids, called hybridomas, produced not only large amounts of antibody to the immunizing antigen, but also maintained the characteristics of the myeloma parent, in that they grew continuously in culture and formed tumors when injected back into syngeneic animals. Further analysis demonstrated that a hybridoma perpetuates the production of the single antibody produced by any one clone of B lymphocyte. From this humble beginning, hybridoma technology has spawned a multimillion-dollar industry worldwide that is dedicated to supplying diagnosticians, clinicians, and researchers with homogeneous and highly specific monoclonal antibodies against antigens of every description.

III. RECOMBINANT-DNA (GENETIC ENGINEERING) TECHNOLOGY

DNA encodes the genetic makeup of virtually all living organisms. The second half of the biotechnology industry is said to have begun when the enzymology of DNA reached the stage where it became possible to clone genes; that is, to insert specific DNA fragments of interest into plasmids or phages by use of restriction endonucleases, polymerases, ligases, and other appropriate enzymes; and introduce the recombinant molecules into microorganisms, and enable the recombinants to multiply, thus producing a colony from which significant quantities of the cloned-DNA fragment could be obtained. The various applications of this recombinant-DNA technology are to clone genes producing specific, useful protein products, to clone genes which can be used in diagnosis, and to clone genes having specific physiological effects for introduction into animals or plants so as to change their phenotypes in a useful way. Examples of the latter include the introduction of disease-resistance genes or genes which increase growth rate in animals and plants.

This section aims to bring "state of the art" methodologies to the attention of applied

* Kohler, G. and Milstein, C., *Nature (London),* 256, 495, 1975.

scientists and development workers involved in the control and management of footrot and foot abscess of ruminants. Three different areas of application of biotechnology are explored. These are the development of an improved vaccine (Chapter 12), the characterization and greater understanding of antigenic variation (Chapter 13), and, finally, the potential use of biotechnology methods and their byproducts in diagnosis (Chapter 14). In each of these areas there is work in progress, albeit achievements to date reflect the perceived commercial potential of the area. The work on vaccine production for footrot using genetic-engineering techniques is the most advanced and most impressive of the three areas. On the other hand, the work on developing gene probes and monoclonal antibodies for diagnosis is just beginning, and understandably there is a dearth of published data. The editors do not, however, apologize for the inclusion of such a chapter in this volume because the development and application of these reagents are opening up exciting possibilities of improved diagnosis in the future.

Chapter 12

THE MOLECULAR BIOLOGY OF THE FIMBRIAE (PILI) OF *BACTEROIDES NODOSUS* AND THE DEVELOPMENT OF A RECOMBINANT-DNA-BASED VACCINE

John S. Mattick

TABLE OF CONTENTS

I. BACKGROUND

The development of a recombinant-DNA-based fimbrial vaccine against footrot has been carried out over the past 4 years by research groups at the CSIRO Divisions of Molecular Biology and Protein Chemistry, in conjunction with those groups having long-term experience with the disease and its essential causative agent *Bacteroides nodosus,* at the CSIRO Division of Animal Health and the University of Sydney Department of Veterinary Clinical Studies. More recently, these groups have been assisted in the commercial development of the vaccine by the research and development facilities of Arthur Webster Pty. Ltd. and Biotechnology Australia Pty. Ltd. The issues involved and progress towards this objective have been briefly reviewed in previous articles.[1,2]

Footrot is a relatively chronic infection of the avascular epidermis of the hoof, an area considered relatively remote from host defense mechanisms. This perspective originally prompted the suggestion that sheep were incapable of generating a systemic response against the infection, a view which for a time inhibited the investigation and application of measures such as vaccination and chemotherapy for the treatment of the disease.[3] The feasibility of such approaches was, however, demonstrated in the late 1960s by Egerton and co-workers who showed that parenteral administration of bactericidal antibiotics was effective against infection[4] and that vaccination with *B. nodosus* cells elicited protective responses.[5,6] although these were largely serotype specific.[7] Following these observations, several companies released mono- or bivalent vaccines, which were unsuccessful, partly because they failed to cover the range of serological variants in the field, and partly because they contained few fimbriae,[8] which are now known to include the relevant antigens (see below). However, a bivalent vaccine is still marketed in Britain by Coopers Animal Health. Extensive serological analyses of Australian field isolates, based on the K-agglutination test,[9] subsequently defined 9 major serogroups of *B. nodosus,* designated A to I, encompassing at least 18 subsidiary serotypes.[9-11] The same range of variants appears to occur in Britain, New Zealand, and other countries, although the relative prevalence of the different serogroups may vary.[12,13] Based on such studies, more complex vaccines containing whole cells of between eight and ten different *B. nodosus* strains have been developed and marketed since 1981, and these appear to be reasonably effective against footrot infection in the field,[8,14,15] although there have been some conflicting reports.[16]

The primary impetus for the development of new recombinant-DNA-based vaccines against footrot was economic. The current vaccines against footrot are relatively costly, up to fivefold higher than other vaccines used in the sheep industry, and a significant proportion of the value of the animal they are designed to protect. Consequently, their use and acceptance by the pastoral community has been limited, and the control of footrot is still largely dependent on the traditional, and laborious, methods of hoof paring and antiseptic bathing.[14,17,18] Apart from their multivalency, the high cost of conventional footrot vaccines appears to be largely related to the fastidious requirements and sparse growth of *B. nodosus* in culture[19] and the difficulty of obtaining stable fimbrial production, especially under the liquid-fermentation conditions[20] required for commercial production. These difficulties may be circumvented by the use of recombinant-DNA technology to direct the production of the appropriate antigens by an alternative host cell. This approach also has the attraction of allowing greater flexibility in the selection of serological variants for inclusion in the vaccine, so that quality and efficacy may be improved, as well as the (longer-term) possibility of its supplementation or replacement by the production of other (normally less prominent) antigens which may elicit broader cross-protective immunity.

II. IDENTIFICATION AND CHARACTERIZATION OF TARGET ANTIGENS

It is now well established that the prominent serological and immunoprotective antigens of *B. nodosus* are the fimbriae (or pili*),[1,2] which are long, surface filaments of about 6 nm in diameter and up to several mm in length.[21,22] In other bacteria, fimbriae have adhesive properties,[23] and although their exact function in *B. nodosus* has not yet been determined, it seems likely that they play an important role in the colonization and penetration by this bacterium of the epithelial matrix of the ovine hoof.[1,22] This view is supported by observations that *B. nodosus* colony variants with few or no fimbriae are relatively benign or noninfectious, respectively.[20,24,25] *B. nodosus* fimbriae have been classified as type 4, sharing several features in common with those of other pathogenic bacteria, including a polar location on the cell, association with a phenomenon known as twitching motility, agar-corroding activity, and certain highly conserved features of the structural protein which comprises the fimbrial strand.[23,26-29] Similar characteristics are observed in fimbriate species classified in a wide range of Gram-negative genera, including *Acinetobacter, Alteromonas, Bacteroides, Eikenella, Moraxella, Neisseria, Pasteurella, Pseudomonas, Vibrio,* and *Wolinella* among others.[23,26,28-33]

The fimbriae of *B. nodosus* are the principal antigenic structures mediating the K-agglutination reaction,[22,34-36] which has been the basis of the serological classification of this organism.[9-11] In addition, sheep vaccinated with purified fimbriae have been shown to be highly resistant to footrot following subsequent challenge,[36-39] although correspondingly, and as with the whole-cell vaccines, the range of protection is largely restricted to the serogroup involved.[38,40,41] Conversely, vaccination with poorly fimbriated cells is much less effective.[21,32,35] Although there have been reports of serogroup-independent cross-protective immunity following whole-cell vaccination, this effect is weak and not universal,[41-43] and the antigens involved are thus far undefined.

On this basis therefore, the fimbriae were selected as the target antigens for the production of a recombinant vaccine against footrot. *B. nodosus* fimbriae are essentially composed of a small, repeated structural subunit of about 17 kDa molecular weight,[35,44] which varies among different serogroups and serotypes (Figure 1A).[45,46] However, isolated fimbrial preparations also contain trace amounts of another polypeptide, of about 80 kDa, which a number of lines of circumstantial evidence suggested might represent a basal protein linking the fimbrial strand to the surface of the cell.[44] This protein also varies among different strains,[44,46] and elicits a substantial antibody response not only in vaccinated animals,[44] but also in those exposed to natural infection[47] (Figure 1B). These observations raised serious questions, especially in the context of a recombinant-DNA approach to vaccine production, about the relative roles of the structural subunit and putative basal protein in the generation of protective immunity against footrot.

The problem was addressed by examining these antigens in the context of the serotypic classification system. Using a simple and rapid procedure,[44] fimbriae were prepared from a large number of strains, including the prototypes of each of the known serogroups and serotypes. Electrophoretic analyses indicated that there was a broad correlation between the apparent size of the fimbrial subunit and the isolate in question, which was not evident with the 80-kDa antigen.[46] Matrix Western transfer analyses were also carried out, in which fimbriae isolated from a range of strains from each of the major serogroups and subtypes were displayed by electrophoresis, transferred to nitrocellulose paper, and challenged with serogroup- and serotype-specific antisera.[48] The results showed that the fimbrial subunits were recognized by antiserum against the homologous serotype and serogroup, but not (generally) by antisera against heterologous groups. There was no relationship between the reactivity of the 80-kDa antigen and serological profile. It was therefore concluded that the

* The terms "fimbriae" and "pili" are both commonly employed to describe nonflagellar, filamentous appendages on the surface of various species of bacteria. The term "pili" is also used to describe surface structures involved in the transfer of bacterial and viral nucleic acids, as well as other structures in plants. Based on historical precedence and to avoid ambiguity, we use the term "fimbriae", as suggested by Ottow.[23]

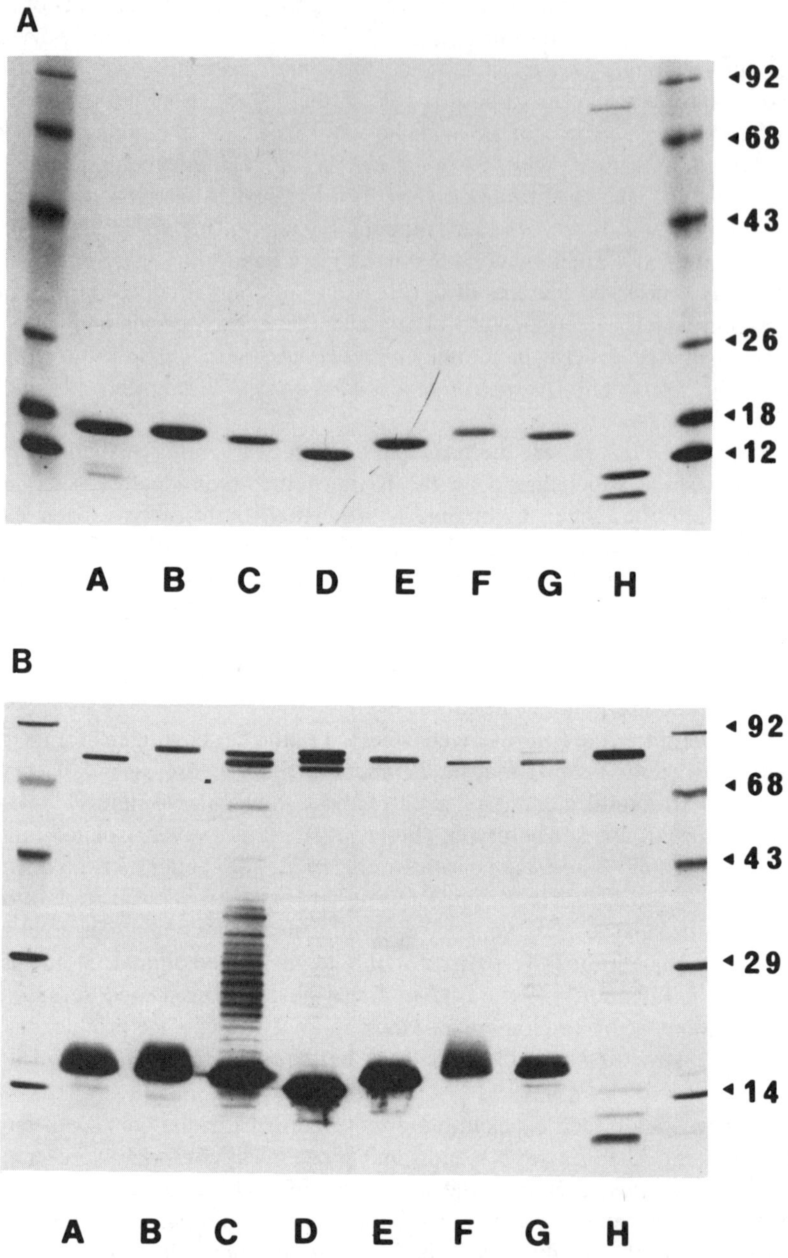

FIGURE 1. Electrophoretic and Western transfer analyses of fimbrial antigens of *B. nodosus*. Fimbriae were isolated from the prototype strains of *B. nodosus* serogroups A to H, displayed on SDS-urea gradient polyacrylamide gels, and (A) stained for protein with Coomassie Blue or (B) challenged with antifimbrial antisera after Western transfer to nitrocellulose paper.[46] (From Anderson, B. J. et al., *J. Bacteriol.*, 166, 453, 1986. With permission.)

primary serotype-specific, and by implication protective, antigen is the structural subunit of the fimbrial strand.[48] This was subsequently confirmed by vaccination trials with fimbriae produced in recombinant cells (see below). However, this does not necessarily preclude immunoprotective contribution by the putative basal antigen, which may, for example, be responsible for a degree of cross protection occasionally observed between different serogroups, depending on whether the vaccinating and challenge strains had this antigen in common.

These studies have also provided a valuable insight into the relationship between the different serogroups and serotypes of *B. nodosus*, and the extent of immunological variance in the field population. The matrix Western transfer analyses suggested that some serogroups are more closely related than others, specifically D and H, C and G, and A, E, F, B, and I,[48] which was subsequently confirmed by sequence analysis of the fimbrial subunit genes (see below). On the other hand, it is also clear that structural and antigenic variation in the fimbrial subunit is widespread in the *B. nodosus* population, even within serotypes, and that different isolates exhibit different patterns of cross-reaction,[46,48] an observation also made in K-agglutination tests.[11,49] This has led to the conclusion that the various serogroups and serotypes of *B. nodosus* comprise a series of overlapping sets of antigenically related strains, with the prototype strains being convenient and/or historical reference points.[48] In terms of vaccine design and antigens selection, there may well be alternative choices providing a more balanced representation of the population as a whole, or a broader or more complementary set of cross-reactions, which would confer superior immunological coverage against infection in the field.

III. CLONING AND EXPRESSION OF *B. NODOSUS* FIMBRIAL SUBUNIT GENES IN *ESCHERICHIA COLI*

As the first step in construction of a genetically engineered vaccine against footrot, the genes encoding the structural subunits of *B. nodosus* fimbriae were cloned into *Escherichia coli*.[50-52] The initial experiments were carried out using *B. nodosus* VCS1001, which is the prototype strain of serogroup A. The corresponding fimbrial gene was isolated by immunological screening of cloned genomic libraries in *E. coli* (Figure 2A), and verified by Western transfer analyses of positive clones[51] and partial sequence data.[52] The fimbrial subunit gene is expressed in *E. coli* from an associated promoter within the cloned sequence.[51,52] The genes encoding the fimbrial subunits representative of all of the other *B. nodosus* serogroups (see below) have since been obtained by analogous procedures.[1,53,54] In these cases, however, more advanced positive selection vector plasmids were employed (see Reference 1), and the cloned gene libraries generally screened with DNA-hybridization probes (Figure 2B) derived from the original cloned serogroup-A fimbrial gene segment.

Despite the fact that *B. nodosus* fimbrial subunit genes are expressed in *E. coli,* analysis of the recombinants showed that these proteins are not assembled into mature fimbriae, but rather remain embedded in the bacterial cell envelope,[51] primarily in the inner membrane.[55] For both immunological and practical reasons this appeared unsuitable for vaccine production,[1] partly because of the likely difficulties and costs of extracting the antigen from this environment on a commercial scale, and partly because isolated denatured fimbrial subunits elicit poor levels of agglutinating antibodies and are not protective.[56] Intact fimbriae, on the other hand, elicit good responses[37-39] and are amenable to simple and industrially applicable procedures for their isolation and purification from the host cell.[44] Overproduction of the fimbrial subunit in *E. coli* by linking the gene to a strong promoter had little impact on the problem. Vaccination trials using sonicated cells or even purified membranes from recombinant *E. coli* containing high levels of the *B. nodosus* fimbrial subunit demonstrated some retardation of the development of foot lesions, but no real reduction in the ultimate severity of the disease.[55]

There are several possible explanations for the lack of formation of mature fimbriae in recombinant *E. coli* cells, the most likely of which is that other genes required for fimbrial biosynthesis are absent from the cloned *B. nodosus* DNA fragments, or that these genes or their products are not properly functional in *E. coli*. Studies on the well-characterized type

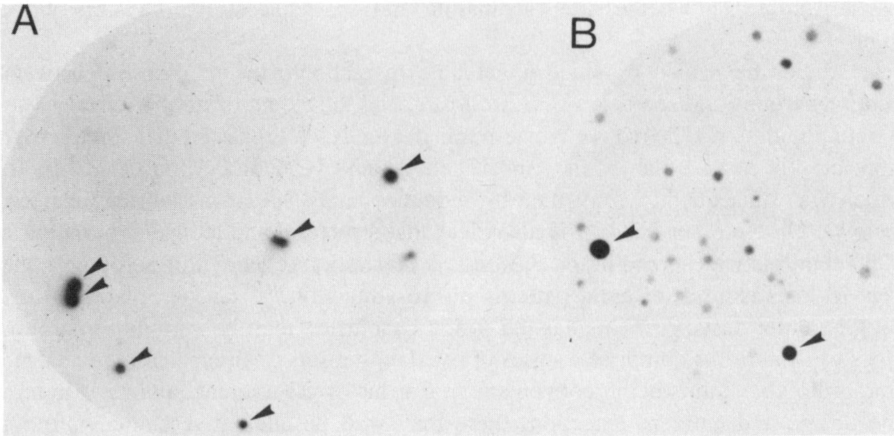

FIGURE 2. Immunological and DNA hybridization screening of cloned *B. nodosus* gene libraries in *E. coli*. Partial *Hind*III digests of *B. nodosus* genomic DNA were ligated into the corresponding site of a positive-selection vector plasmid, and transformed into *E. coli*. The resulting recombinant clones were replica plated onto nitrocellulose filters, lysed *in situ*, and screened (filter A) with serotype-specific anti-fimbrial antiserum[51] or (filter B) with a DNA-hybridization probe derived from the highly conserved region of the fimbrial subunit gene.[54,62] The arrowheads indicate positive colonies.

1 and related fimbrial systems found in enterotoxigenic and uropathogenic strains of *E. coli* itself have shown that fimbrial biosynthesis involves several genes, which are normally clustered.[57-60] Apart from the structural subunit, these encode a basal protein located in the outer membrane and an assembly factor located in the periplasmic space, as well as other proteins related to the subunit, which function as adhesions or influence the length and distribution of the fimbriae on the cell.[57-61] However, these fimbriae are clearly distinct from those occurring in *B. nodosus*, and, not surprisingly, provision of *E. coli* fimbrial-assembly genes did not result in the assembly of *B. nodosus* subunits into fimbrial structures on recombinant cells.[51] The problem was ultimately solved by cloning the *B. nodosus* fimbrial subunit genes into an alterantive host possessing a complementary type 4 fimbrial system, as described in Section V below.

IV. MOLECULAR GENETIC ANALYSIS OF *B. NODOSUS* FIMBRIAL GENES

Studies of *B. nodosus* fimbrial genes cloned into *E. coli* have nonetheless yielded considerable information about the organization of the fimbrial gene region in this bacterium, as well as the structure of the fimbrial subunit itself, and its variation among different serogroups and serotypes. It has recently become evident that there are in fact two distinct classes of fimbrial genes in the *B. nodosus* population[54,62] (Figure 3), although the peripheral flanking sequences are the same and overall genomic DNA-restriction digest patterns are essentially identical. Class I, the larger group, comprises serogroups A, B, C, E, F, G, and I, and class II, serogroups D and H.[54,62]

Upstream of the fimbrial subunit gene in strains of both classes is an open reading frame which appears to encode 5′-enoylpyruvyl shikimate 3′-phosphate synthase,[62] an enzyme involved in the biosynthesis of aromatic amino acids. This gene, termed *aroA*, begins approximately 200 nucleotides upstream of the start codon of the fimbrial subunit gene and reads in the opposite transcriptional orientation. It was identified on the basis of its homology with the known sequences of the equivalent enzymes from *E. coli, Salmonella typhimurium*, and *Aspergillus nidulans,* and confirmed by direct complementation to prototrophy of an *aroA* mutant of *E. coli*.[62] It would, therefore, also appear that this gene is not related to fimbrial biosynthesis.

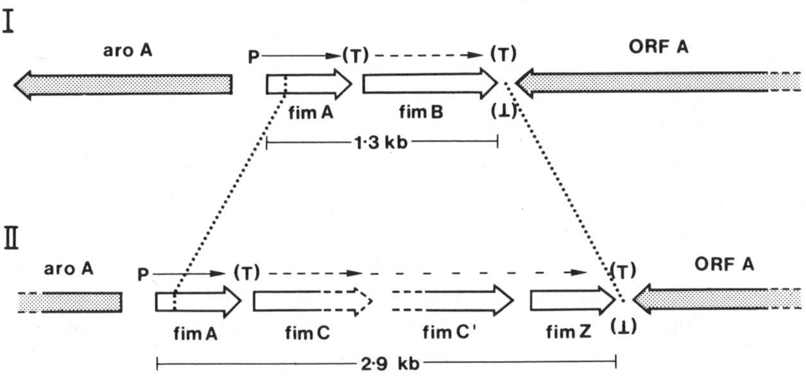

FIGURE 3. Genetic organization of the fimbrial gene region of class I and class II strains of *Bacteroides nodosus*. The boxes indicate open reading frames and their orientation. Shaded boxes indicate flanking genes conserved between class I and class II strains, which do not appear to be involved in fimbrial biosynthesis. For gene designations see text. P indicates the position of the putative *fim*A promoter and T the positions of transcription-termination signals. The direction and probable extent of transcription across the fimbrial gene region is indicated by the arrows.

The fimbrial subunit gene and following sequences are quite different between class I and class II strains, but homology is restored further downstream (see Figure 3). The latter region encodes a large open reading frame of at least 800 codons (ORF A) which shows a close relationship with unidentified genes from *E. coli* (ORF BG, located upstream of the ribosomal *rrnG* operon) and *Rhodopseudomonas blastica* (URF 2, located upstream of the *atp* operon).[62] This gene in *B. nodosus* also reads in the opposite transcriptional orientation to the fimbrial genes from which it is separated, in both class I and class II strains, by a probable bidirectional transcription termination signal.[62] Although its function is unknown, its transcriptional orientation and its presence in diverse bacterial species, including *E. coli* (which does not produce fimbriae equivalent to those of *B. nodosus,*) suggests that this gene, like *aroA*, is unlikely to be involved in fimbrial biosynthesis.

In class I strains, there is only one other gene, aside from the fimbrial subunit gene (*fimA*), located between *aroA* and ORF A (Figure 3). This gene has been termed *fimB* and reads in the same transcriptional orientation as *fimA*, from which it is separated by a short sequence which includes a characteristic transcription-termination signal.[63] In some strains, the distance between the start of *fimB* and this signal is so short as to preclude a separate promoter, and it appears that this gene is probably co-transcribed with the fimbrial subunit, but at an attenuated level determined by the strength of the termination signal in the intergenic region.[63] The *fimB* coding sequence is closely followed by the probable bi-directional transcription-termination signal, which separates *fimB* from ORF A. The sequence of *fimB* predicts a protein of 257 amino acids which is highly conserved between strains and which shows certain features in common with the fimbrial subunit, including a similar short, positively charged sequence preceding a phenylalanine residue at the amino-terminal end of the protein. The *fimB* gene product also contains three typical transmembrane domains and is possibly an inner-membrane protein involved in fimbrial assembly,[63] although its exact location and function remain to be determined.

Class II strains have a more extended region of DNA between the fimbrial subunit gene and ORF A (Figure 3). Immediately downstream of the subunit gene in much the same arrangement as class I is the beginning of an open reading frame (*fimC*)[64] which has a similar amino-terminal sequence to *fimB* (and *fimA*) but no other obvious homology, although its sequence remains to be completed. The available sequence does, however, reveal possible transmembrane domains in similar positions to those of *fimB*, and *fimC* may encode an analogous if not homologous protein. Further downstream is the end of an open reading frame (*fimC′*)[65] which may be part of *fimC*, or an additional gene. This open reading frame is followed closely by a variant fimbrial subunit gene termed *fimZ*.[65] The *fimZ* genes of serogroups D and H are highly homologous, with predicted protein products closely resembling those of their more divergent *fimA* counterparts (see below). The *fimZ*-encoded subunits are, however, unusual in that they contain an amino-terminal leader sequence different from that common to the class I and class II *fimA* subunits, and from those described thus far in other type-4 fimbriate bacteria. *FimZ* is followed by a likely bidirectional transcription-termination signal, at which point homology between class I and class II strains is abruptly restored. The evidence all suggests that *fimZ* represents a duplicate, possibly redundant, fimbrial subunit gene closely related to the progenitor of *fimA* in class II strains, and that the class II-specific region of *B. nodosus* (from the end of the conserved 5′ region of *fimA* to *fimZ*) was orignally derived by recombination from another type-4 species with a different evolutionary history (Figure 3).[65]

The 5′ coding and noncoding regions of the fimbrial subunit genes of class I and class II are virtually identical.[53,54] Probable promoter sequences have been identified in the non-coding region which show homology to equivalent sequences upstream of other type-4 fimbrial subunit genes and to the consensus sequence of *ntrA*-dependent promoters.[66] The latter are recognized by a modified form of RNA polymerase containing the *ntrA* gene product,[67] and are subject to transcriptional activation by transacting factors such as the *xylR* gene product of *Pseudomonas putida tol* plasmids[68] or the *nifA* and *ntrC* gene products involved in nitrogen fixation and nitrogen metabolism in *Klebsiella pneumoniae*.[69,70] There is now direct evidence that the fimbrial subunit genes of *Neisseria gonorrhoeae* and *P. aeruginosa*, which are closely related to those of *B. nodosus*, are also subject to transcriptional activation.[71,72] Moreover, the fimbrial subunit gene of *Vibrio cholerae*, which also appears to be type 4, has recently been shown to be under the transcriptional regulation of the *toxR* gene product, as part of the coordinate regulation of a range of virulence determinants.[32] These observations, taken together, suggest that analogous mechanisms may control the expression of *B. nodosus* fimbrial genes, possibly in concert with those encoding other pathogenic factors, such as extracellular proteases.[73,74]

The nucleotide-coding sequences of the fimbrial subunit genes representative of all *B. nodosus* serogroups and some serotypes have now been determined.[53,54,73-75] Earlier protein-sequence analyses had shown that the fimbrial subunit contains an unusual modified amino acid *N*-methylphenylalanine (MeF), as the first residue in the mature protein, followed by a conserved hydrophobic amino-terminal domain.[76] Comparison with the corresponding gene sequence revealed that the primary translation product contains a short, positively charged leader sequence,[73] which is cleaved from the protein at some point prior to its assembly into the fimbrial strand. These studies also revealed striking similarities between the fimbrial subunits of *B. nodosus* and those expressed by other type-4 fimbriate bacteria, such as *Moraxella bovis*,[77] *M. nonliquefaciens*,[78] *N. gonorrhoeae*,[79,80] *N. meningitidis*,[79,81] and *P. aeruginosa*,[66,82] especially in their amino-terminal sequences (Figure 4). In different species, the fimbrial subunit varies from about 145 to 160 amino acids in length, with both inter- and intraspecific variation occurring primarily in the more hydrophilic carboxy-terminal two thirds of the protein (Figure 5).[29,54,74,77,82-84] This variation includes not only amino acid substitutions, but also insertions and deletions, as well as changes in the location of internal disulfide loops, although some conserved elements are also discernible, notably near the carboxy-terminus of the molecule.[29,53]

The derived protein sequences of the fimbrial subunits prototypal of *B. nodosus* serogroups A to I[54,73] are shown in Figure 6. Such data together with those obtained by direct protein sequencing[84,85] have revealed the molecular basis and extent of variation of this antigen in

FIGURE 4. Comparison of the amino-terminal sequences of the fimbrial subunits of *B. nodosus* class I (VCS1006, serogroup B)[54] and class II (VCS1172, serogroup D),[54] with those of *P. aeruginosa* (PAK),[82] *M. bovis* (EPP63β),[77] and *N. gonorrhoeae* (MS11).[80] Amino acids are indicated by the standard single-letter code. The arrowhead indicates the point of cleavage of the short leader sequence, preceding the phenylalanine residue, which is methylated in the mature protein.

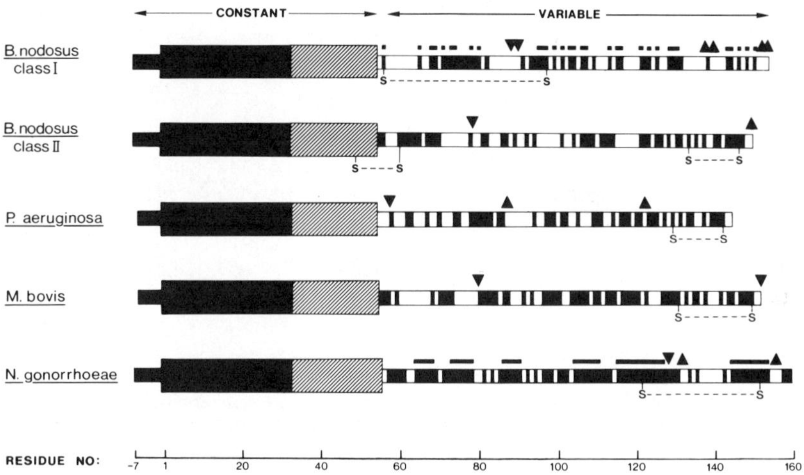

FIGURE 5. Diagrammatic representation and comparison of the primary sequences of the fimbrial subunits of *B. nodosus* (class I and class II), *P. aeruginosa, M. bovis,* and *N. gonorrhoeae.* In the constant region (see Figure 4), the thinner dark block represents the short 6- to 7-amino-acid leader sequence, and the thicker dark block is the highly conserved hydrophobic 32-amino-acid sequence, beginning with the modified phenylalanine residue at position 1 in the mature protein. The shaded area from residue 33 to the glycine at position 54/55 indicates a secondary region which is conserved within species or classes, but which shows divergence between them. Serotypic differences occur in the variable region, wherein little homology is observed between species or classes, except at the carboxy-terminal end.[29,54] The dark and light areas of this region indicate conserved or different amino acids, respectively, between two serological variants from within the species in question, with the sites of small insertions (▼) and deletions (▲) indicated accordingly. Cysteine (disulfide) bridges are marked S---S. Strains compared are *B. nodosus* (class I) A/C,[54] *B. nodosus* (class II) D/H,[54] *P. aeruginosa* PAK/PAO,[82] *M. bovis* β/α,[77] and *N. gonorrhoeae* MS11/variant 3C.[83] All serogroups of *B. nodosus* class I[54] and several variants of *N. gonorrhoeae* have been sequenced, and the bars above the variable region indicate residues conserved in all cases. The two variants selected for direct comparison here were those which were the least alike.

the *B. nodosus* population, a topic which is reviewed in more detail in an accompanying article.[74] Essentially, however, the same dichotomy is observed between class I and class II strains in the structure of the fimbrial subunit as is observed in the organization of the fimbrial gene region. Both classes contain an identical leader sequence and hydrophobic amino-terminal region up to residue 18, wherein the nucleotide-coding sequence is also very highly conserved.[54] The close amino-acid-sequence homology observed among all type-4 fimbrial subunits extends up to residue 32, but thereafter the classes diverge dramatically

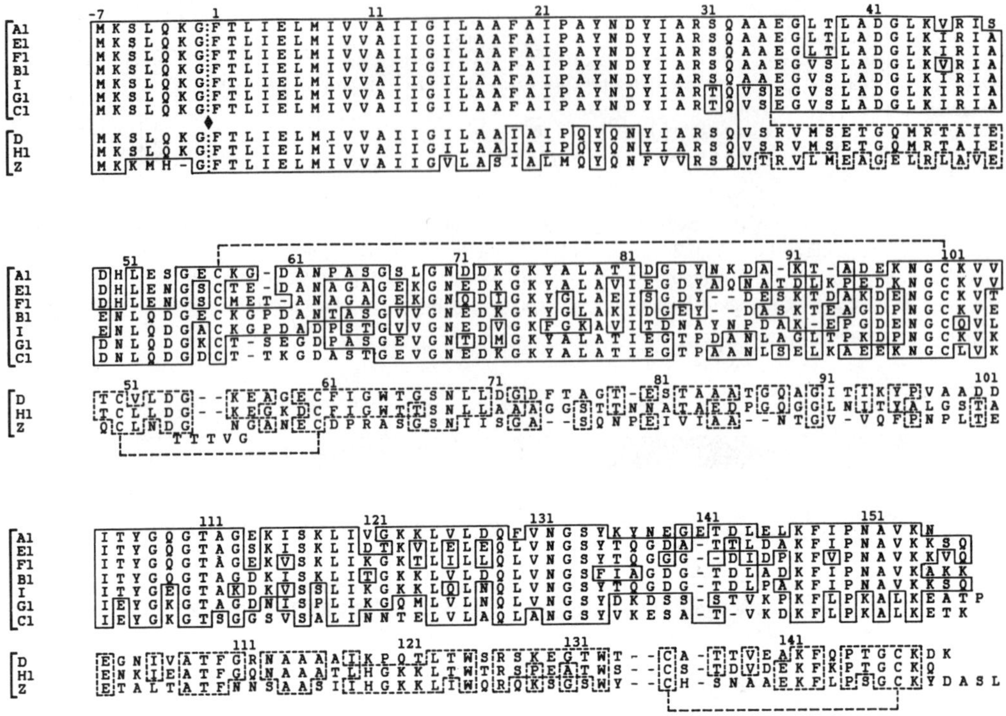

FIGURE 6. The predicted protein sequences, derived from gene sequences, of the fimbrial subunits representative of *B. nodosus* serogroups A to I.[54] Fimbrial subunit gene sequences were determined from the prototype strains of each serogroup as follows: VCS1001 (A), VCS1006 (B), VCS1008 (C), VCS1172 (D), VCS1137 (E), VCS1017 (F), VCS1220 (G), VCS1215 (H), and VCS1636 (I).[54] The sequences have been grouped into class I and class II and ordered according to their similarity. The predicted sequence of *fimZ* (from VCS1172) is also shown for comparison. Boxed residues indicate homologies in the conserved amino-terminal regions of the proteins and between serotypes in class I. Dotted residues indicate homologies within class II. The dashed lines indicate the position of putative disulfide loops within the subunits and the diamond indicates the point of cleavage of the leader sequences.

with only some loose similarities evident up to the conserved glycine at residue 54, and near the carboxy-terminus, both regions in which class II subunits actually appear to be more closely related to those of other type-4 fimbriate species than to those of class I.[53,54]

Class I subunits (serogroups A, B, C, E, F, G, and I) are characterized by a large disulfide loop spanning the central region of the protein. The mature proteins vary from 151 to 155 amino acids in length, and contain three major clusters of sequence variation, one at each end within the disulfide loop, and one near the carboxy-terminus (Figure 6). Each of these clusters requires small insertions or deletions to align flanking conserved motifs. There is also scattered variation evident at other positions in the protein (Figure 6). Based on sequence differences and similarities in conserved and variable regions of the subunits, the class I subunits may be divided into subgroups comprising serogroups ([A, E, F][B,I]) and serogroups (G, C),[54] in general agreement with serological cross-reaction patterns.[48]

Comparison of fimbrial subunit gene sequences representative of different serotypes within serogroups A,[74] C, and E[75] shows more limited variation of between 8 and 15 amino acid changes dispersed throughout the carboxy-terminal two thirds of the protein, and not confined to the hypervariable clusters, suggesting that there is no discrete serotype-specific domain, and that these variants have arisen via genetic drift.[75]

Class II subunits are characterized by two small, putative disulfide loops, one at the end of the conserved amino-terminal region, and the other near the carboxy-terminus (Figure

6).[53,54] The same arrangement is observed in the predicted products of the *fimZ* genes of these strains (Figure 6).[65] The carboxy-terminal disulfide loop is similar to those observed in other characterized type-4 fimbrial subunits, but not those of class II. The mature subunits prototypal of serogroups D and H (*fimA*) are both 149 amino acids in length, although an insertion and compensating deletion is again necessary to align conserved residues. With only two representatives for comparison, the significance of similarities and differences between these subunits is difficult to ascertain. However, the major differences occur in the central portion of the protein, with significant variation also observed in the carboxy-terminal region. The serogroup H subunit is unusual in that it appears to be proteolytically cleaved *in vivo*.[46] This cleavage occurs between the pair of alanine residues at positions 72 and 73,[53] which are not present in the serogroup D subunit.[54,74] Serological cross-reaction between these subunits involves the amino-terminal fragment which is largely conserved between them.[48]

V. MORPHOGENETIC EXPRESSION OF *B. NODOSUS* FIMBRIAE IN *P. AERUGINOSA*

It is now evident that the type-4 fimbriae expressed by different bacteria are structurally, functionally, and evolutionarily related.[29] Studies of the fimbrial subunits of *N. gonorrhoeae*[86] and *P. aeruginosa*[87] had indicated that the hydrophobic amino-terminal domain is involved in subunit-subunit interactions within the fimbrial strand, and this may well be the case. However, the extremely strong conservation of this sequence, including the small leader peptide, across the wide range of species and genera which possess type-4 fimbriae, suggested that this region may also contain important signals for fimbrial morphogenesis,[1,88] in terms of the interaction of the subunit with other factors involved in the assembly pathway. If so, the fimbrial subunits containing these signals might be interchangeable throughout the type-4 group, and thereby provide a means for the production of *B. nodosus*-type fimbriae from suitably engineered cloned genes.[88] *P. aeruginosa* (Figure 7) was selected from among the known type-4 fimbriate bacteria as the recipient host, on the basis that this species is a genetically well-characterized and easily grown aerobe, highly suitable for industrial fermentation, and for which a range of plasmid shuttle vectors are available for gene cloning and transfer.

The prototype studies were carried out using the fimbrial subunit gene from *B. nodosus* serogroup A (VCS1001). This gene however was found to be expressed very poorly in *P. aeruginosa*,[28] a problem which was subsequently overcome by placing it under the transcriptional control of a strong promoter, either P_L or P_R from bacteriophage lambda.[28,88,89] An example of the type of construction employed is shown in Figure 8. In this case, a 576 base pair *Dra*I restriction fragment of cloned *B. nodosus* genomic DNA, containing the complete structural subunit gene-coding sequence (from 30 bases upstream of the initiation codon to 69 bases downstream of the termination codon), as well as the 5′ ribosome-binding sequence and 3′ transcription-termination signal,[54,73] was subcloned into the *Hpa*I site downstream from the P_L promoter in the plasmid pPL-1. The entire gene-promoter construction was then transferred as a *Bam*HI cartridge into the corresponding site of the broad-host range vector plasmid pKT240.[28]

Analysis of *P. aeruginosa* transformants containing this type of construction showed that not only were large amounts of the *B. nodosus* subunit expressed, but also that these were assembled into fimbriae on the surface of the cell.[28,29] Although *B. nodosus* fimbriae are physically very similar to those naturally produced by *P. aeruginosa*, they may be distinguished both immunologically and electrophoretically.[28] The apparent molecular weight of the *B. nodosus* fimbrial subunit in SDS-PAGE is about 17 kDa, whereas that of the host *P. aeruginosa* subunit is about 16kDa; their actual molecular weights calculated from sequence are closer to 16 kDa (Table 1) and 15 kDa,[82] respectively.

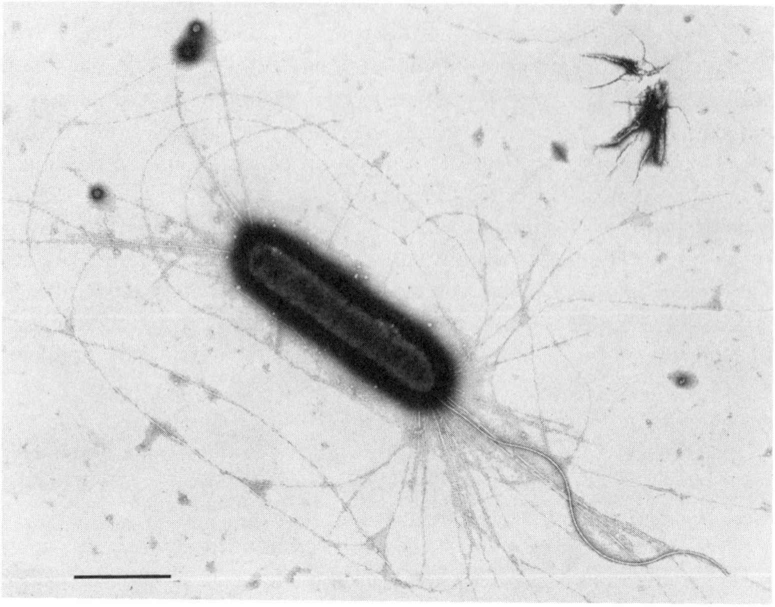

FIGURE 7. *Pseudomonas aeruginosa*. The fine surface filaments are the fimbriae, and the single thick filament the flagellum. The bar represents 1 μm. (Photomicrogaph courtesy of M. Mott.)

High yields of fimbriae were obtained from recombinant *P. aeruginosa* cells using either of two standard purification procedures, $MgCl_2$ precipitation or isoelectric precipitation at pH 4.5 (Figure 9).[28,88] These fimbriae were found to be composed of a structural subunit with the same electrophoretic mobility as that of *B. nodosus* (Figure 9, 10A), which was confirmed by Western transfer analysis with anti-*B. nodosus* fimbrial antiserum (Figure 10C). Quite unexpectedly, however, there was no trace at all in these fimbriae of the subunit characteristic of the host strain itself (Figure 10A), even in relatively sensitive Western transfer analyses using anti*P. aeruginosa* fimbrial antiserum (Figure 10B).[28]

The production of *B. nodosus*-type fimbriae in recombinant *P. aeruginosa* cells was independently verified by electron microscopy with immunogold labeling, as shown in Figure 11. In these experiments, antibody binding is marked by the presence of (electron-dense) spherical gold particles, as well as by the clumping of the fimbrial strands. As expected, the fimbriae normally produced by the host *P. aeruginosa* strain reacted strongly with homologous antiserum (Figure 11A), but not with antiserum, against *B. nodosus* fimbriae (Figure 11B). This pattern is completely reversed in the recombinant *P. aeruginosa* cells, whose fimbriae were not recognized by antibodies directed against *P. aeruginosa* fimbriae (Figure 11C), but rather reacted with those against *B. nodosus* fimbriae (Figure 11D).[28] This is the same profile exhibited by *B. nodosus* itself.

These and other studies, including agglutination tests, have demonstrated that recombinant *P. aeruginosa* cells containing the *B. nodosus* fimbrial subunit gene under the control of a strong promoter produce mature fimbrial structures (Figure 12) which are physically, structurally, and antigenically indistinguishable from those produced by the *B. nodosus* strain from which the fimbrial subunit gene was originally derived.[28] The complete substitution of the *B. nodosus* fimbrial subunit for that of the host in the recombinant cells appears to be the result of the very high level of expression of the cloned gene. This also implies that the *P. aeruginosa* host cell has little, if any preference for the assembly of the homologous over the heterologous subunit into fimbrial strands.

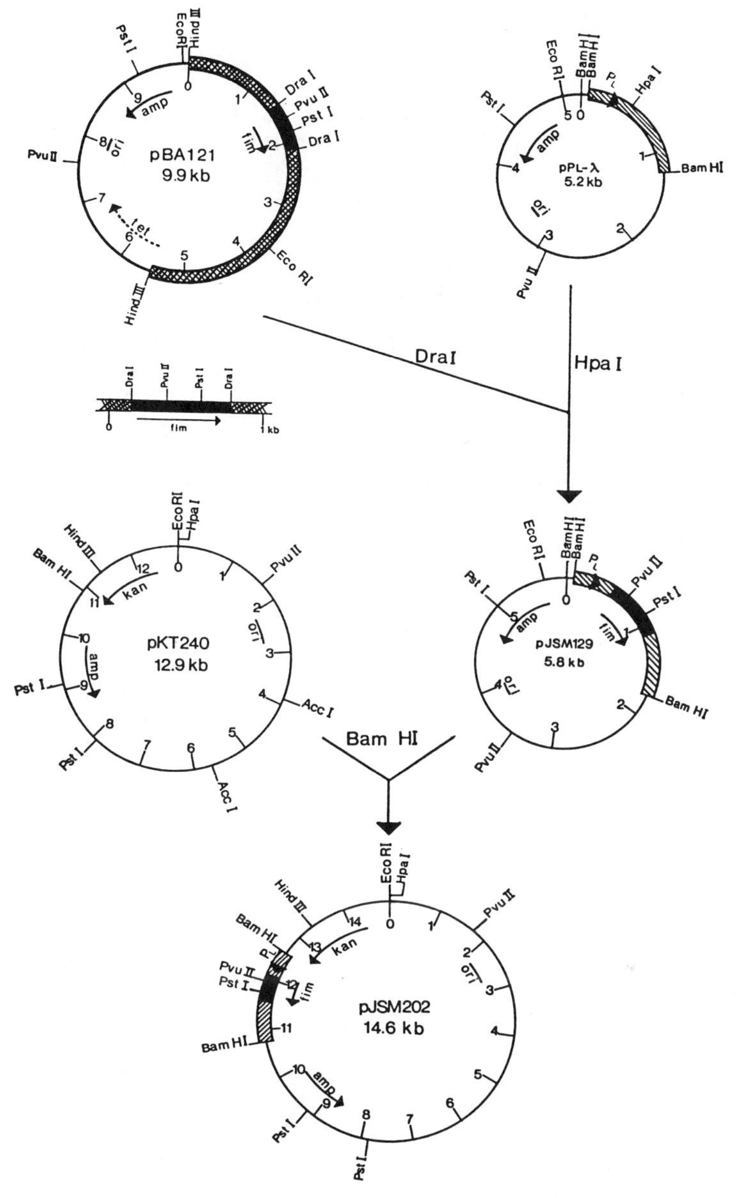

FIGURE 8. Engineering of the *B. nodosus* fimbrial subunit gene for high level expression in a broad host-range vector.[28] The cross-hatched sequences in pBA121 indicate those originally cloned from *B. nodosus* VCS1001,[51] with the internal dark segment delineating the *Dra*I fragment which spans the fimbrial subunit gene. The hatched sequences in pPL-1, pJSM129, and pJSM202 indicate the *Bam*HI segment (originally derived from bacteriophage I) which contains the P_L promoter and into which the *Dra*I fragment was cloned (at the *Hpa*I site). The orientation of the fimbrial subunit gene with respect to the P_L promoter in the *Bam*H1 segment was determined by reference to the internal *Pst*I and *Pvu*II sites and verified by overproduction of the protein under appropriate conditions. The thin line in pJSM202 represents sequences derived from the broad host range vector pKT240 (see Reference 28). Various restriction endonuclease sites are indicated. The numbers refer to the distance in kilobases (kb) around each plasmid. The arrows indicate the direction of transcription and approximate boundaries of the structural genes in question. Abbreviations: fim, fimbrial subunit gene; amp, ampicillin (or carbenicillin) resistance gene; tet, tetracycline-resistance gene (nonfunctional in pBA121); kan, kanamycin-resistance gene; ori, origin of replication. (From Anderson, B. J. et al., *J. Bacteriol.*, 166, 453, 1986. With permission.)

Table 1
COMPARATIVE AMINO ACID CHAIN LENGTHS AND CALCULATED MOLECULAR WEIGHTS OF THE FIMBRIAL SUBUNITS OF *B. NODOSUS* SEROGROUPS A TO I

Serotype strain	Preprotein (number aa)	Mature subunit[a] (mol wt, Da[b])
A1 (VCS1001)	158	16,104
B1 (VCS1006)	160	15,907
C1 (VCS1008)	160	16,054
D (VCS1172)	156	15,785
E1 (VCS1137)	161	16,280
F1 (VCS1017)	158	15,895
G1 (VCS1220)	162	16,243
H1 (VCS1215)	156	15,864
I (VCS1636)	161	16,127

Note: All protein sequences were derived from the DNA sequence of the corresponding genes[54] (see Figure 6).

[a] Cleavage occurs after the seventh reside, leaving a (modified) phenylalanine as the first residue in the mature subunit.[73,76]

[b] Including *N*-methylation of the terminal phenylalanine residue.

These results have broad biological and biotechnological significance. The morphogenetic expression of *B. nodosus* fimbriae in *P. aeruginosa* provides strong support for the proposition that type-4 fimbrial subunits from different bacterial species are assembled by a common mechanism and may be interchanged among compatible hosts.[28] This is certainly the case for *B. nodosus* and *P. aeruginosa*, and should be equally true for other type 4 species retaining the conserved amino-terminal region of the fimbrial subunit. In this respect, *B. nodosus* and *P. aeruginosa* appear to be no more or less closely related than other characterized members of the group, and indeed in the remainder of the protein, the class I subunits of *B. nodosus* appear to have less in common than the rest, although there are similarities in overall structure.[29,54] The absence of strict morphogenetic barriers to the assembly of heterologous subunits suggests that genetic exchange is one possible route by which variation may be generated in natural populations,[28] and the likely origin of the class II fimbrial genes in *B. nodosus* itself.[62,65]

VI. VACCINATION TRIALS

Vaccination trials have been carried out to examine the immunoprotective potential of the *B. nodosus* fimbrial antigens produced by recombinant *P. aeruginosa* cells.[90,91] Represent-

1 2 3 4 5 6

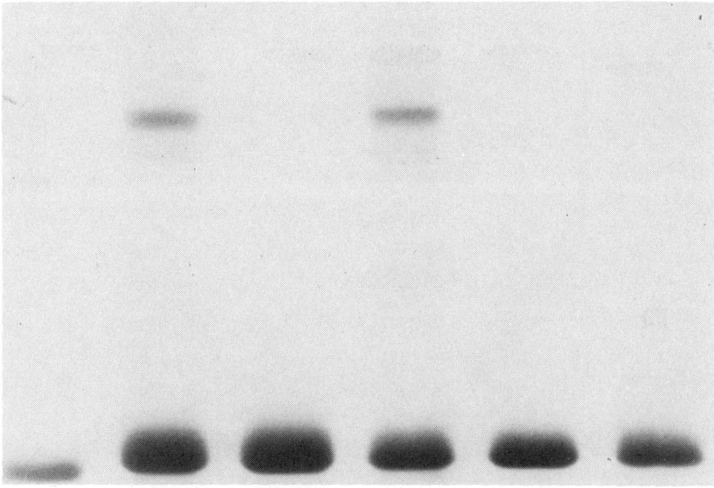

FIGURE 9. Isolation of fimbriae from cells containing pJSM202.[28] Recombinant cells were cultured on nutrient agar plates containing carbenicillin (1 mg/ml) and harvested in phosphate-buffered saline (PBS), as described previously.[44] The cells were removed by centrifugation, and the supernatant fraction was reserved. The cells were resuspended in PBS, subjected to mechanical blending, and again pelleted by centrifugation to yield a blended-supernatant fraction. Fimbriae were recovered from these fractions by isoelectric precipitation with sodium acetate at pH 4.5 or by $MgCl_2$ precipitation.[44] Samples were then displayed on SDS-urea gradient polyacrylamide gels, and precipitation, respectively. Lanes 4 and 5 contain fimbriae from the blended-supernatant fraction by isoelectric or $MgCl_2$ precipitation, respectively. Lanes 2 to 5 all contain an amount of material equivalent to 3.5% of the isolated from a single standard (85 mm) petri dish culture of recombinant *P. aeruginosa* cells. Lane 1 contains purified *P. aeruginosa* PAK/2Pfs fimbriae (about 5 μm) and lane 6 contains purified *B. nodosus* VCS1001 fimbriae (about 10 μg). (From Mattick, J. S., *J. Bacteriol.*, 169, 33, 1987. With permission).

ative data are shown in Table 2. In this case, sheep were vaccinated with either 250 or 500 μg of fimbriae purified from recombinant cells containing pJSM202, and compared with positive control groups vaccinated with whole cells or purified fimbriae from the source *B. nodosus* strain VCS1001 (serogroup A), as well as with negative control groups which were either left untreated or were vaccinated with normal *P. aeruginosa* fimbriae. Two doses were given 1 month apart, a protocol which approximates normal field usage. No local or systemic toxic reactions were observed in vaccines containing recombinant fimbriae, even at the higher dose.

The results (Table 2) show that groups vaccinated with recombinant fimbriae developed average serum-agglutinating titres against *B. nodosus* comparable to those vaccinated with *B. nodosus* whole cells or fimbriae. These groups also displayed the *same* level of protective immunity against homologous footrot challenge, although this was not absolute. Complete immunity is rarely observed in such trials, especially when a highly virulent strain such as VCS1001 is involved. The response of individual animals to vaccination and challenge is also quite variable.[90] Nervetheless, there was a large difference between the vaccinated groups and negative controls. All animals in the latter category became affected with severe footrot, 65% of their feet suffering underrunning lesions, with an average lesion score of 2.4 (Table 2). Such lesions were relatively rare in the groups vaccinated with *B. nodosus*

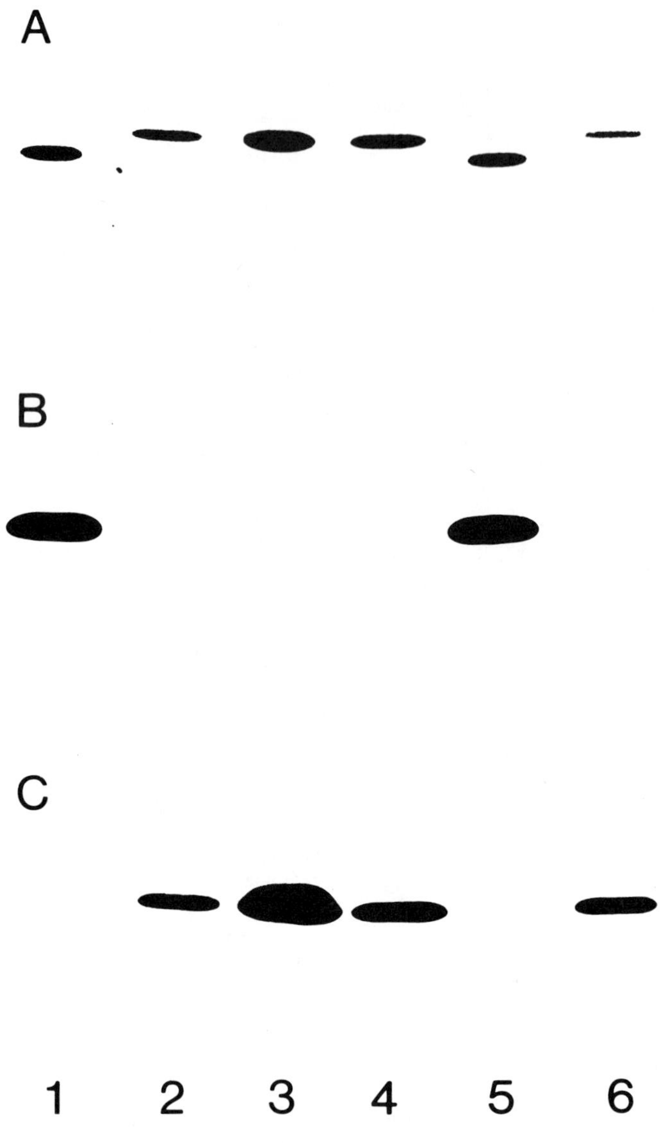

FIGURE 10. Electrophoretic and immunological analysis of fimbriae produced by *P. aeruginosa* PAK/2Pfs (pJSM202).[28] Fimbriae were purified (by two rounds of isoelectric precipitation) from two independent primary clones of *P. aeruginosa* PAK/2Pfs transformed with pJSM202 (lanes 3 and 4) and compared with fimbriae obtained from the host *P. aeruginosa* PAK/2Pfs (lanes 1 and 5) and from *B. nodosus* VCS1001 (lanes 2 and 6). (A) These fimbriae (approximately 2 μg) were displayed on SDS-urea gradient polyacrylamide gels and stained for protein with Coomassie blue R250. (B) Duplicate (unstained) displays were subjected to Western transfer analysis with either anti-*P. aeruginosa* PAK/2Pfs fimbrial antiserum or (C) anti-*B. nodosus* VCS1001 fimbrial antiserum. In each case only the lower portion of the gel (containing the fimbrial subunit) is shown. (From Mattick, J. S., *J. Bacteriol.*, 169, 33, 1987. With permission.)

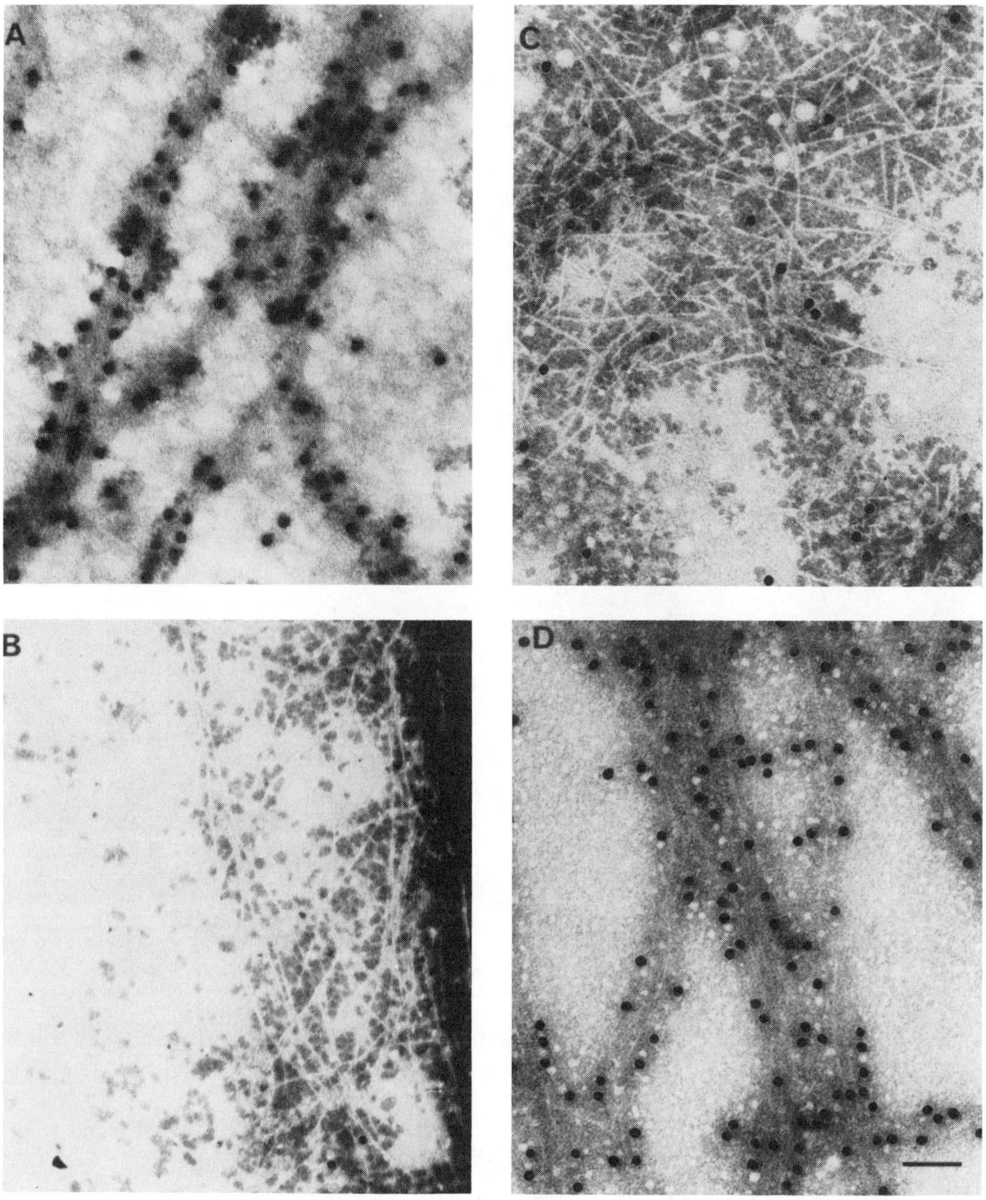

FIGURE 11. Immunogold labeling of host and recombinant fimbriae.[28] (A,B) Whole cells of the *P. aeruginosa* PAK/2Pfs or (CD) the same strain containing pJSM202 were placed onto parlodion-carbon-coated gold grids and treated with (A and C) either anti-*P. aeruginosa* PAK/2Pfs fimbrial antiserum or (B and D) anti-*B. nodosus* fimbrial antiserum and then with protein A labeled with 15 nm colloidal gold particles. The grids were then negatively stained and examined in an electron microscope at 60 kV. Each micrograph is shown at the same magnification (bar = 0.1 μm). (From Mattick, J. S., *J. Bacteriol.*, 169, 133, 1987. With permission.)

cells or fimbriae, or the fimbriae produced by the recombinant *P. aeruginosa* cells, severe clinical signs being observed in only 15 to 20% of all feet. The average lesion score among these groups ranged from 0.7 to 1.1; this effectively means the difference between sheep which are debilitated by the disease, and those which are mildly affected, albeit with some individual exceptions. This practical difference is important in the field. These results also

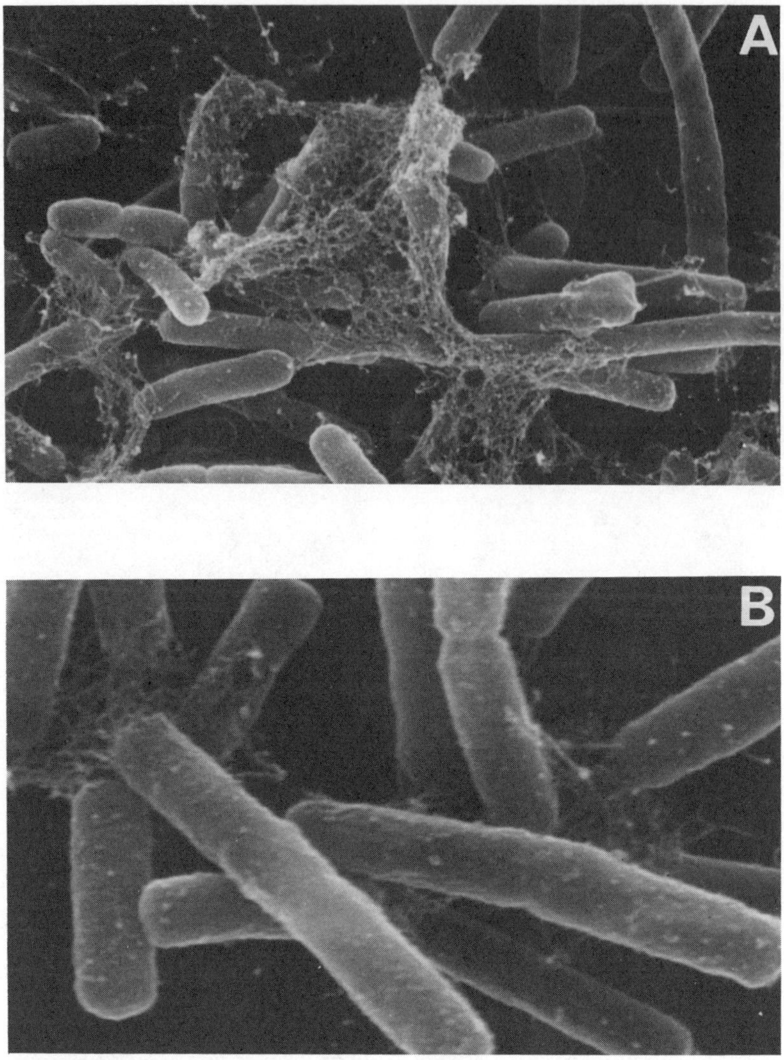

FIGURE 12. Scanning electron micrographs of recombinant *P. aeruginosa* cells containing pJSM202 and expressing *B. nodosus*-type fimbriae. The cells were fixed with glutaraldehyde subjected to chemical dehydration followed by critical point drying with liquid CO_2, then gold coated and examined in a JEDL JSM840 scanning electron microscope. The bar at the base of each micrograph = 1 μm. (A) Broad field, (B) higher magnification. Note the fimbriae emerging from the poles of the cells (Photomicrographs by kind courtesy of Sue Doyle, Macquarie University.)

provide unequivocal evidence that the fimbrial subunit of *B. nodosus* is the primary protective antigen against footrot.

Extensions of these studies have shown that the recombinant fimbriae are also effective in accelerating the rate of recovery from *B. nodosus* infection.[90,91] Immunotherapy is an important adjunct to immunoprophylaxis in the management of chronic infections, and is particularly relevant for footrot, as well as other superficial infections affecting epithelial tissues, such as bovine keratoconjunctivitis (caused by *Moroxella bovis*), where antibiotic therapy may be ineffective or impractical under field conditions.

Table 2
EFFECTIVENESS OF A RECOMBINANT-DNA-BASED FIMBRIAL VACCINE AGAINST FOOTROT

Group	1	2	3	4	5	6
Vaccine	Nil	*B. nodosus* cells (10^{10})	*B. nodosus* fimbriae (250 μg)	*P. aeruginosa* fimbriae (250 μg)	Recombinant fimbriae (250 μg)	Recombinant fimbriae (500 μg)
Underrunning infections	25/40	7/44	6/40	29/44	9/44	6/40
GMFS	2.2	0.7	1.1	2.5	1.0	0.75
GMAT	200	7000	3400	200	8000	6800

Note: Groups contained either 10 or 11 animals, and were vaccinated 28 d apart with two doses of the indicated antigen preparations, followed by immediate challenge.[90] The clinical and serological responses shown were those measured 3 weeks after challenge. Footrot-lesion scores were assessed on a standard scale of 0 to 4,[6] as follows: 0, no symptoms or mild interdigital dermatitis (see Reference 90); 2, obvious interdigital dermatitis; 3, underrunning lesions of the soft horn of the heel of the hoof; and 4, underrunning lesions of the hard horn of the hoof. GMFS is the group mean footscore, and was calculated as the sum of the lesion scores of all animals in the group. GMAT is the group mean agglutinin titre, measured as the reciprocal of the maximum dilution at which a visible K-agglutination reaction was observed.[9] The data are expressed as the geometric mean value for the group, rounded off to the nearest 100th.

VII. OUTLOOK

The vaccination trials described above have shown that fimbriae produced by recombinant *P. aeruginosa* cells are equally as effective as either whole cells or isolated fimbriae of *B. nodosus* itself in inducing protective immunity against footrot. This, then, represents the prototype for the development of a multivalent vaccine for use in the field. Fimbriae representative of other serogroups or serotypes may be produced simply by substituting the relevant subunit genes into the expression system described herein. These representatives may be selected solely on their immunological profile, rather than the growth-yield properties of the original host strain. The further development of the vaccine will involve assessment of various host-vector-promoter expression systems to obtaine optimum yields and stable production of the fimbriae under commercial fermentation conditions, followed by the normal requirements of veterinary-product registration.

This system should be equally applicable to the production of similar vaccines against other type-4 fimbriate pathogens, such as *N. gonorrhoeae* and *M. bovis*.[28] Available evidence suggests that fimbriae are also immunoprotective against these infections.[92-94] and it seems that the main limitation to the development of recombinant-DNA-based vaccines against such pathogens may be the availability of serological classification data upon which to base a multivalent formulation. Recent studies have identified six or seven serogroups in the *M. bovis* population,[95,96] suggesting a situation analogous to that in *B. nodosus*, although some additional groups may be uncovered by more extensive surveys.

Footrot itself is an unusual disease in that the infection is relatively remote from host immune surveillance, and does not generally provoke immunoprotective responses, although such responses can be aroused by appropriate vaccination. In the absence of intervention, the condition is chronic or semichronic, depending on environmental factors,[3] although natural healing or recovery has been observed.[97] Intervention by vaccination may occur either in advance of or during infection, and in this context it is unlikely that there are any real differences between prophylactic and therapeutic responses. In any event, it is one-way traffic, and it seems that *B. nodosus* infection does not stimulate an amnestic response, even though already primed by vaccination. Hence, the duration of protection is more limited than usual. Since the means for efficient and economic production of the protective fimbrial antigens are now available via recombinant-DNA technology, the contemporary issues in footrot-vaccine development would now seem to be to identify the immunological mechanisms involved in dealing with *B. nodosus* infection, and to design strategies for vaccine formulation or delivery which will increase the intensity and duration of the relevant responses. There is also the possibility that other antigens, not normally prominent in whole-cell preparations, may be capable of eliciting broad cross-protective responses if presented in sufficient quantity or purity. Potential candidates are the proteases secreted by *B. nodosus*,[98] which appear to play an important role in its pathogenicity,[1] and which may, theoretically at least, be also produced from genetically engineered host cells.[99]

ACKNOWLEDGMENTS

I would like to thank my colleagues John Egerton and Brian Dalrymple for their helpful suggestions and critical reading of this article. I would also like to thank the other members of my laboratory for their contribution to the work and the results reviewed herein. We have been most grateful for the support of the Australian Meat and Livestock Research and Development Corporation and the Australian Wool Corporation throughout this project.

REFERENCES

1. **Mattick, J. S., Anderson, B. J., and Egerton, J. R.,** Molecular biology and footrot of sheep, in *Reviews in Rural Science,* Vol. 6, Leng, R. A., Barker, J. S. F., Adams, D. B., and Hutchinson, K. J., Eds., University of New England, Armidale, N.S.W., 1985, 79.
2. **Mattick, J. S., Anderson, B. J., Bills, M. M., Cox, P. T., Dalrymple, B., Egerton, J. R., Elleman, T. C., Finney, K., Hoyne, P. A., Lehrbach, P., McKern, N. M., McNicholas, P., Schwartzkoff, C. L., Stewart, D. J., and Wright, D.,** Anaerobe to aerobe: the development of a recombinant DNA-based vaccine against ovine footrot, in Proc. 7th Aust. Biotechnol. Conf., Melbourne, 1986, 91.
3. **Egerton, J. R.,** Footrot of sheep — pathogenesis and immunity, *Prog. Immunol.,* 3, 645, 1977.
4. **Egerton, J. R., Parsonson, I. M., and Graham, N. P. H.,** Parenteral chemotherapy of ovine foot-rot, *Aust. Vet. J.,* 44, 275, 1968.
5. **Egerton, J. R. and Burrell, D. J.,** Prophylactic and therapeutic vaccination against ovine foot-rot, *Aust. Vet. J.,* 46, 517, 1976.
6. **Egerton, J. R. and Roberts, D. S.,** Vaccination against ovine foot-rot, *J. Comp. Pathol.,* 81, 179, 1971.
7. **Egerton, J. R.,** Significance of *Fusiformis nodosus* serotypes in resistance of vaccinated sheep to experimental foot-rot, *Aust. Vet. J.,* 50, 59, 1974.
8. **Reed, G. A.,** The role of footrot vaccines in Australia, in *Footrot in Ruminants,* Proc. Workshop, Melbourne, 1985, Stewart, D. J., Peterson, J. E., McKern, N. M., and Emery, D. L., Eds., CSIRO Division Animal Health/Australian Wool Corp., Glebe, N.S.W., 1986, 173.
9. **Egerton, J. R.,** Surface and somatic antigens of *Fusiformis nodosus, J. Comp. Pathol.,* 83, 151, 1973.
10. **Claxton, P. D., Ribeiro, L. A., and Egerton, J. R.,** Classification of *Bacteroides nodosus* by agglutination tests, *Aust. Vet. J.,* 60, 331, 1983.
11. **Claxton, P. D.,** Serogrouping of *Bacteroides nodosus* isolates, in *Footrot in Ruminants,* Proc. Workshop, Melbourne, 1985, Stewart, D. J., Peterson, J. E., McKern, N. M., and Emery, D. L., Eds., CSIRO Division Animal Health/Australian Wool Corp., Glebe, N.S.W., 1986, 131.
12. **Hindmarsh, F. and Fraser, J.,** Serogroups of *Bacteroides nodosus* isolated from ovine footrot in Britain, *Vet. Rec.,* 116, 187, 1985.
13. **Kingsley, D. F., Hindmarsh, F. H., Liardet, D. M., and Chetwin, D. H.,** Distribution of serogroups of *Bacteroides nodosus* with particular reference to New Zealand and the United Kingdom, in *Footrot in Ruminants,* Proc. Workshop, Melbourne, 1985, Stewart, D. J., Peterson, J. E., McKern, N. M., and Emery, D. L., Eds., CSIRO Division Animal Health/Australian Wool Corp., Glebe, N.S.W., 1986, 143.

14. **Plant, J. W. and Claxton, P. D.,** Efficacy of paring, footbathing and vaccination in the treatment of footrot, in *Footrot in Ruminants,* Proc. Workshop, Melbourne, 1985, Stewart, D. J., Peterson, J. E., McKern, N. M., and Emery, D. L., Eds., CSIRO Division Animal Health/Australian Wool Corp., Glebe, N.S.W., 1986, 57.
15. **Liardet, D. M., Chetwin, D. H., Kingsley, D. F., and Hindmarsh, F. H.,** Results of field trials in New Zealand to confirm the protective and curative effects of a 10-strain ovine footrot vaccine, in *Footrot in Ruminants,* Proc. Workshop, Melbourne, 1985, Stewart, D. J., Peterson, J. E., McKern, N. M., and Emery, D. L., Eds., CSIRO Division Animal Health/Australian Wool Corp., Glebe, N.S.W., 1986, 181.
16. **Mulvaney, C. J., Jackson, R., and Jopp, A. J.,** Field trials with a killed nine-strain, oil adjuvanted *Bacteroides nodosus* footrot vaccine in sheep, *N.Z. Vet. J.,* 32, 137, 1984.
17. **Malecki, J. C. and McCausland, I. P.,** *In vitro* penetration and adsorption of chemicals into the ovine hoof, *Res. Vet. Sci.,* 33, 192, 1982.
18. **Mulvaney, C. J., Jackson, R., and Jopp, A. J.,** A revised concept of ovine footrot control, in *Footrot in Ruminants,* Proc. Workshop Melbourne, 1985, Stewart, D. J., Peterson, J. E., McKern, N. M., and Emery, D. L., Eds., CSIRO Division Animal Health/Australian Wool Corp., Glebe, N.S.W., 1986, 63.
19. **Skerman, T. M.,** Determination of some *in vitro* growth requirements of *Bacteroides nodosus, J. Gen. Microbiol.,* 87, 107, 1975.
20. **Skerman, T. M., Erasmuson, S. K., and Every, D.,** Differentiation of *Bacteroides nodosus* biotypes and colony variants in relation to their virulence and immunoprotective properties in sheep, *Infect. Immun.,* 32, 788, 1981.
21. **Stewart, D. J.,** An electron microscope study of *Fusiformis nodosus, Res. Vet. Sci.,* 14, 132, 1973.
22. **Walker, P. D., Short, J., Thompson, R. O., and Roberts, D. S.,** The fine structure of *Fusiformis nodosus* with special reference to the antigens associated with immunogenicity, *J. Gen. Microbiol.,* 77, 351, 1973.
23. **Ottow, J. C. G.,** Ecology, physiology and genetics of fimbriae and pili, *Annu. Rev. Microbiol.,* 29, 79, 1975.
24. **Short, J. A., Thorley, C. M., and Walker, P. D.,** An electron microscope study of *Bacteroides nodosus:* ultrastructure of organisms from primary isolates and different colony types, *J. Appl. Bacteriol.,* 40, 311, 1976.
25. **Every, D. and Skerman, T. M.,** Surface structure of *Bacteroides nodosus* in relation to virulence and immunoprotection in sheep, *J. Gen. Microbiol.,* 129, 225, 1983.
26. **Henrichsen, J.,** Twitching motility, *Annu. Rev. Microbiol.,* 37, 81, 1983.
27. **Depiazzi, L. J. and Richards, R. B.,** Motility in relation to virulence of *Bacteroides nodosus, Vet. Microbiol.,* 10, 107, 1985.
28. **Mattick, J. S., Bills, M. M., Anderson, B. J., Dalrymple, B., Mott, M. R., and Egerton, J. R.,** Morphogenetic expression of *Bacteroides nodosus* fimbriae in *Pseudomonas aeruginosa, J. Bacteriol.,* 169, 33, 1987.
29. **Dalrymple, B. and Mattick, J. S.,** The structure and evolution of type 4 (MePhe) fimbrial subunit proteins, *J. Mol. Evol.,* 25, 261, 1987.
30. **Fuerst, J. A. and Hayward, A. C.,** Surface appendages similar to fimbriae (pili) on *Pseudomonas* species, *J. Gen. Microbiol.,* 58, 227, 1969.
31. **Henrichsen, J.,** The occurrence of twitching motility among gram-negative bacteria, *Acta Pathol. Microbiol. Scand. B,* 83, 171, 1975.
32. **Taylor, R. L., Miller, V. L., Furlong, D. B., and Mekalanos, J. J.,** Use of *phoA* gene fusions to identify a pilus colonization factor coordinately regulated with cholera toxin *Proc. Natl. Acad. Sci. U.S.A.,* 84, 2833, 1987.
33. **Dalrymple, B. and Mattick, J. S.,** Type 4 fimbriae among the pseudomonads, Manuscript in preparation.
34. **Stewart, D. J.,** Studies on the antigenic structure of *Bacteroides nodosus, Res. Vet. Sci.,* 24, 293, 1978.
35. **Every, D.,** Purification of pili from *Bacteroides nodosus* and an examination of their chemical, physical and serological properties, *J. Gen. Microbiol.,* 115, 309, 1979.
36. **Lee, S. W., Alexander, B., and McGowan, B.,** Purification, characterization, and serologic characteristics of *Bacteroides nodosus* pili and use of a purified pili vaccine in sheep, *Am. J. Vet. Res.,* 44, 1676, 1983.
37. **Stewart, D. J.,** The role of various antigenic fractions of *Bacteroides nodosus* in eliciting protection against foot-rot in vaccinated sheep, *Res. Vet. Sci.,* 24, 14, 1978.
38. **Every, D. and Skerman, T. M.,** Protection of sheep against experimental foot-rot by vaccination with pili purified from *Bacteroides nodosus, N.Z. Vet. J.,* 30, 156, 1982.
39. **Stewart, D. J., Clark, B. L., Emery, D. L., Peterson, J. E., and Fahey, K. J.,** Importance of pilus-associated antigen in *Bacteroides nodosus* vaccines, *Res. Vet. Sci.,* 32, 140, 1982.
40. **Claxton, P. D.,** Studies on *Bacteroides nodosus* Vaccines with Particular Reference to Factors Associated with their Efficacy in Protecting Against Ovine Virulent Footrot, Ph.D. thesis, University of Sydney, Sydney, 1981.

41. **Stewart, D. J., Clark, B. L., Peterson, J. E., Emery, D. L., Smith, E. F., Griffiths, D. A., and O'Donnell, I. J.**, The protection given by pilus and whole cell vaccines of *Bacteroides nodosus* strain 198 against foot-rot induced by strains of different serogroups, *Aust. Vet. J.*, 62, 153, 1985.
42. **Thorley, C. M. and Egerton, J. R.**, Comparison of alum-absorbed or nonalum-absorbed oil emulsion vaccines containing either pilate or non-pilate *Bacteroides nodosus* cells in inducing and maintaining resistance of sheep to experimental foot-rot, *Res. Vet. Sci.*, 30, 32, 1981.
43. **Stewart, D. J., Clark, B. L., Emery, D. L., Peterson, J. E., and Fahey, K. J.**, A *Bacteroides nodosus* immunogen, distinct from the pilus, which induces cross-protective immunity in sheep vaccinated against footrot, *Aust. Vet. J.*, 60, 83, 1983.
44. **Mattick, J. S., Anderson, B. J., Mott, M. R., and Egerton, J. R.**, Isolation and characterization of *Bacteroides nodosus* fimbriae: structural subunit and basal protein antigens, *J. Bacteriol.*, 160, 740, 1984.
45. **Handley, B. G., Ng, M. L., and Schwartzkoff, C. L.**, Electrophoretic mobility of *Bacteroides nodosus* pilus antigens, *Aust. Vet. J.*, 63, 62, 1986.
46. **Anderson, B. J., Kristo, C. L., Egerton, J. R., and Mattick, J. S.**, Variation in the structural subunit and basal protein antigens of *Bacteroides nodosus* fimbriae, *J. Bacteriol.*, 166, 453, 1986.
47. **O'Donnell, I. J., Stewart, D. J., and Clark, B. L.**, Serological identification of pilus antigen and other protein antigens of *Bacteroides nodosus* using elctro-blot radioimmunoassay after electrophoretic fractionation of the proteins on sodium dodecyl sulphate polyacrylamide gels, *Aust. J. Biol. Sci.*, 36, 15, 1983.
48. **Anderson, B. J., Mattick, J. S., Cox, P. T., Kristo, C. L., and Egerton, J. R.**, Matrix Western transfer analysis of the fimbrial antigens of *Bacteroides nodosus, J. Bacteriol.*, In press
49. **Day, S. E. J., Thorley, C. M., and Beesley, J. E.**, Serotyping of *Bacteroides nodosus:* proposal for 9 further serotypes (J-R) and a study of the antigenic complexity of *B. nodosus* pili, in *Footrot in Ruminants*, Proc. Workshop, Melbourne, 1985, Stewart, D. J. Peterson, J. E., McKern, N. M., and Emery, D. L., Eds., CSIRO Division Animal Health/Australian Wool Corp., Glebe, N.S.W., 1986, 147.
50. **Mattick, J. S. and Anderson, B. J.**, Australian Patent Application 34979/84, 1984; New Zealand Patent Application 210067, 1984.
51. **Anderson, B. J., Bills, M. M., Egerton, J. R., and Mattick, J. S.**, Cloning and expression in *Escherichia coli* of the gene encoding the structural subunit of *Bacteroides nodosus* fimbriae, *J. Bacteriol.*, 160, 748, 1984.
52. **Elleman, T. C., Hoyne, P. A., Emery, D. L., Stewart, D. J., and Clark, B. L.**, Isolation of the gene encoding pilin of *Bacteroides nodosus* (strain 198), the causal organism of ovine footrot, *FEBS Lett.*, 173, 103, 1984.
53. **Elleman, T. C., Hoyne, P. A., McKern, N. M., and Stewart, D. J.**, Nucleotide sequence of the gene encoding the two-subunit pilin of *Bacteroides nodosus* 265, *J. Bacteriol.*, 167, 243, 1986.
54. **Mattick, J. S., Anderson, B. J., Cox, P. T., Bills, M. M., Dalrymple, B., and Egerton, J. R.**, Gene sequences and comparison of the fimbrial subunits representative of *Bacteroides nodosus* serogroups A to I: class I and class II strains, *J. Bacteriol.*, Submitted for publication.
55. **Elleman, T. C., Hoyne, P. A., Emery, D. L., Stewart, D. J., and Clark, B. L.**, Expression of the pilin gene from *Bacteroides nodosus* in *Escherichia coli, Infect. Immun.*, 51, 187, 1986.
56. **Emery, D. L., Stewart, D. J., and Clark, B. L.**, The structural integrity of pili from *Bacteroides nodosus* is required to elicit protective immunity against foot-rot in sheep, *Aust. Vet. J.*, 61, 237, 1984.
57. **Orndorff, P. E. and Falkow, S.**, Organization and expression of genes responsible for type 1 piliation in *Escherichia coli, J. Bacteriol.*, 159, 736, 1984.
58. **Klemm, P.**, Fimbrial ahesions of *Escherichia coli, Rev. Infect. Dis.*, 7, 321, 1985.
59. **Mooi, F. R. and de Graaf, F. K.**, Molecular biology of fimbriae of enterotoxigenic *Escherichia coli, Curr. Top. Microbiol. Immunol.*, 118, 119, 1985.
60. **Uhlin, B. E., Baga, M., Goransson, M., Lindberg, F. P., Lund, B., Norgren, M., and Normark, S.**, Genes determining adhesin formation in uropathogenic *Escherichia coli, Curr. Top. Microbiol. Immunol.*, 118, 163, 1985.
61. **Baga, M., Norgren, M., and Normark, S.**, Biogenesis of *E. coli* Pap pili: PapH, a minor pilin subunit involved in cell anchoring and length modulation, *Cell*, 49, 241, 1987.
62. **Dalrymple, B., Livingston, S., Cox, P. T., and Mattick, J. S.**, Organization of the fimbrial gene region of *Bacteroides nodosus:* class I and class II strains, *J. Bacteriol.*, Submitted for publication.
63. **Dalrymple, B. and Mattick, J. S.**, The identification of a putative accessory gene for fimbrial morphogenesis in class I strains of *Bacteroides nodosus*, Manuscript in preparation.
64. **Mattick, J. S. and Dalrymple, B.**, Unpublished DNA sequence data from *Bacteroides nodosus* serogroups D and H, 1987.
65. **Hobbs, M., Dalrymple, B., Delaney, S. F., and Mattick, J. S.**, The identification of a variant fimbrial subunit gene in class II strains of *Bacteroides nodosus, Mol. Gen. Genet.*, Submitted for publication.
66. **Johnson, K., Parker, M. L., and Lory, S.**, Nucleotide sequence and transcriptional initiation site of two *Pseudomonas aeruginosa* pilin genes, *J. Biol. Chem.*, 261, 15703, 1986.

67. **Hirschmann, J., Wong, P.-K., Sei, K., Keener, J., and Kustu, S.,** Products of nitrogen regulatory genes *ntrA* and *ntrC* activate *glnA* transcription *in vitro* evidence that the *ntrA* product is a sigma factor, *Proc. Natl. Acad. Sci. U.S.A.*, 82, 7525, 1985.
68. **Inouye, S., Nakazawa, A., and Nakazawa, T.,** Determination of the transcription initiation site and identification of the protein product of the regulatory gene *xylR* for *xyl* operons on the TOL plasmid, *J. Bacteriol.*, 163, 863, 1985.
69. **Ausubel, F. M.,** Regulation of nitrogen fixation genes, *Cell*, 37, 5, 1984.
70. **Drummond, M., Whitty, P., and Wooten, J.,** Sequence and domain relationships of *ntrC* and *nifA* from *Klebsiella pneumoniae:* homologies to other regulatory proteins, *EMBO J.*, 13, 547, 1986.
71. **So, M., Seifert, H. S., Deal, C. D., Stromberg, N., Karlsson, K. A., Nyberg, G., and Normark, S.,** Genetics of pilus expression and function, *J. Cell. Biochem.*, Suppl. 11B, 110, 1987.
72. **Hobbs, M., Dalrymple, B., Delaney, S. F., and Mattick, J. S.,** Transcription of the fimbrial subunit gene and an associated transfer RNA gene of *Pseudomonas aeruginosa, Gene,* Accepted for publication.
73. **Elleman, T. C. and Hoyne, P. A.,** Nucleotide sequence of the gene encoding pilin of *Bacteroides nodosus,* the causal organism of ovine footrot, *J. Bacteriol.*, 160, 1184, 1984.
74. **McKern, N. M., Elleman, T. C., and Hoyne, P. A.,** Molecular basis of antigenic variation in the pilus of *Bacterioides nodosus,* — application to footrot vaccines, this volume.
75. **Cox, P. T., Mattick, J. S., Dalrymple, B., and Egerton, J. R.,** Sequence and comparison of the fimbrial subunit genes representative of *Bacteroides nodosus* serotypes C2 and E2: comparison with sister serotypes and other serogroups, Manuscript in preparation.
76. **McKern, N. M., O'Donnell, I. J., Inglis, A. S., Stewart, D. J., and Clark, B. L.,** Amino-acid sequence of pilin from *Bacteroides nodosus* (strain 198), the causative organism of ovine foot-rot, *FEBS Lett.*, 164, 149, 1983.
77. **Marrs, C. F., Schoolnik, G., Koomey, J. M., Hardy, J., Rothbard, J., and Falkow, S.,** Cloning and sequencing of a *Moraxella bovis* pilin gene, *J. Bacteriol.*, 163, 132, 1985.
78. **Froholm, L. O. and Sletten, K.,** Purification and N-terminal sequence of a fimbrial protein from *Moraxella nonliquefaciens, FEBS Lett.*, 73, 29, 1977.
79. **Hermodsen, M. A., Chen, K. C. S., and Buchanan, T. M.,** *Neisseria* pili proteins: amino-terminal amino-acid sequences and identification of an unusual amino-acid, *Biochemistry,* 17, 442, 1978.
80. **Meyer, T. F., Billyard, E., Haas, R., Storzbach, S., and So, M.,** Pilus genes of *Neisseria gonorrhoeae:* chromosomal organization and DNA sequence, *Proc. Natl. Acad. Sci. U.S.A.*, 81, 6110, 1984.
81. **Olafson, R. F., McCarthy, P. J., Bhatti, A. R., Dooley, J. S. G., Heckels, J. E., and Trust, T. J.,** Structural and antigenic analysis of meningococcal piliation, *Infect. Immun.*, 48, 336, 1985.
82. **Sastry, P. A., Finlay, B. B., Paloske, B. L., Paranchych, W., Pearlstone, J. R., and Smillie, L. B.,** Comparative studies of the amino acid and nucleotide sequences of pilin derived from *Pseudomonas aeruginosa* PAK and PAO, *J. Bacteriol.*, 164, 571, 1985.
83. **Hagblom, P., Segal, E., Billyard, E., and So, M.,** Intragenic recombination leads to pilus antigenic variation in *Neisseria gonorrhoeae, Nature,* 314, 156, 1985.
84. **McKern, N. M., O'Donnell, I. J., Stewart, D. J., and Clark, B. L.,** Primary structure of pilin protein from *Bacteroides nodosus* strain 216: comparison with the corresponding protein from strain 198, *J. Gen. Microbiol.*, 131, 1, 1985.
85. **McKern, N. M., Stewart, D. J., and Strike, P. M.,** Unpublished protein sequence data from representatives of *Bacteroides nodosus* serogroups B and C, 1987.
86. **Schoolnik, G. K., Fernandez, R., Tai, J. Y., Rothbard, J., and Gotschlich, E. C.,** Gonococcal pili: primary structure and receptor binding domain, *J. Exp. Med.*, 159, 1351, 1984.
87. **Watts, T. H. and Kay, C. M.,** Dissociation and characterization of pilin isolated from *Pseudomonas aeruginosa* strains PAK and PAO, *Can. J. Biochem.*, 60, 867, 1982.
88. **Mattick, J. S., Anderson, B. J., and Elleman, T. C.,** Australian Patent Application 50154/85; New Zealand Patent Application of 214017; European Patent Application 85905494.2; Patent Corporation Treaty Application W086/02557, 1985.
89. **Elleman, T. C., Hoyne, P. A., Stewart, D. J., McKern, N. M., and Peterson, J. E.,** Expression of pili from *Bacteriodes nodosus in Pseudomonas aeruginosa, J. Bacteriol.*, 168, 574, 1986.
90. **Egerton, J. R., Cox, P. T., Anderson, B. J., Kristo, C., Norman, M., and Mattick, J. S.,** Protection of sheep against footrot with a recombinant DNA-based fimbrial vaccine, *Vet. Microbiol.*, 14, 393, 1987.
91. **Stewart, D. J. and Elleman, T. C.,** A *Bacteroides nodosus* pili vaccine produced by recombinant DNA for the prevention and treatment of foot-rot in sheep, *Aust. Vet. J.*, 64, 79, 1987.
92. **Pugh, G. W., Hughes, D. E., and Booth, G. D.,** Experimentally induced infectious bovine keratoconjunctivitis: effectiveness of a pilus vaccine against exposure to homologous strains of *Moraxella bovis, Am. J. Vet. Res.*, 38, 1519, 1977.

93. **Brinton, C. C., Bryan, J., Dillon, J.-A., Guerina, N., Jacobsen, L. T., Labik, A., Lee, S., Levine, A., Lim, S., McMichael, J., Polen, S., Rogers, K., To, A. C.-C., and To, S. C.-M.,** Uses of pili in gonorrhea control: role of bacterial pili in disease, purification and properties of gonococcal pili, and progress in the development of a gonococcal pilus vaccine for gonorrhea, in *Immunology of Neisseria gonorrhoeae,* Brooks, G. F., Jr., Gotschlich, E. C., Holmes, K. K., Sawyer, W. D., and Young, F. E., Eds., American Society for Microbiology, Washington, D.C., 1978, 155.
94. **Lehr, C., Jayappa, H. G., and Goodnow, R. A.,** Serologic and protective characterization of *Moraxella bovis* pili, *Cornell Vet.,* 75, 484, 1985.
95. **Lepper, A. W. D. and Hermans, L. R.,** Characterization and quantitation of pilus antigens of *Moraxella bovis* by ELISA, *Aust. Vet. J.,* 63, 401, 1986.
96. **Moore, L. J. and Rutter, J. M.,** Antigenic analysis of fimbrial proteins from *Moraxella bovis, J. Clin. Microbiol.,* In press.
97. **Egerton, J. R., Ribeiro, L. A., Kieran, P. J., and Thorley, C. M.,** Onset and remission of ovine footrot, *Aust. Vet. J.,* 60, 334, 1983.
98. **Stewart, D. J., Clark, B. L., Emery, D. L., Peterson, J. E., and Kortt, A. A.,** The phenomenon of cross protection against footrot induced by vaccination of sheep with *Bacteroides nodosus* vaccines, in *Footrot in Ruminants,* Proc. Workshop, Melbourne, 1985, Stewart, D. J., Peterson, J. E., McKern, N. M., and Emery, D. L., Eds., CSIRO Division Animal Health/Australian Wool Corp., Glebe, N.S.W., 1986, 185.
99. **Moses, E. K., Yong, W. K., Mattick, J. S., and Riffkin, G. G.,** Expression of *Bacteroides nosodus* genes encoding extracellular proteins in *Escherichia coli,* in *Footrot in Ruminants,* Proc. Workshop, Melbourne, 1985, Stewart, D. J., Peterson, J. E., McKern, N. M., and Emery, D. L., Eds., CSIRO Division Animal Health/Australian Wool Corp., Glebe, N.S.W., 1986, 245.

Chapter 13

MOLECULAR BASIS OF ANTIGENIC VARIATION IN PILI (FIMBRIAE) OF *BACTEROIDES NODOSUS* — APPLICATION TO FOOTROT VACCINES

Neil M. McKern, Thomas C. Elleman, and Peter A. Hoyne

TABLE OF CONTENTS

I. INTRODUCTION

Common pili are filamentous appendages frequently occurring on Gram-negative bacteria.[1] Pili of *Bacteroides nodosus* (Figure 1) have a polar location and are responsible for a primitive type of locomotion termed "twitching motility".[2,3] Several hundred pili are present on each cell, each pilus being approximately 5 nm in diameter and of variable length up to 25 μm.[4] In common with other pili, those of *B. nodosus* are composed of numerous identical protein subunits. Pili are excellent immunogens and those from *B. nodosus* are as effective as whole *B. nodosus* cells in providing protection against footrot in sheep challenged by the same strain of the organism.[5] In contrast, depiliated cells provide little protection against challenge.[6] Based on immunological cross-reactivity of pilus antigens, the numerous strains of *B. nodosus* can be divided into a number of serogroups,[7] across which little protection is afforded following vaccination against footrot infection.[8] Initial commercial vaccines failed to provide adequate protection against footrot under Australian field conditions, due to poor piliation of cells and the omission of key serogroups.[9]

A better understanding of the serological relationships among strains of *B. nodosus* resulted in improved commercial vaccines, but some deficiencies in these have prevented their widespread use in the Australian sheep industry. These deficiencies included the high cost of production due to poor growth characteristics of *B. nodosus*,[10] the occurrence of vaccinal lesions at the site of inoculation,[11] and the low proportion of effective antigens in whole-cell preparations. Application of new technologies based on pilus antigens, may permit the production of an effective footrot vaccine which overcomes the deficiencies of contemporary formulations. To this end, genetic and structural studies aimed at understanding the pilus of *B. nodosus* at a molecular level have been undertaken to produce an inexpensive and more efficacious pilus-based vaccine against footrot.

II. ANTIGENIC RELATIONSHIP OF *B. NODOSUS* STRAINS

Strains of *B. nodosus* may be divided into a number of separate groups based on agglutination reactions with antisera against prototype strains of the organism.[7] Pili play a central role in these *in vitro* agglutination reactions, and mediate protection through induction of agglutinating antibody *in vivo*. Consequently, the classification system based on agglutination correlates well with the pattern of cross protection observed among strains.[12] On the basis of agglutination reactions, *B. nodosus* strains have been classified into at least eight noncross-reacting serogroups,[7] which have been further divided into 16 subgroups (Table 1), the latter confirmed by the incomplete removal of agglutinating activity following absorption of antisera by antigens of the heterologous subgroup.

Vaccination studies using killed cells have shown that little protection against ovine footrot is afforded across different serogroups. However, the limited data available suggest that cross protection often occurs among subgroups within a serogroup.[12] Hence, vaccination with a single strain from a serogroup may afford protection against challenge by other strains from that serogroup. Similar protection studies using pili show that the same relationships of cross protection occur as were observed with killed cells.[13,14]

III. PILUS CHARACTERIZATION

Bacterial pili can be divided into sex pili, which are involved in transfer of genetic information,[1] and nonsex pili, such as those which mediate attachment to and/or locomotion over host cell surfaces.[15] These latter pili are generally more numerous on the cell and may have a peritrichous or polar arrangement. Those that are of a more polar location are frequently involved in twitching motility.[16,17] Bacterial pili of all types are essentially com-

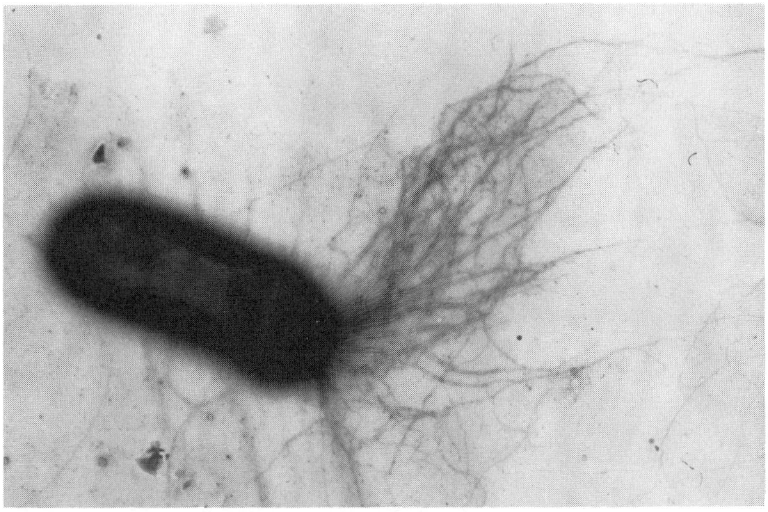

FIGURE 1. Electron micrograph of *Bacteroides nodosus*. Size bar represents 1 μm. (Courtesy of J. Peterson.)

posed of multiple copies of a subunit protein termed pilin, whose sizes have been shown to vary from 7 to 43 kDa.[18,19]

Protein-sequence studies of pilins generally support this subdivision of pili. In particular, there is a striking conservation of amino-terminal sequence from pilins[20] of those bacterial species showing twitching motility (Figure 2).[21] The first 30 amino acid residues of pilins from *Bacteroides nodosus*,[22] *Moraxella bovis*,[23] *M. nonliquefaciens*,[20] *Neisseria gonorrhoeae*,[24] *N. meningitidis*,[25] and *Pseudomonas aeruginosa*[26] are virtually identical. Additionally, the following 25 residues are highly conserved.

The unusual amino acid *N*-methylphenylalanine is generally the amino-terminal residue of this group of pilins, although a proportion of the pilin molecules may lack this residue.[24] These pili are collectively termed ''*N*-methylphenylalanine pili'' (abbreviated ''mePhe pili'').[23,27] Furthermore, pilins of this group show a relatively small range in molecular size (16 to 18 kDa) and exhibit a degree of sequence similarity in the carboxy-terminal regions.[21] The well-studied pilins of the Enterobacteriaceae show no sequence homology to the mePhe group but may be divided into several groups, some of which are interrelated (Figure 3).[28-35] These pilins have a molecular size within the range 7.0 to 29.5 kDa.[19,28] Overall, these data indicate that mePhe pili form a class quite distinct from other bacterial pili.

The N-terminal sequence conservation of mePhe pilins from separate bacterial families is quite remarkable, considering the diversity of tissues colonized (Table 2) and the variability of sequences of bacterial pilins seen within the family *Enterobacteriaceae* (Figure 3). This sequence conservation may reflect the need to conserve a topogenic signal necessary for translocation of pilin to its ultimate destination within the pilus.[36] An additional function of this region may be the nucleation of protein folding necessary to produce the defined subunit structure required for pilus formation. In this respect, it is of interest to note that secondary structure analysis strongly predicts an α-helical domain for these sequences. Furthermore, the highly hydrophobic nature of this region may promote intersubunit interactions for the maintenance of structural integrity in the mature pilus.[20] Titration and hydrodynamic studies in detergent solutions of *P. aeruginosa* pili have suggested that the primary interactions between subunits are hydrophobic in nature and involve the N-terminal regions of the molecule.[37]

Table 1
AGGLUTINATION TITERS AMONG STRAINS OF *B. NODOSUS* AND THEIR ANTISERA PREPARED IN RABBITS[a]

Antigen serogroup[b]	Agglutination titer[c]															
	A(1)	A(2)	B(1)	B(2)	B(3)	B(4)	C	D	E(1)	E(2)	F(1)	F(2)	G(1)	G(2)	H(1)	H(2)
A(1)	10240	2560														
(2)	2560	10240														
B(1)			10240	10240	1280	1280										
(2)			5120	10240	2560	320										
(3)			5120	10240	20840	5120										
(4)			5120	1280	5120	20480										
C							10240									
D								10240								
E(1)									10240	2580						
(2)									1280	20480						
F(1)											20480	320				
(2)											320	5120				
G(1)													5120	1280		
(2)													640	10240		
H(1)															5120	640
(2)															320	5120

[a] Data taken from Claxton et al.[7]
[b] Figures in parenthesis designate subgroups.
[c] Titers less than 80 not recorded.

N-terminal Sequences

	1 10 20 30 40 50
B.nodosus (strain 198)	mF T L I E L M I V V A I I G I L A A F A I P A Y N D Y I A R S Q A A E G L T L A D G L K V R I S D H - L E S G E
B.nodosus (strain 265)	mF T L I E L M I V V A I I G I L A A I A I P Q Y Q N Y I A R S Q V S R V M S E T G Q M R T A I E T C - L L D G K
M.bovis (EPP63 β)	mF T L I E L M I V I A I I G I L A A I A L P A Y Q D Y I S K S Q T T R V V G E L A A G K T A V D A A - L F E G K
N.gonorrhoeae (MS11)	mF T L I E L M I V I A I V G I L A A V A L P A Y Q D Y T A R A Q V S E A I L L A E G Q K S A V T E Y Y L N H G K
P.aeruginosa (PAK)	mF T L I E L M I V V A I I G I L A A I A I P Q Y Q N Y V A R S E G A S A L A S V N P L K T T V E E A - L S R G W

FIGURE 2. Amino-terminal amino acid sequences of bacterial mePhe-type pilins. Amino acid residues are denoted by the standard single-letter code, except for the uncommon amino acid *N*-methylphenylalanine, which is denoted mF. Amino acid residues showing a majority occupancy are shaded.

E.coli IC	V T T V N G G T V H F K G E V V N A A C A V N T N S F D Q T V N L
E.coli fimA	A A T T V N G G T V H F K G E V V N A A C A V D A G S V D Q T V Q L
E.coli $F7_2$	A P T I P Q G Q G K V T F N G T V V D A P C G I D A Q S A D Q S I D F
E.coli PapA	A P T I P Q G Q G K V T F N G T V V D A P C S I S Q K S A D Q S I D F
Klebsiella pneumoniae 1	N T T T V N G G T V A F K G E V V D A A S A V D A
Salmonella enteritidis	A G F V G N K A V V Q A A V T I A A Q N T T S A N W S
Salmonella typhimurium	A D P T P T S V S G G T I H F E G K L V N A A X A V S T
E.coli K99	N T G T I N F N G K I T S A T C T I E P E V N G N R T S T
E.coli K88	W M T G D F N G S V D I G G S I T A D D Y R Q K W
E.coli F	A G S S G Q D L M A S G N T T V K A T F G K D S S V V K W V V L A
E.coli CFA1	V E K N I T V T A S V D P V I D L L Q A D G N A L P S A V K L A Y

FIGURE 3. Amino-terminal amino acid sequences of pilins of Enterobacteriaceae aligned to show homology.

Table 2
ORGANISMS BEARING PILI COMPOSED OF mePhe-TYPE PILINS, SHOWING THE VARIETY OF HOSTS AND AFFECTED TISSUES

Organism	Important host	Tissue affected	Disease
B. nodosus	Goats/sheep	Epithelium of hoof	Footrot
M. bovis	Cattle	Cornea	Infectious keratoconjunctivitis (Pinkeye)
M. nonliquefaciens	Man	Upper-respiratory tract	Bronchitis
N. gonorrhoeae	Man	Urogenital tract	Gonorrhea
N. meningitidis	Man	Meninges/cerebrospinal fluid	Cerebrospinal meningitis
P. aeruginosa	Wide range	Epithelium and traumatized tissue	Opportunistic pathogen

In contrast to the highly conserved nature of the N-terminal regions of mePhe pilins, there is little similarity between sequences of the central regions of the molecules (Figure 4). These central regions may constitute the different receptor-binding sites of pili from various organisms, since studies on gonococcal pili have implicated part of this region in the attachment of pili to host epithelial cells.[38] No conserved features of predicted secondary structure are evident between these pilins in this central region.

As with the N-terminal sequences, homology is evident between the C-terminal regions of mePhe pilins, although the extent of sequence similarity is considerably less than that

Central Region Sequences

B.nodosus (strain 198)	$_{56}$C K G D A N P A S G S L G N D D K G K Y A L A T I D G D Y N K D A K T A D E K N G C K$_{98}$
B.nodosus (strain 265)	$_{56}$E G K D C F I G W T T S N L L A A A G G S T T N N A T A A D P G Q G G L N I T Y A L E S T A E N K$_{104}$
M.bovis (EPP63 β)	$_{56}$T P K L G K A A N D T E E D I G L T T T G G T A R S N L M S S V N I G G G A F A T G A G T$_{100}$
N.gonorrhoeae (MS11)	$_{57}$W P E N N T S A G V A S P P S D I K G K Y V K E V E V K N G V$_{87}$
P.aeruginosa (PAK)	$_{56}$W S V K S G T G T E D A T K K E V P L G V A A D A N K L G T I A L K P D P A D G T A D$_{98}$

FIGURE 4. Amino acid sequences of the central region from mePhe pilins.

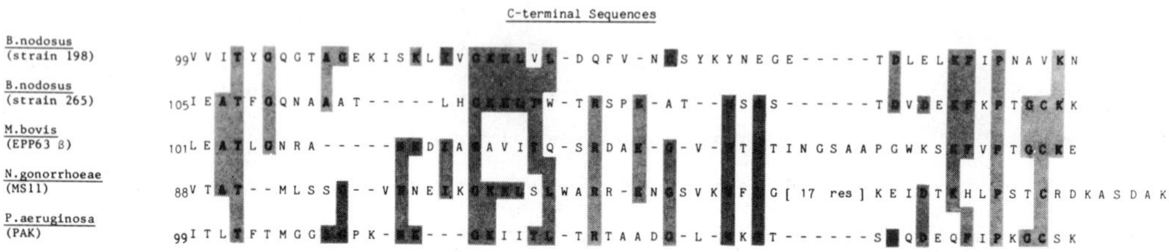

FIGURE 5. Comparison of amino acid sequences from the carboxy-terminal region of pilins of the mePhe type. Amino acid residues showing a majority occupancy are shaded. A 17-residue sequence in pilin of *N. gonorrhoeae* is abbreviated as [17 res].

observed between the N-termini (Figure 5). A feature of the predicted secondary structure in the C-terminal region is the consistent occurrence of β-sheet, which may indicate a conserved functional domain. Studies on *P. aeruginosa* pili have shown approximately 40% β-sheet content, in accordance with the predicted structural features of the C-terminal region. The predicted interspersion of β-turns between β-sheets suggests an antiparallel arrangement of the β-sheets. The conservation of both the N-terminal and C-terminal domains suggests a common architecture for all mePhe pilins, thereby permitting a similar tertiary and quaternary structure to be attained, which is reflected in their common morphology.

IV. COMPARISON OF PILIN SEQUENCES FROM ANTIGENIC VARIANTS OF *B. NODOSUS*

Sequences of pilins from representative strains from each of the eight recognized serogroups of *B. nodosus* have been determined directly by protein sequencing or indirectly by gene sequencing.[21,22,36,39-43] Comparison of these sequences shows extensive conservation in the N-terminal region (Figure 6). The first 22 residues show absolute identity and the following 32 residues are well conserved. Comparison of the remainder of the pilin sequences indicates that *B. nodosus* pilins may be grouped into two distinct sets,* comprising pilins from serogroups A, B, C, E, F, and G (''A-set'' or Class I pilins), and those from serogroups D and H (''D-set'' or Class II pilins). Within each set, aligned sequences show about 60% or more identity, whereas less than 40% identity is found in comparisons of sequences from different sets (Table 3). Representatives from the separate sets bear little sequence similarity in the central region of the molecules, while comparison of the C-terminal regions shows a degree of sequence conservation which is no better than that found between mePhe pilins in general.

* Referred to as class I and class II in Chapter 12.

```
Serogroup   Strain   1                 10                  20                  30                  40

A.          198      mF T L I E L M I V V A I I G I L A A F A I P A Y N D Y I A R S Q A A E G L T L A D G L K V
B.          234      mF T L I E L M I V V A I I G I L A A F A I P A Y N D Y I A R S Q A A E G V S L A D G L K V
C.          217      mF T L I E L M I V V A I I G I L A A F A I P A Y N D Y I A R T Q V S E G V S L A D G L K I
E.          216      mF T L I E L M I V V A I I G I L A A F A I P A Y N D Y I A R S Q A A E G L T L A D G L K I
F.          1017     mF T L I E L M I V V A I I G I L A A F A I P A Y N D Y I A R S Q A A E G L T L A D G L K I
G.          1220     mF T L I E L M I V V A I I G I L A A F A I P A Y N D Y I A R T Q V S E G V S L A D G L K I

D.          340      mF T L I E L M I V V A I I G I L A A I A I P Q Y Q N Y I A R S Q V S R V M S E T G Q M R T
H.          265      mF T L I E L M I V V A I I G I L A A I A I P Q Y Q N Y I A R S Q V S R V M S E T G Q M R T


            50                  60                  70                  80                  90                  100

A.  R I S D H L E S G E C K G - D A N P A S G S L G N D D K G K Y A L A T I D G D Y N K D A K T A D - - E K N G C K
B.  R I A E N L Q N G E C K G P D A N T A S G V V G N E D K G K Y G L A K I D G E Y D A S K T E A G - - D P N G C K
C.  R I A D N L Q D G D C T T - K G D A S T G E V G N E D K G K Y A L A T I E G T P A A N L S E L K A E E K N G C L
E.  R I A D H L E N G S C T E - D A N A G A G E K G N E D K G K Y A L A V I E G D Y A Q N A T D L K - - P E N G C K
F.  R I A D H L E N G S C M E - T A N A G A G E K G N Q D I G K Y G L A E I S G D Y D E S K T D A K - - D E N G C K
G.  R I A D N L Q D G K C T S - E G D P A S G E V G N T D M G K Y A L A T I E G T P D A N L A G L T P K D P N G C K

D.  A I E T C V L D G K E A D K C F I G W T G S N L L D G E F T A G T E - S T A A A T G Q T G I T I - - K Y P V A A
H.  A I E T C L L D G K E G K D C F I G W T T S N L L A A A G G S T T N N A T A A D P G Q G G L N I - - T Y A L E S


                  110                 120                 130                 140                 150

A.  V V I T Y G Q G T A G E K I S K L I V G K K L V L D Q F V N G S Y K Y N E G E T D L E L K F I P N A V K N
B.  V E I T Y G Q G T A G D K I S K L I T G K K L V L D Q L V N G S F - I A G D G T D L A D K F I P N A V K A K K
C.  V K I E Y G K G T S G G S V S A L I N N T E L V L A Q L A N G S Y - - V K E S A T V K D K F L P K A L K E T K
E.  V V I T Y G Q G T A G D K I S K L I D T K V L E L E Q L V N G S Y - - T Q G D A T L D A K F I P N A V K K S Q
F.  V T I T Y G Q G T A G E K V S K L I K G K T L I L L Q L V N G S Y - - T Q G G G D I D P K F V P N A V K K V Q
G.  V K I E Y G K G T A G D N I S P L I K G Q M L V L N Q L V N G S Y - D K D S S S T V K P K F L P K A L K E A T P

D.  D D E G N I V A T F G R N A A A A I K P Q T L T W S R S K E G T W - - - T C A T T V E A K F Q P T G C K D G K
H.  T A E N K I E A T F G Q N A A A T L H G K K L T W T R S P E A T W - - - S C S T D V D E K F K P T G C K K
```

FIGURE 6. Comparison of amino acid sequences of pilins from *B. nodosus*. A, B, C, E, F, and G are boxed. Alignment of pilins was maximized by the insertion of gaps. In protein sequences derived from gene sequence studies, it is assumed that the N-terminal residue would be *N*-methylphenylalanine in mature pilins.

Hydrophilicity profiles of pilin sequences reinforce the clear division of serogroups into two sets. Profiles of serogroups D and H (D-set) pilins are very similar to each other, but bear little relationship to those of the A-set pilins, apart from the conserved amino-terminal regions (Figure 7). Likewise, the A-set pilins display similar profiles to each other.

Within pilins of the A-set, a single, intramolecular disulfide bridge spans about 40 residues of the central region of the molecule. The location of this bridge is a characteristic of A-set pilins and it is absent from D-set and other mePhe pilins (Figure 8). Three regions of hypervariability can be discerned in comparisons of A-set sequences. These encompass alignment positions 57 to 65, 84 to 97, and 134 to 145 (Figure 6). Each of these regions is hydrophilic in nature (Figure 7) and hence has a potential surface location. They may, therefore, represent antigenic determinants. The first two regions lie within the disulfide loop, which may effectively segregate these regions of high variability from those regions of the molecule which are required for functional integrity of the pilus. This potential segregation into domains by a disulfide loop is reminiscent of the isolation of antigenic determinants of gonococcal pili by a disulfide bridge.[38]

In contrast to pilins of the A-set, D-set pilins possess two intramolecular disulfide bridges spanning 11 and 14 residues, respectively (Figure 8). Sequence analysis of the regions surrounding these bridges demonstrates their unrelatedness to the disulfide bridge of A-set pilins. However, the C-terminal disulfide loop and surrounding sequences of the D-set pilins are closely related to the corresponding regions of mPhe pilins from other bacterial species (see Figure 5).

Pilins from serogroup H differ from all other mePhe pilins studied, in that they are composed of two chains of approximately equal size. The combined molecular weight of these chains (16 kDa) approximates the size of other *B. nodosus* pilins. Gene-sequence studies have shown that these subunits are encoded in a single chain, which is subsequently

Table 3
PERCENTAGE IDENTITY IN SEQUENCE COMPARISONS OF PILINS FROM *B. NODOSUS* STRAINS FROM DIFFERENT SEROGROUPS

Serogroup (strain)	A (198)	B (234)	C (217)	E (216)	F (1017)	G (1220)	D (340)	H (265)
A (198)	100	76	59	74	71	64	33	36
B (234)		100	64	72	72	65	33	37
C (217)			100	65	59	78	36	37
E (216)				100	81	64	35	37
F (1017)					100	62	37	34
G (1220)						100	36	38
D (340)							100	68
H (265)								100

Note: The number of identities in each comparison is expressed as a percentage of the maximum possible number of aligned residues in a comparison. Gaps have been introduced in alignments providing that the total number of identical residues aligned increases by more than two for each gap.

cleaved between residues at positions 72 and 73.[21] Cleavage at this site may have resulted either from the action of a proteolytic enzyme unique to serogroup H of *B. nodosus,* or from the susceptibility of the particular amino acid sequence in pilin from this serogroup to a protease common to all serogroups.

With only two serogroups comprising the D-set pilins, it is not possible to define the location of hypervariable regions and hence potential antigenic determinants, as was possible with the A-set pilins, except to say that the location of these determinants is limited to the c-terminal two thirds of the molecules.

An indication of the basis of serotypic variation of pilins can be gained from comparisons of their amino acid sequences. Within most serogroups, a number of subgroups have been distinguished on the basis of serological cross-reactivity.[7] For example, in the A serogroup, two subgroups have been identified, which show considerable cross protection in vaccination studies.[12,13] Gene-sequence studies on representatives from these two subgroups show that the predicted protein sequences differ at eight amino acid positions. Six of these are the result of single base changes, while the remaining two are the result of double base changes. These changes are a consequence of genetic drift and have only a minor serological effect. Nevertheless, it is of interest to note that the majority (75%) of these changes are located within the disulfide loop, and of these changes, most (five of six) occur within hypervariable regions of the A-set pilin sequences.

The lack of cross-reactivity that is observed between serogroups reflects the greater sequence variability that occurs between pilins from different serogroups when compared to pilins from the same serogroup. Within the A-set pilins, approximately half of the sequence positions are unchanged across the group, although comparison of any two members of the group shows greater than 59% identity (Figure 6 and Table 3). On the basis of sequence comparisons, the A-set pilins can be further subdivided into two subsets (Table 3), comprising

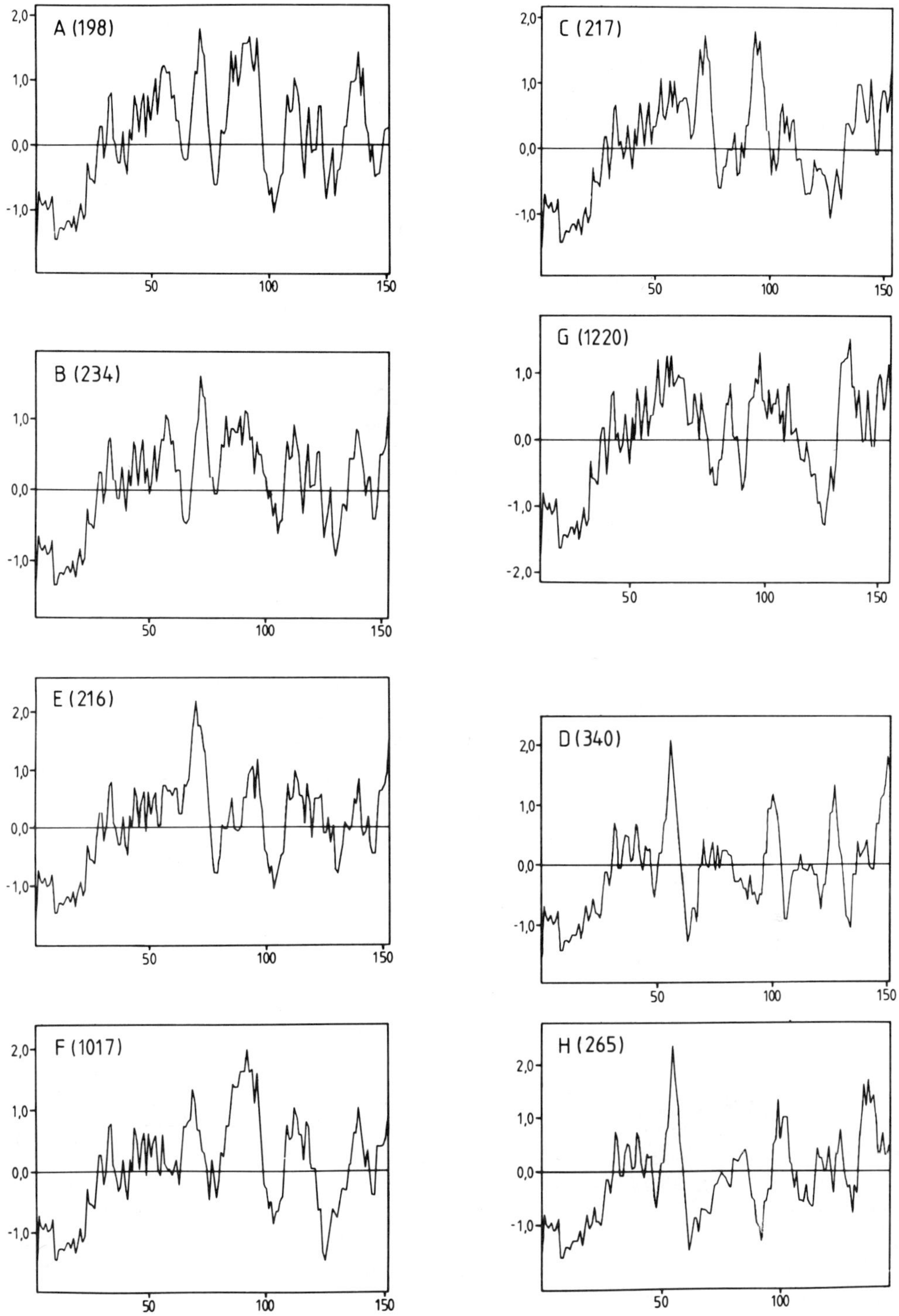

FIGURE 7. Hydrophilicity profiles (using a seven-residue running average) of amino acid sequences of *B. nodosus* pilins. Hydrophilicity index values are shown on the vertical axes, residue numbers along the horizontal axes. The serogroup designation and strain number (in parenthesis) are shown in the upper left corner of each panel.

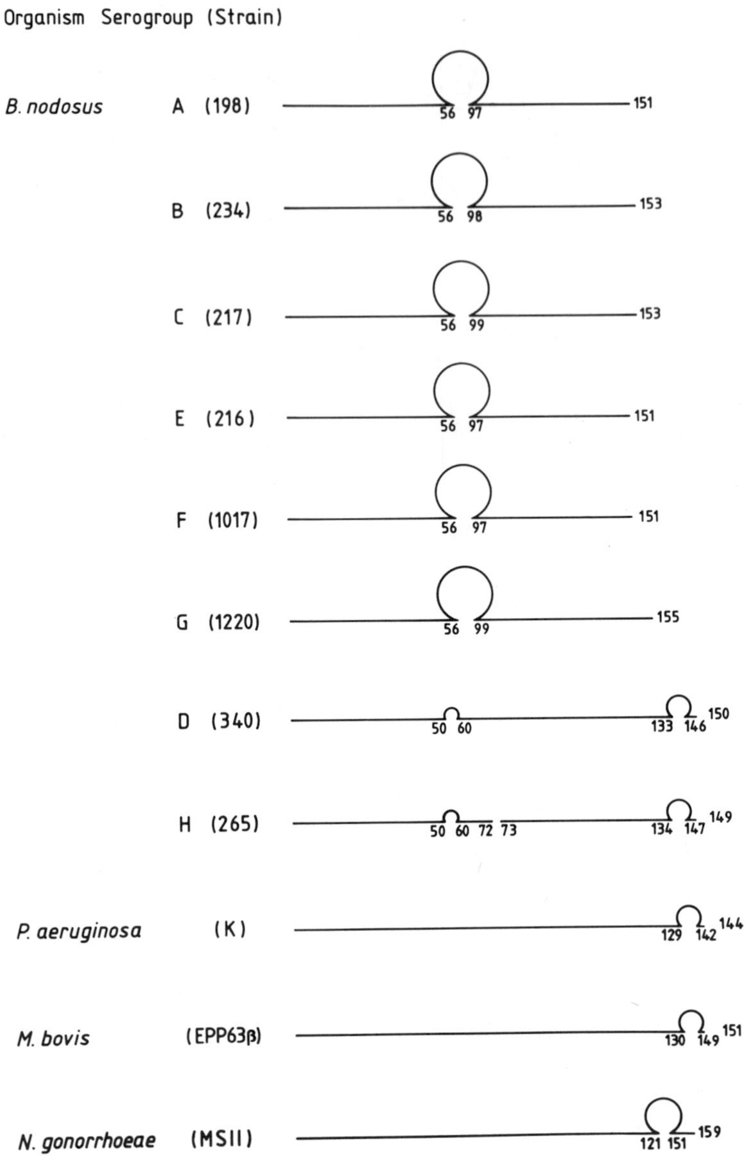

FIGURE 8. Line diagram representations of mePhe pilin sequences, illustrating the location of disulfide bridges. The residue length of each pilin is also shown.

serogroups A, B, E, and F, and serogroups C and G. This subdivision is supported by observed differences in hydrophilicity profiles of the subsets (Figure 7). The majority of sequence changes within A-set pilins are amino acid substitutions rather than insertions or deletions and this observation indicates that these serogroups may have arisen simply by an extension of the process of genetic drift, as was evidenced within the A serogroup. Similarly, sequence studies of pilin genes from representatives of serogroups D and H show that all changes are amino acid substitutions, with the exception of a single residue insertion/deletion at residue position 80. This sequence comparison of D-set pilins also shows that there is a similar level of identity (68%) to that found between any pair of A-set pilins.

The simple relationship observed within both the A-set and D-set pilin sequences is lacking when these sets are compared to each other. Despite the near identity of the first 30 residues

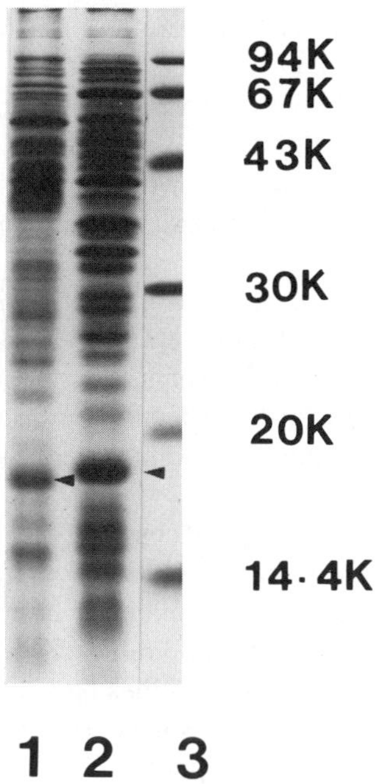

FIGURE 9. SDS-polyacrylamide gel electrophoresis of cellular products from bacteria harboring the pilin gene of *B. nodosus*. Lane 1, *P. aeruginosa* PAODB2 (pTE22) grown at 37°C overnight; lane 2, *E. coli* C600 (pTE22) induced at 42°C for 2.5 h; lane 3, molecular weight markers. Arrowheads indicate the positions of *B. nodosus* pilin and prepilin in lanes 1 and 2, respectively.

of all *B. nodosus* pilin sequences determined to date, the remaining regions are very different between sets, and the evolutionary relationship of the sets to one another is difficult to discern. Furthermore, no better insight into this evolutionary relationship can be gained by analysis of the gene sequences, other than the distant relationship of the C-terminal regions of pilins discerned from protein sequence studies. Despite the uncertainty of the evolutionary relationship, data are nevertheless compatible with the assumption that all *B. nodosus* pilins have evolved from a common ancestor.

V. APPLICATION OF STRUCTURAL DATA TO VACCINES

The poor growth characteristics and fastidious nature of *B. nodosus* in culture makes this organism unsuitable as either a killed, whole-cell vaccine, or as a source of pili for a cost-effective, pilus-based vaccine. The use of surrogate hosts in conjunction with recombinant-DNA technology offers a means of overcoming the deficiencies associated with the use of *B. nodosus* as an industrial microorganism. To this end, the pilin gene from *B. nodosus* was inserted into the surrogate hosts — *Escherichia coli* and *P. aeruginosa*, and was expressed at high yields in both (Figure 9).[27,45,46] In the widely used and well characterized *E. coli* organism, no morphological *B. nodosus* pili were produced, although high levels of an unprocessed prepilin were present. This product failed to protect sheep against footrot infection when challenged with the same strain of *B. nodosus*.

In contrast to the results obtained with *E. coli,* expression of the *B. nodosus* pilin gene in *P. aeruginosa* produced morphological pili indistinguishable from those produced by *B.*

nodosus itself.[27] These pili protected sheep against footrot infection in homologous challenge. The successful morphological expression of pili in *P. aeruginosa* contrasts with the failure to produce pili in *E. coli*. This may reflect the recognized difference in the types of pili naturally produced by *E. coli,* and those of the mePhe group of which *P. aeruginosa* and *B. nodosus* pili are members (see Section III. Pilus Characterization). Thus, the failure of *E. coli* to produce morphological pili may indicate the absence of appropriate processing and assembly mechanisms required for mePhe pilus construction, whereas these mechanisms are obviously present in *P. aeruginosa* and are capable of fabricating *B. nodosus* prepilins into mature pili.

Using this recombinant-DNA approach with *P. aeruginosa,* representative pili have been produced from both the A-set and the D-set. These pili have been efficiently expressed in high yield and protect against footrot in homologous challenge.[27] A new generation pilus vaccine based on recombinant-DNA technology is therefore feasible, provided that pili from sufficient serogroups of *B. nodosus* are included to confer an adequate spectrum of protection in the field.

An alternative to a pilus-based vaccine is a synthetic-peptide vaccine which would incorporate peptides corresponding to sections of pilin sequences. This synthetic-peptide approach has been investigated for a number of infectious diseases, including foot and mouth disease, hepatitis A, poliomyelitis, and rabies.[47-50] In the case of both foot and mouth disease and hepatitis A, synthetic-peptide vaccines have been shown to provide host protection.[47,48] Advantages of synthetic-peptide vaccines compared to conventional vaccines are their biological safety, stability, and high purity. Conventional vaccines suffer from possible contamination by the infectious agent used in their production or by the reversion of an attenuated to a virulent strain. Furthermore, even stable, attenuated strains may infect a small proportion of individuals who have a low threshold of susceptibility. Because of the biological method of their manufacture, conventional vaccines may also contain toxins and pyrogens.

Synthetic peptides which elicit antibodies to antigenic regions of pathogens are likely candidates for use in a peptide vaccine. Antigenic regions may be defined by the identification of binding sites of neutralizing antibodies to the pathogen. In the absence of this information, the location of hypervariable sequences between serotypes of the pathogen may reflect antigenic determinants and so indicate potential amino acid sequences suitable as peptide vaccines. Alternatively, surface-located antigenic determinants may be recognized from hydrophilicity profiles.[51] Both these inferential methods require a knowledge of amino acid sequence and are dependent on the antigenic determinant being composed of contiguous amino acid residues.

In diseases where the pathogen shows substantial serotypic variation (such as footrot), a multivalent vaccine may be necessary if the component peptides are modeled on the sequences of hypervariable regions. In these instances, a more efficient approach may be the choice of an invariant region of the molecules as a candidate peptide vaccine, even though this region may be only weakly immunogenic in the intact protein. This novel approach has been investigated with gonococcal pili,[38] where antibodies to peptides corresponding to a conserved, weakly immunogenic region inhibited heterologous gonococcal strains from binding to eukaryotic cells.

Preliminary studies with peptides based on various regions of a pilin from a serogroup A representative (strain 198) of *B. nodosus* have been carried out.[52] Four synthetic peptides of 12 to 14 residues (Figure 10), encompassing constant or variable regions were each cross linked with glutaraldehyde both in the presence and absence of a carrier protein (ovine serum albumin). In addition, the N-terminal 30-residue tryptic peptide from pilin was isolated. Vaccines incorporating these peptides were tested in sheep for their ability to elicit antibody. Each peptide failed to elicit a substantial response as measured by agglutination and ELISA.

The inability of those peptides so far tested to elicit a worthwhile antigenic response may

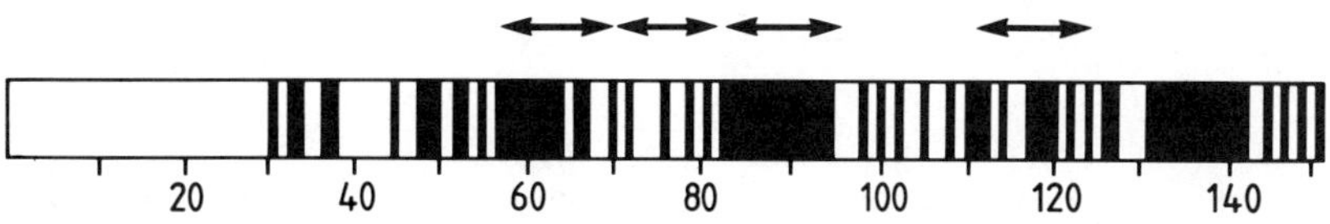

FIGURE 10. Block diagram showing regions of variability (in black) between A-set pilins of *B. nodosus* and the residue positions of synthetic peptides modelled on the sequence of strain 198 (serogroup A).

result from the possible discontinuous nature of the residues comprising the antigenic determinants, or the failure of the peptides to assume the appropriate conformation. Alternatively, the limited size and choice of peptides so far examined may have been inadequate to span an antigenic determinant. Furthermore, the cross-linking agent used (glutaraldehyde) binds to side-chain amino groups of lysine residues which occur internally in the sequences of three of the four tested peptides. Potential determinants may therefore be modified by the cross-linked lysine residues and be unable to assume a conformation capable of eliciting the production of effective antibodies. Further investigations, using peptides of suitable length and specifically linked to a carrier via a terminal residue may prove more worthwhile. Only peptides from those two hypervariable regions contained within the disulfide loop were included in this study. Synthetic peptides associated with the third region near the C-terminus of the molecule have yet to be examined for their ability to elicit a measurable antigenic response.

Attempts to stimulate an effective antigenic response to synthetic peptides usually rely on the direct stimulation of antibody production by B lymphocytes. An alternative, supplementary approach is to synthesize peptides which initiate the stimulation of helper T cells which in turn augment the production of specific antibody in B cells. Epitopes recognized by T cells are usually different chemically or functionally from those recognized by antibody-producing B cells.[53] One advantage of the use of peptides to stimulate T cells is that the latter recognize antigenic determinants which have been "processed" (unfolded or degraded), thus overcoming the potential difficulty of conformational requirements often associated with B-cell antigens. Screening of T-cell clones with synthetic peptides from the pilin sequence of *B. nodosus* strain 198 has been initiated and preliminary investigations are encouraging.[54]

VI. CONCLUSIONS AND PROSPECTS

The major thrust towards development of improved footrot vaccines has occurred as a result of the ability of *P. aeruginosa* to produce *B. nodosus* pili.[27,46] The identification of common N-terminal sequences in pilins of these bacterial species was fundamental to this development. Additionally, application of similar technology to other members of the mePhe class of pilins will permit the production of vaccines against other diseases whose causative agents possess such pili. In this regard, the availability of an inexpensive source of pilus-based vaccines from recombinant-DNA technology might offer an inexpensive means for control of several diseases of medical and veterinary importance, since pili from pathogens such as *N. gonorrhoeae* and *M. bovis* are highly immunogenic and provide protection against infection.

Organisms possessing mePhe pili generally show extensive serotypic variation. Eight non-cross-reacting serogroups of *B. nodosus* have been identified and at least one representative strain from each of these serogroups would be required for an effective vaccine, whether based on whole cells or pili. However, by genetic manipulation of the pilin gene, it may be possible to incorporate antigenic determinants from a number of serogroups within the same pilin molecule, thereby reducing the number of components required in a vaccine. The feasibility of this approach is supported by the occurrence of a small number of strains which cross-react with antiserum from more than one serogroup.[55]

Although less well-advanced than the development of pilus-based vaccines, vaccines based on synthetic peptides offer distinct advantages such as purity and biological safety which makes research into their development a worthwhile endeavor. In addition, recent advances in large-scale methods of peptide synthesis make this approach economically viable.[56] An intimate knowledge of the tertiary and quaternary structure of pilin will permit the identification of surface structural features, while X-ray crystallographic studies of pili-antibody complexes may enable antigenic determinants to be defined, thereby allowing the identification of sequences suitable for use as candidate peptide vaccines. In an approach analogous to that proposed for pilus vaccines based on recombinant DNA, hybrid peptides which encompass antigens from more than one serogroup may be feasible, and this would reduce the complexity of a comprehensive vaccine. An understanding of the molecular conformation of the pilus may provide information against footrot.

The development of a new generation of cost-effective peptide or pilus vaccines offers the possiblity of improved prophylactic control of footrot infection.

REFERENCES

1. **Brinton, C. C., Jr.,** The structure, function, synthesis and genetic control of bacterial pili and a molecular model for DNA transport in Gram negative bacteria, *Ann. N.Y. Acad. Sci.*, 27, 1003, 1965.
2. **Every, D.,** Purification of pili from *Bacteroides nodosus* and an examination of their chemical, physical and serological properties, *J. Gen. Microbiol.*, 115, 309, 1979.
3. **Depiazzi, L. J. and Richards, R. B.,** Motility in relation to virulence of *Bacteroides nodosus, Vet. Microbiol.*, 10, 107, 1985.
4. **Stewart, D. J.,** An electron microscopic study of *Fusiformis nodosus, Res. Vet. Sci.*, 14, 132, 973.
5. **Stewart, D. J.,** The role of various antigenic fractions of *Bacteroides nodosus* in eliciting protection against foot-rot in vaccinated sheep, *Res. Vet. Sci.*, 24, 14, 1978.
6. **Stewart, D. J., Clarke, B. L., Peterson, J. E., Griffiths, D. A., and Smith, A. E.,** Importance of pilus-associated antigen in *Bacteroides nodosus* vaccines, *Res. Vet. Sci.*, 32, 140, 1982.
7. **Claxton, P. D., Ribeiro, L. A., and Egerton, J. R.,** Classification of *Bacteroides nodosus* by agglutination tests, *Aust. Vet. J.*, 60, 331, 1983.
8. **Stewart, D. J., Clark, B. L., Peterson, J. E., Emery, D. L., Smith, E. F., Griffiths, D. A., and O'Donnell, I. J.,** The protection given by pilus and whole cell vaccines of *Bacteroides nodosus* strain 198 against ovine footrot induced by strains of different serogroups, *Aust. Vet. J.*, 62, 153, 1985.
9. **Keogh, J.,** Footrot vaccine trials, *N. S. W. Vet. Proc.*, 12, 35, 1976.
10. **Mattick, J. S., Anderson, B. J., and Egerton, J. R.,** Molecular biology and footrot of sheep, *Rev. Rural Sci.*, 6, 79, 1985.
11. **Stewart, D. J., Clark, B. L., Peterson, J. E., Griffiths, D. A., Smith, E. F., and O'Donnell, I. J.,** Effect of pilus dose and type of Freund's adjuvant on the antibody and protective responses of vaccinated sheep to *Bacteroides nodosus, Res. Vet. Sci.*, 35, 130, 1983.
12. **Claxton, P. D.,** Serogrouping of *Bacteroides nodosus* in relation to protection by vaccination, in Ovine Footrot, Rep. Workshop, University of Sydney, May 3 to 6, 1981, 16.
13. **Stewart, D. J., Clark, B. L., Peterson, J. E., Emery, D. L., Smith, E. F., Griffiths, D. A., and O'Donnell, I. J.,** The protection given by pilus and whole cell vaccines of *Bacteroides nodosus* strain 198 against ovine foot-rot induced by strains of different serogroups, *Aust. Vet. J.*, 62, 153, 1985.
14. **Stewart, D. J. and Elleman, T. C.,** A *Bacteroides nodasus* pili vaccine produced by recombinant DNA for the prevention and treatment of footrot in sheep, *Aust. Vet. J.*, 64, 79, 1987.
15. **Ottow, J. C. G.,** Ecology, physiology and genetics of fimbriae and pili, *Annu. Rev. Microbiol.*, 29, 79, 1975.
16. **Henrichsen, J.,** The occurrence of twitching motility among Gram-negative bacteria, *Acta Pathol. Microbiol. Scand. Sect. B*, 83, 171, 1975.
17. **Henrichsen, J.,** Twitching motility, *Annu. Rev. Microbiol.*, 37, 81, 1983.

18. **Yoshimura, F., Takasawa, T., Yoneyama, M., Yamaguchi, T., Shiokawa, H., and Suzuki, T.,** Fimbriae from the oral anaerobe *Bacteroides gingivalis:* physical, chemical and immunological properties, *J. Bacteriol.*, 163, 730, 1985.
19. **Finlay, B. B., Frost, L. S., and Paranchych, W.,** Localization, cloning and sequence determination of the conjugative plasmid ColB2 pili gene, *J. Bacteriol.*, 160, 402, 1984.
20. **Fröholm, L. O. and Sletten, K.,** Purification and N-terminal sequence of a fimbrial protein from *Moraxella nonliquefaciens, FEBS Lett.*, 73, 29, 1977.
21. **Elleman, T. C., Hoyne, P. A., McKern, N. M., and Stewart, D. J.,** Nucleotide sequence of the gene encoding the two-subunit pilin of *Bacteroides nodosus* 265, *J. Bacteriol.*, 167, 243, 1986.
22. **McKern, N. M., O'Donnell, I. J., Inglis, A. S., Stewart, D. J., and Clark, B. L.,** Amino acid sequence of pilin from *Bacteroides nodosus* (strain 198), the causative organism of ovine footrot, *FEBS Lett.*, 164, 149, 1983.
23. **Marrs, C. F., Schoolnik, G., Koomey, J. M., Hardy, J., Rothbard, J., and Falkow, S.,** Cloning and sequencing of a *Moraxella bovis* pili gene, *J. Bacteriol.*, 163, 132, 1985.
24. **Hermodson, M. A., Chen, K. C. S., and Buchanan, T. M.,** *Neisseria* pili proteins: amino terminal amino acid sequences and identification of an unusual amino acid, *Biochemistry,* 17, 442, 1978.
25. **Olafson, R. W., McCarthy, P. J., Bhatti, A. R., Dooley, J. S. G., Heckels, J. E. and Trust, T. J.,** Structural and antigenic analysis of meningococcal piliation, *Infect. Immun.*, 48, 336, 1985.
26. **Sastry, P. A., Pearlstone, J. R., Smillie, L. B., and Paranchych, W.,** Amino acid sequence of pilin isolated from *Pseudomonas aeruginosa* PAK, *FEBS Lett.*, 151, 253, 1983.
27. **Elleman, T. C., Hoyne, P. A., Stewart, D. J., McKern, N. M., and Peterson, J. E.,** Expression of pili from *Bacteroides nodosus* in *Pseudomonas aeruginosa, J. Bacteriol.*, 168, 574, 1986.
28. **Klemm, P.,** Fimbrial adhesins of *Escherichia coli, Rev. Infect. Dis.*, 7, 321, 1985.
29. **Van Die, I., Van Geffen, B., Hockstra, W., and Bergmans, H.,** Type 1C fimbriae of a uropathogenic *Escherichia coli* strain: cloning and characterization of the genes involved in the expression of the 1C antigen and nucleotide sequence of the subunit gene, *Gene,* 34, 187, 1984.
30. **Van Die, I. and Bergmans, H.,** Nucleotide sequence of the gene encoding the $F7_2$ fibrial subunit of a uropathogenic *Escherichia coli* strain, *Gene,* 32, 83, 1984.
31. **Klemm, P.,** The complete amino-acid sequence of the K88 antigen, a fimbrial protein from *Escherichia coli, Eur. J. Biochem.*, 117, 617, 1981.
32. **Frost, L. S., Paranchych, W., and Willetts, N. S.,** DNA sequence of the F tra ALE region that includes the gene for F pilin, *J. Bacteriol.*, 160, 395, 1984.
33. **Fader, R. C., Duffy, L. K., Davis, C. P., and Kurosky, A.,** Purification and chemical characterization of type 1 pili isolated from *Klebsiella pneumoniae, J. Biol. Chem.*, 257, 3301, 1982.
34. **Klemm, P.,** Primary structure of the CFA1 fimbrial protein from human enterotoxigenic *Escherichia coli* strains, *Eur. J. Biochem.*, 124, 339, 1982.
35. **Feutrier, J., Kay, W. W., and Trust, T. J.,** Purification and characterization of fimbriae from *Salmonella enteritidis, J. Bacteriol.*, 168, 221, 1986.
36. **Elleman, T. C. and Hoyne, P. A.,** Nucleotide sequence of the gene encoding pilin of *Bacteroides nodosus,* the causal organism of ovine footrot, *J. Bacteriol.*, 160, 1184, 1984.
37. **Watts, T. H., Kay, C. M., and Paranchych, W.,** Spectral properties of three quaternary arrangements of *Pseudomonas aeruginosa* pilin, *Biochemistry,* 22, 3640, 1983.
38. **Rothbard, J. B., Fernandez, R., Wang, L., Teng, N. N. H., and Schoolnik, G. K.,** Antibodies to peptides corresponding to a conserved sequence of gonococcal pilins block bacterial adhesion, *Proc. Natl. Acad. Sci. U.S.A.* 82, 915, 1985.
39. **Mattick, J. S.,** Personal communication of pilin sequences from serogroup B and C representatives, 1987.
40. **Anderson, B. J. and Mattick, J. S.,** Personal communication of pilin sequences from serogroup F and G representatives, 1987.
41. **McKern, N. M., Stewart, D. J., and Strike, P. M.,** Amino acid sequences of piliris from serologically distinct strains of *Bacteroides nodosus, J. Protein Chem.*, 7, 157, 1988.
42. **McKern, N. M., O'Donnell, I. J., Stewart, D. J., and Clark, B. L.,** Primary structure of pilin protein from Bacteroides nodosus strain 216: comparison with the corresponding protein from strain 198, *J. Gen. Microbiol.*, 131, 1, 1985.
43. **Finney, K. G., Elleman, T. C., and Stewart, D. J.,** Nucleotide sequence of the pilin gene of *Bacteroides nodosus* 340 (Serogroup D) and implications for the relatedness of serogroups, *J. Gen. Microbiol.*, 134, 575, 1988.
44. **Rothbard, J. B. and Schoolnik, G. K.,** Immunochemistry of gonococcal pili: approaches toward a gonococcal vaccine, in *Vaccines 85,* Lerner, R. A., Chanock, R. M., and Brown, F., Eds., Cold Spring Harbor Laboratory, Cold Spring Harbor, NY, 1985, 85.
45. **Elleman, T. C., Hoyne, P. A., Emery, D. L., Stewart, D. J., and Clark, B. L.,** Expression of the pilin gene from *Bacteroides nodosus* in *Escherichia coli, Infect. Immun.*, 51, 187, 1986.

46. **Mattick, J. S., Bills, M. M., Anderson, B. J., Dalrymple, B., Mott, M. R., and Egerton, J. R.,** Morphogenetic expression of *Bacteroides nodosus* fimbriae in *Pseudomonas aeruginosa, J. Bacteriol.,* 169, 33, 1987.
47. **DiMarchi, R., Brooke, G., Gale, C., Cracknell, V., Doel, T., and Mowat, N.,** Protection of cattle against foot and mouth disease by a synthetic peptide, *Science,* 232, 639, 1986.
48. **Emini, E. A., Jameson, B. A., and Wimmer, E.,** Priming for and induction of anti-poliovirus neutralising antibodies by synthetic peptides, *Nature,* 304, 699, 1983.
49. **Emini, E. A., Berger, J., Hughes, J. V., Mitra, S. W., and Linemeyer, D. L.,** Priming of an anti-hepatitis-A virus antibody response by poliovirus-specific synthetic peptides: localisation of potential antigenic sites of hepatitis-A virus neutralisation, in *Vaccines 85,* Lerner, R. A., Chanock, R. M., and Brown, F., Eds., Cold Spring Harbor Laboratory, Cold Spring Harbor, NY, 1985, 217.
50. **Koprowski, H., Reagan, K. J., MacFarlan, R. I., Dietzschold, B., and Wiktor, T. J.,** New generation of rabies vaccines: rabies glycoprotein gene recombinants, anti-idiotypic antibodies, and synthetic peptides, in *Vaccines 85,* Lerner, R. A., Chanock, R. M., and Brown, F., Eds., Cold Spring Harbor Laboratory, Cold Spring Harbor, NY, 1985, 151.
51. **Hopp, T. P. and Woods, K. R.,** Prediction of protein antigenic determinants from amino acid sequences, *Proc. Natl. Acad. Sci. U.S.A.,* 78, 3824, 1981.
52. **Kirkpatrick, A., Maclaren, J. A., McKern, N. M., and Stewart, D. J.,** Unpublished data, 1987.
53. **Schwartz, R. H.,** T-lymphocyte recognition of antigen in association with gene products of the major histocompatibility complex, *Annu. Rev. Immunol.,* 3, 237, 1985.
54. **Emery, D. L.,** Personal communication, 1987.
55. **Claxton, P. D.,** Serogrouping of *Bacteroides nodosus* isolates, in *Footrot in Ruminants,* Stewart, D. J., Peterson, J. E., McKern, N. M., and Emery, D. L., Eds., Proc. Workshop, Melbourne, 1985, CSIRO Division of Animal Health/Australian Wool Corp., Glebe, N.S.W., Melbourne, 1986, 131.
56. **Jakubke, H. D., Kuhl, P., and Könnecke, A.,** Basic principles of protease-catalysed peptide bond formation, *Angew. Chem. Int. Ed. Eng.,* 24, 85, 1985.

Chapter 14

APPLICATION OF BIOTECHNOLOGY TO THE DIAGNOSIS OF FOOTROT

Julian I. Rood and W. K. Yong

TABLE OF CONTENTS

I. INTRODUCTION

Clinical diagnosis of footrot is limited in value because it cannot distinguish among early cases of the virulent, intermediate, and benign forms of the disease. Laboratory methods are employed because it is important for quarantine and treatment purposes that such a distinction is made.[1] Laboratory diagnosis is a two-step procedure which requires the isolation and identification of *Bacteroides nodosus* from the lesion. For confirmation of a case of virulent footrot, it is then necessary to demonstrate that the isolate has the properties typical of a virulent strain. These properties include the ability to produce relatively heat-stable proteases,[2,3] elastase activity,[4] a typical protease isoenzyme-activity pattern after gel electrophoresis,[5-7] and typical levels of twitching motility.[8,9] These tests are tedious, time consuming, expensive, and are complicated by the stringent growth requirements and slow growth of *B. nodosus* in culture. There is, therefore, a well-established need for more efficient and specific laboratory diagnostic methods which can rapidly identify and differentiate virulent, subvirulent, and benign *B. nodosus* isolates.[10]

Another major problem in the diagnosis of footrot is the identification of infected animals which do not show any clinical signs of disease. At present, there is no practical method of detecting such sheep except by physically tipping each animal over, and carefully paring and examining its feet. Because of the size of the individual sheep flocks in countries such as Australia and New Zealand, this method is laborious and generally impractical under normal farm-management procedures. Footrot-carrier sheep have been one of the major sources of transmission of ovine footrot from one district or farm to another, particularly when farmers buy apparently healthy animals either for improvement or replacement of their stock. The breakdowns in the footrot-control program in Southwestern Victoria (Australia) between 1983 and 1984 were attributed to the introduction of such animals at the end of the 1982 drought in this district.[11] The failure to identify carrier sheep has also placed serious constraints on the success of footrot-eradication programs, as these animals remain a constant threat to susceptible and uninfected animals in the flock.

In recent years, many important advances in clinical diagnosis have been made by the application of contemporary molecular biology techniques such as monoclonal antibody (MAb) and recombinant-DNA technologies. It is now possible to produce MAbs specific for almost any epitope or cell-surface determinant, to use these antibodies to screen genomic libraries, and to isolate the genes encoding important antigens of pathogens, tumor cells, receptors, and other functional biomolecules. Using recombinant-DNA technology, specific gene sequences can be cloned and analyzed in considerable detail. The resultant DNA fragments can then be used as sensitive and specific gene probes in diagnostic assays. Both technologies have been used either directly on clinical specimens to detect pathogenic bacteria[12-15] or on bacterial cultures to identify the causative organism,[16] and to detect specific virulence-related gene products or their structural genes.[17-31]

There is considerable potential for the application of both MAb and gene-probe technologies to the diagnosis of ovine footrot. If the ideal MAb or gene probe could be constructed, it would be possible to diagnose accurately the early stages of virulent ovine footrot directly from lesion materials, within 2 to 24 h, without the need to culture the causative microorganism. Semiautomated screening procedures could also be developed to identify footrot-carrier sheep in any flock. This should enable both the farmer and the appropriate government agency to make better decisions as to whether costly and disruptive quarantine and control programs should be implemented.

In this chapter, the principles used in the application of each of these technologies will be briefly discussed. The resultant development of new approaches to *B. nodosus* identification and differentiation, and their applications to the diagnosis of footrot-carrier sheep will then be explored. In addition, we will briefly describe potential applications in the selection and breeding of sheep which are innately resistant to footrot infections.

II. PRINCIPLES OF HYBRIDOMA AND RECOMBINANT-DNA TECHNOLOGIES

A. Monoclonal-Antibody Production

Hybridoma technology has changed the scope of diagnostic and investigative serology. Previously, the preparation of high-titer monospecific antibodies required the purification of antigen, the careful immunization of multiple animals, and the exhaustive absorption of antisera. It was more of an art than a science to produce good reagents. However, it is now relatively easy to generate hybridomas that secrete antibody molecules against impure and even uncharacterized antigens. These MAbs bind to antigens uniformly and reproducibly and can be used to mediate a diverse array of biological activities. In thc diagnostic laboratory, monoclonal hybridoma antibodies are now used routinely to enumerate lymphocyte subsets for tissue typing, to identify and classify pathogens in both body fluids and tissue sections, and in many other practical applications which utilize the unique specificity offered by monoclonal reagents.

Even though only a decade has passed since hybridoma technology was first described,[32] numerous reviews,[33-37] monographs,[38-41] and books[42,43] devoted specifically to its theory, application, and practice, have been published. Various refinements have been made to the methodology, but the basic principles have not altered (Figure 1). Mice are immunized with the antigen of interest and antibody-forming cells from the immunized mouse spleen are then fused with tissue-culture-adapted murine myeloma cells by means of somatic cell-hybridization techniques. The resultant fused cells (termed hybridomas) will not only grow continuously in culture, but each cloned hybridoma produces antibody molecules which are specific for a single antigenic determinant.

B. Gene-Probe Production

In the early 1970s, studies on the mechanism of action of restriction endonuclease and DNA ligase enzymes led to the development of techniques whereby DNA molecules could be specifically digested and the resultant DNA fragments ligated to suitable cloning vectors, usually multicopy plasmid DNA molecules.[44] Since the basic structure of DNA is essentially the same irrespective of its source, it is therefore possible to carry out test-tube manipulations that enable the recombination of DNA molecules from different organisms. With the development of reliable transformation methods, these recombinant-DNA plasmid molecules can be introduced efficiently into suitable host strains, such as the genetically well-defined bacterium, *Escherichia coli*. By careful selection of the *E. coli*-derived vector molecule, the recombinant-DNA fragment can be incorporated into the host cell's genome so that it is stably maintained in an extrachromosomal state and results in high levels of gene expression of the cloned genes.[45] Recombinant-DNA, or gene-cloning, technology is now routinely used as the first step in the detailed molecular analysis of procaryotic genes.[46] However, of more relevance to this chapter is the use of easily purified recombinant plasmids as a readily available source of gene-specific probes. These probes can be used in DNA-hybridization assays to diagnose microbial infections.[47-51]

DNA-DNA-hybridization assays rely upon the presence of complementary DNA sequences on different single stranded (SS) DNA molecules. The test DNA is immobilized on a nitrocellulose filter, denatured to form SS molecules, and is baked irreversibly onto the filter. After blocking of nonspecific binding sites with control DNA, the hybridization reaction is carried out by adding a ^{32}P-labeled or biotin-labeled reference SS probe. If the test DNA contains nucleotide sequences that are complementary to those of the probe DNA,

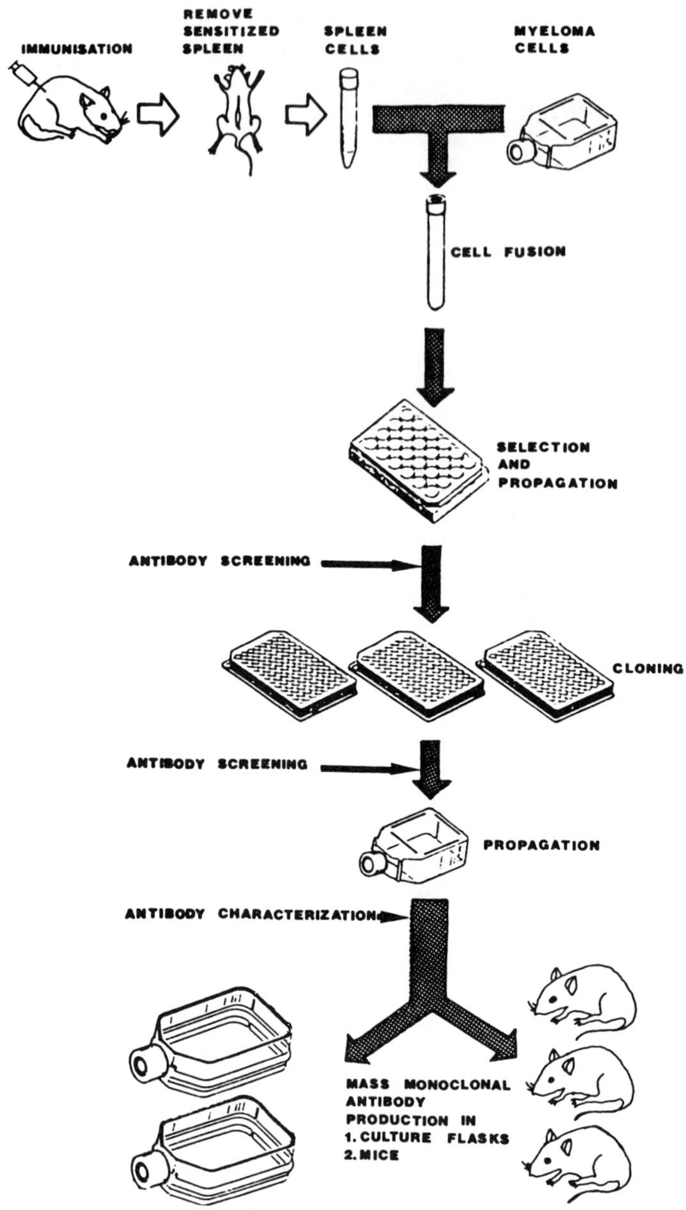

FIGURE 1. Schematic representation of the basic protocol for the derivation of monoclonal antibodies.

then the SS-probe DNA will form hydrogen bonds with the test DNA and will therefore become bound to the filter. Alternatively, if no homologous sequences are present, then the probe will not bind to the filter. After incubation with the hybridization mixture, the filter is washed to remove unbound probe DNA. Depending on the method used to label the probe, either autoradiography or an enzymic method is used to detect the probe DNA that is bound specifically to the filter.

III. PATHOGEN IDENTIFICATION AND DIFFERENTIATION OF VIRULENT STRAINS

There are two major classes of specific footrot MAbs or gene probes that could be constructed. The first type would be species specific and used for the detection of *B. nodosus* in clinical specimens and for the identification of *B. nodosus* cultures. The second type would be virulence specific and used primarily for the differential diagnosis of the virulent and benign forms of ovine footrot. Although the development of either type would lead to significant improvements in footrot diagnosis, the most useful advances would be made with the development of virulence-specific tests.

A. Species-Specific Diagnostic Tests

There are several candidates for the construction of species-specific *B. nodosus* detection systems. It should be possible to construct MAbs specific for common epitopes present on components of the fimbrial system, proteases, or common antigens that are constituents of the *B. nodosus* cell wall. Gene probes could be constructed by cloning the genes coding for these antigens. Alternatively, it should be possible to construct specific gene probes by random cloning of chromosomal DNA fragments from *B. nodosus*, as has been done for other *Bacteroides* spp.[52-54] Irrespective of the source of the common antigen or DNA segment that is to be used, the new detection system must be both specific and sensitive. It will be essential to show that the MAb or gene probe reacts with all of the clinical *B. nodosus* isolates (from all of the major serological groups) that can realistically be tested, and that it does not react with closely related bacteria or with other microorganisms that may be present in lesions or exudates obtained from the hooves of either clinically or subclinically infected animals.

Major advances would be made by use of a species-specific diagnostic test if the test were sensitive enough for the direct examination of clinical specimens, thus eliminating the necessity to culture the causative *B. nodosus* strain, or to use tedious and insensitive Gram stains. It is uncertain whether gene probes based on single chromosomal genes, even if highly specific for *B. nodosus*, will be sensitive enough for direct tests on clinical material. However, one solution to this problem may be a species-specific gene probe based on the detection of rRNA molecules. It should be possible, by using purified *B. nodosus* 16S-rRNA as an initial probe, to clone the structural genes coding for *B. nodosus* 16S-rRNA.

These genes, or species-specific fragments of them, would be ideal for use as probes in DNA-rRNA-hybridization tests on material obtained from footrot lesions. These probes would be more sensitive than those based on the detection of chromosomal genes because of the large number of rRNA molecules present in the bacterial cell.[55] Therefore, gene probes based on rRNA genes would appear to have the greatest potential for the development of a species-specific diagnostic test suitable for the direct examination of clinical specimens. However, since these probes will be unable to distinguish between virulent and benign isolates, their use will not solve the problems of the footrot diagnostic laboratory. It will still be necessary to develop additional virulence-specific diagnostic tests.

B. Virulence-Specific Diagnostic Tests

All pathogenic bacteria have an enormous array of antigens. Many of these will be shared with other bacterial species, while there will be some which will be unique to the particular pathogen. By utilizing the reaction of these unique antigenic molecules or their structural genes with specific MAbs or gene probes, respectively, it is possible to diagnose many infectious diseases by directly identifying the pathogen either in smears, tissue sections, or bacterial cultures. The key to the development of MAb- or gene-probe-based diagnostic tests for the rapid differentiation of *B. nodosus* isolates is therefore the identification of such unique antigens.

The best-characterized antigenic determinants of *B. nodosus* are the fimbriae.[56-61] However, the role of fimbriae as virulence determinants requires further clarification.[57,62,63] Nevertheless, the fimbriae should not be discarded as a possible source of antigenic determinants which reflect the degree of virulence of a particular strain. Recent studies have shown that although both virulent and benign strains of *B. nodosus* may contain an equivalent number of fimbriae,[8,64,65] virulent isolates clearly exhibit a higher degree of twitching motility than do benign isolates.[8,9] The differences in twitching motility must reflect functional differences in either the fimbrial subunit genes or other genes coding for components of the fimbrial-assembly system. Recently, it has become evident that there are in fact two distinct classes (I and II) of fimbrial genes in the *B. nodosus* population.[67,68] Nucleotide-sequence analyses also showed that the fimbrial subunit genes in Class I and Class II strains are quite different.[67] In addition, studies on the primary structure of fimbrial subunit proteins from two *B. nodosus* strains have already revealed regions of conservation and variability.[69,70] It remains to be determined whether the regions of variability are associated with antigenic determinants and whether they reflect actual functional differences, such as motility and virulence, between strains. If so, fimbrial subunit proteins or their structural genes will be important targets for the development of virulence-specific diagnostic systems.

The other well-studied *B. nodosus* antigens are the proteases. These enzymes[71] form the basis of current laboratory tests for differentiating isolates which cause virulent and benign footrot. Antigenic differences between the proteases elaborated by virulent and benign strains were recently reported.[72] Antibodies were successfully raised in rabbits against various electrophoretically separated proteases from virulent and benign *B. nodosus* isolates. Using immunoelectrophoresis, antibodies from rabbits immunized with a protease from a virulent strain specifically reacted with the antigen from the homologous strain. Similarly, antibodies from rabbits immunized with a protease from the benign strain specifically reacted with only the homologous antigen. However, when these antibodies were conjugated with horseradish peroxidase and used in an enzyme-linked immunosorbent assay (ELISA) system, extensive cross-reactions were demonstrated between whole cells from virulent and benign strains.[64] Cross-reactions of antibodies from sheep immunized with purified proteases from either virulent or benign *B. nodosus* strains were also reported.[73] These results suggest that antigenic differences between proteases from virulent and benign strains are probably confined to a few epitopes of each enzyme. Hybridoma techniques whereby MAbs reactive against only a single antigenic site are selected, are therefore ideal for exploring these restricted differences between proteases, and should overcome many of the problems associated with the use of polyclonal antisera in the development of rapid differential diagnostic tests for footrot.

Other antigens which are present in the extracellular protein complex of *B. nodosus* have also been reported.[72] The identity of these antigens is not clear. They are neither of fimbrial nor of outer-cell membrane origin, as rabbit antibodies raised against both fimbriae and outer-membrane proteins do not react against these antigens. Virulent *B. nodosus* isolates have at least three of these unique extracellular protein antigens, whereas benign *B. nodosus* have four such proteins. These proteins and other *B. nodosus* antigens may be suitable targets for the production of virulence-specific MAbs, or their structural genes may have potential value in the development of gene probes. These antigens include components of the outer-membrane complex[74-78] and the external superficial layers.[79,80]

Two classes of antigenic molecules, lipopolysaccharides (LPS) and proteins, have been identified as components of the outer membrane of *B. nodosus*.[74,76-78] It remains to be established whether any of these molecules are specifically associated with either virulent or benign *B. nodosus* isolates. Stewart[78] reported that the heat-stable LPS contained at least two common carbohydrate O-antigenic determinants but no type-specific O antigens. However, this concept of the specificity and function of antibodies that react with *B. nodosus* LPS is based on studies done with polyclonal antibodies. The availability of MAbs, with their exquisite specificity for single epitopes within the LPS molecule may reveal type- (or strain-) specific O antigens in different strains. Such epitopes have been recently demonstrated in other Gram-negative bacteria, such as *Pseudomonas aeruginosa*.[81,82] Six major *B. nodosus* outer-membrane proteins have been described,[74] and antigenic differences noted when proteins from benign strains were compared with those from virulent and subvirulent strains.[83] These results suggest that some outer-membrane proteins may be good antigen targets for the development of rapid differential diagnostic tests for the identification of virulent, subvirulent, and benign *B. nodosus* isolates.

IV. DIAGNOSIS OF FOOTROT-CARRIER SHEEP

A. Use of Monoclonal Antibodies

Since serology became established as a discipline in clinical laboratories, the detection of circulating antibodies has been of key importance in the immunodiagnosis of infectious diseases. An indication that hybridoma-derived antibodies would be useful in the development of new immunodiagnostic tests based on the detection of circulating serum antibodies was initially provided in a model parasite system.[84] An IgG1 MAb was produced from the fusion of cells of mice infected with the proliferating larval cestode, *Mesocestoides corti*, and used in a competitive binding radioimmunoassay (RIA) with a crude parasite homogenate as antigen. The binding of the labeled MAb to the antigen mixture on the RIA plate was inhibited by sera from *M. corti*-infected mice but not by sera from mice infected with other helminth parasites. With no false positive reactions, no false negatives, and positive reactions occurring as early as 7 d after infection, the test clearly had a high specificity and adequate sensitivity for diagnostic use. Similar MAb-based competitive immunological assays have been developed for the diagnosis of a variety of infectious diseases.[85]

Although the bacterial invasion in footrot infections is restricted to the avascular epidermis, it has been established that naturally affected animals develop increased antibody titers to a number of *B. nodosus* antigens.[86-88] The analysis of antibody responses has played a key role in the identification of the protective antigen and subsequent development of effective vaccines. Determination of specific circulating antibodies may be one method by which footrot-infected sheep will be identified, even those sheep which do not show any clinical manifestations of the disease.

The control of ovine footrot by the widespread use of multivalent fimbrial vaccines seems inevitable in the near future, since methods for the efficient and economic production of the protective fimbrial antigens are now available via recombinant-DNA technology.[67] One consequence of vaccination programs for any large-scale, immunological, diagnostic, screening program for footrot in sheep (and for that matter any diseases[89,90]) is the difficulty in separating naturally infected animals from those which are serologically positive because they have been immunized. This problem is further complicated by large quantitative and qualitative differences in the antibody responses in individual animals, both within naturally infected and vaccinated groups. Despite these inherent variations, Ferrier et al.,[91] using ELISA serology based on a crude whole-cell antigen preparation, showed that sheep from a footrot-endemic flock could be categorized into either vaccinated or infected status according to their antibody levels. However, these promising results must be confirmed by follow-up studies. MAbs, with their high specificity, may be the tools required for differentiating vaccination from true-infection responses, especially if separate antigenic epitopes are responsible for the induction of the respective response in the host. Recently, a competitive-binding ELISA for the detection of circulating bovine antibodies to *Brucella abortus* has been developed using murine MAb raised against *B. abortus* cell-surface antigens.[92] The results from a panel of 30 sera from animals with a known status of vaccination or infection led the authors to claim that the test was able to discriminate infected from uninfected or immunized animals. Unfortunately, this promising finding was not confirmed in a follow-up study.[93] The interpretation of serological results in the diagnosis of *B. nodosus* infection in sheep will face similar problems once large-scale vaccination is introduced. Nevertheless, it is clear that improved specificity and sensitivity for the immunodiagnosis of footrot infection is linked to a greater understanding of the different immunological responses mounted by the host to different *B. nodosus* antigens and to the development of specific MAbs for these antigens.

Another approach to the immunological diagnosis of chronic infectious diseases is the detection of specific antigens present in tissues or body fluids of the animal. The antigens in body fluids are usually in the form of soluble antigens which exist either as free molecules or attached to other immunological elements to form immune complexes.[94] Both specific polyclonal antibodies[95,96] and MAbs[97] have been used successfully to identify such circulating antigens. However, the detection of circulating antigens as a means of diagnosing *B. nodosus* infection in sheep has never been attempted. Given the localized nature of this infection, as well as the special situation of the carrier sheep, it seems unlikely that circulating antigens will be available in abundance to allow easy detection.

B. Use of Gene Probes

Gene probes may also play a role in the diagnosis of footrot-infected but clinically and immunologically nonresponsive animals. The gene-probe method relies on the presence of complementary DNA fragments. Therefore, the whole pathogen or relevant gene fragments from it must be present either in the clinical specimen or DNA prepared from the specimen. The sensitivity of gene-probe assays may be a problem for the diagnosis of carrier animals. However, by careful selection of the type of gene probe, such as a 16S-rRNA probe, the presence of genetic material from only a few cells will provide sufficient material for a DNA-RNA-hybridization reaction to occur. It remains to be seen whether enough cells will be present for a sensitive assay to be developed.[54] The prospective use of gene probes for the identification of footrot-carrier sheep may, therefore, be contraindicated because of the low *B. nodosus* load in these animals.

V. IDENTIFICATION OF SHEEP WITH INNATE RESISTANCE TO FOOTROT

There has always been an interest in breeding footrot-resistant sheep as a long-term strategy to control this disease.[98,99] Between breed as well as within breed differences in the susceptibility of sheep to footrot are well recognized.[100-104] However, this knowledge has rarely been exploited. Without a reliable diagnostic test, breeding programs have to rely on time consuming and subjective methods for the selection of resistant sheep. These methods involve deliberate exposure of sheep to footrot, followed by selecting and breeding sheep which are able to withstand such a challenge. Diagnostic difficulties notwithstanding, some success with this approach has been reported in New Zealand.[98,105] The combination of long delays in achieving the desired results and the difficulty in identifying and selecting the resistant progeny probably explain why such breeding programs are not vigorously pursued.

The application of MAb or gene-probe technology may alleviate at least one of the problems involved in the selective breeding of resistant sheep. If the genetic markers responsible for innate resistance to footrot can be identified, then using either of these techniques, a more rapid, reliable, and objective laboratory diagnostic test for identifying and selecting the resistant sheep can be developed. Recently, a presumptive marker associated with a class-I-type major histocompatibility gene which appeared to code for the liability of sheep to footrot infection was identified.[106] These preliminary findings offer encouraging prospects for the development and application of MAb- or gene-probe-based diagnostic tests which can be used in breeding programs. The availability of such specific probes will allow the isolation of the resistance gene from the host's genome,[107] and may facilitate its rapid dissemination within the sheep population by gene-transfer techniques.[108]

VI. CONCLUSION

The use of hybridoma technology in footrot research will certainly accelerate. In principle, MAbs can resolve a single protein from a complex mixture, or indeed a single epitope responsible for a specific function of a complex macromolecule. Therefore, MAbs are ideal tools for in-depths studies of the function of different *B. nodosus* antigens and for use in antigen purification. Considerable activity can be expected in the next few years in both the production of MAbs and in their application in a variety of different footrot research and diagnostic programs. However, there is still a need for fundamental research on the different antigenic moieties which characterize the various virulence groups. Apart from some preliminary data on the extracellular proteases,[72] there has been no attempt to immunologically identify and characterize virulence antigens. Although whole cells and other crude antigen mixtures can be used for immunization for hybridoma production, the isolation of virulence-specific MAbs requires the appropriate purified antigen for screening the hybridomas. The rapid progress in molecular and immunological techniques has not been matched by progress in our understanding of the structure, occurrence, and functional relationships of the important antigens which differentiate *B. nodosus* strains.

There is also a need for fundamental research on the molecular genetics of *B. nodosus* virulence antigens. Such studies are an essential component of attempts to develop diagnostic gene probes and will contribute greatly to our understanding of the pathogenesis of ovine footrot. We conclude that the aim of future research on *B. nodosus* gene probes should be to develop gene-probe methods that are of such sensitivity and specificity that they can be used for the direct examination of clinical specimens. The ideal gene probe would be a nonradioactively labeled, virulence-specific DNA sequence, bearing a nonradioactive label, which could readily be used in small regional veterinary diagnostic laboratories. However, it may be possible to develop field methods whereby clinical specimens are applied directly to nitrocellulose filters which are then transported to a central reference laboratory for analysis. Such a method has been successfully used for the study of enterotoxigenic *Escherichia coli* strains from developing countries. In these studies, clinical specimens were swabbed onto nitrocellulose filters and mailed to the U.S., where the DNA-hybridization reactions with specific gene probes were done.[109-113] In this stituation, there would be no need for the development of nonradioactive-DNA probes, since the larger reference laboratory would have both the facilities and the expertise to carry out diagnostic autoradiography on a routine basis.

The development of more rapid and efficient laboratory methods for the diagnosis of ovine footrot is an important step in efforts to both control and eradicate this economically significant disease. In this chapter we have shown that both MAb and gene-probe technologies have considerable potential for the development of these methods. It appears that the most suitable initial targets for research are the extracellular protease antigenic determinants and their structural genes. Collaborative research in our laboratories has therefore concentrated on these proteins and genes. However, research on the application of both MAb and gene-probe technologies for differentiation of footrot organisms is at an extremely early stage. Moreover, the use of these technologies for the identification of footrot-carrier sheep and the identification of sheep which are innately resistant to footrot infection is yet to be explored.

In conclusion, we have identified and attempted to discuss the application of MAb and gene probes to three different aspects of footrot diagnosis. At the time of writing, there have been no published papers on the application of either of these contemporary technologies to any of these areas. Therefore, it is not possible at this time to predict which approach is likely to be the most successful or which antigens or genes will prove to be of most value. As a result, it is important that both researchers and granting bodies continue to support varied and innovative approaches to this problem.

ACKNOWLEDGMENTS

We acknowledge the research support that has been given to both of our laboratories by grants from the Wool Research Trust Fund of the Australian Wool Corporation and the Rural Credits Development Fund of the Reserve Bank of Australia.

REFERENCES

1. **Egerton, J. R.,** Control and eradication of ovine footrot, in *Footrot in Ruminants,* Proceedings of a Workshop, Melbourne, 1985, Stewart, D. J., Peterson, J. E., McKern, N. M., and Emery, D. L., Eds., CSIRO Division Animal Health and Australian Wool Corp., Sydney, 1986, 35.
2. **Depiazzi, L. J. and Richards, R. B.,** A degrading proteinase test to distinguish benign and virulent ovine isolates of *Bacteroides nodosus, Aust. Vet. J.,* 55, 25, 1979.
3. **Depiazzi, L. J. and Rood, J. I.,** The thermostability of proteases from virulent and benign strains of *Bacteroides nodosus, Vet. Microbiol.,* 9, 227, 1984.
4. **Stewart, D. J.,** The role of elastase in the differentiation of *Bacteroides nodosus* infections in sheep and cattle, *Res. Vet. Sci.,* 27, 99, 1979.
5. **Every, D.,** Proteinase isoenzyme patterns of *Bacteroides nodosus:* distinction between ovine virulent isolates, ovine benign isolates and bovine isolates, *J. Gen. Microbiol.,* 129, 809, 1982.
6. **Kortt, A. A., Burns, J. E., and Stewart, D. J.,** Detection of the extracellular proteases of *Bacteroides nodosus* in polyacrylamide gels: a rapid method of distinguishing virulent and benign ovine isolates, *Res. Vet. Sci.,* 35, 171, 1983.
7. **Gordon, L. M., Yong, W. K., and Woodward, C. A. M.,** The temporal relationships and characterization of extracellular proteases from benign and virulent strains of *Bacteroides nodosus* as detected in zymogram gels, *Res. Vet. Sci.,* 39, 165, 1985.
8. **Depiazzi, L. J. and Richards, R. B.,** Motility in relation to virulence of *Bacteroides nodosus, Vet. Microbiol.,* 10, 107, 1985.
9. **Depiazzi, L. J., Richards, R. B., McQuade, N. C., and Rood, J. I.,** Virulence, protease, pilus and motility characteristics of *Bacteroides nodosus,* in *Footrot in Ruminants,* Proceedings of a Workshop, Melbourne 1985, Stewart, D. J., Peterson, J. E., McKern, N. M., and Emery, D. L., Eds., CSIRO Division Animal Health and Australian Wool Corp., Sydney, 1986, 185.
10. **Anon.,** Recommendations from the workshop on footrot, in *Footrot in Ruminants,* Proceedings of a Workshop, Melbourne, 1985, Stewart, D. J., Peterson, J. E., McKern, N. M., and Emery, D. L., Eds., CSIRO Division Animal Health and Australian Wool Corp., Sydney, 1986, 281.
11. **Napthine, D. V.,** Future directions for the control of footrot in Victoria, in *Footrot in Ruminants,* Proceedings of a Workshop, Melbourne, 1985, Stewart, D. J., Peterson, J. E., McKern, N. M., and Emery, D. L., Eds., CSIRO Division Animal Health and Australian Wool Corp., Sydney, 1986, 21.
12. **Tam, M. R., Stamm, W. E., Handsfield, H. H., Stephens, R., Kuo, C. C., Holmes, K. K., Ditzenberger, K., Krieger, M., and Nowinski, R. C.,** Culture independent diagnosis of *Chlamydia trachomatis* using monoclonal antibodies, *N. Engl. J. Med.,* 310, 1146, 1984.
13. **Madsen, R. D., Saeed, F. A., Gray, O., Fendly, B. M., and Coates, S. R.,** Species-specific monoclonal antibody to a 43,000 molecular weight membrane protein of *Mycoplasma pneumoniae, J. Clin. Microbiol.,* 24, 680, 1986.
14. **Dorman, M. A., Blair, C. D., Collins, J. K., and Beaty, B. J.,** Detection of bovine herpesvirus 1 DNA immobilised on nitrocellulose by hybridization with biotinylated DNA probes, *J. Clin. Microbiol.,* 22, 990, 1985.
15. **Collins, J. K., Butcher, A. C., Teramoto, Y. A., and Winston, S.,** Rapid detection of bovine herpesvirus type 1 antigens in nasal swab specimens with an antigen capture enzyme-linked immunosorbent assay, *J. Clin. Microbiol.,* 21, 375, 1985.

16. **Fitts, R., Diamond, M., Hamilton, C., and Neri, M.,** DNA-DNA hybridization assay for detection of *Salmonella* spp. in foods, *Appl. Environ. Microbiol.*, 46, 1146, 1983.
17. **Miner, K. M., Manyak, C. L., Williams, E., Jackson, J., Jewell, M., Gammon, M. T., Ehrenfreund, C., Hayes, E., Callahan, L. T., III, Zweerink, H., and Sigal, N. H.,** Characterization of murine monoclonal antibodies to *Escherichia coli* J5, *Infect. Immun.*, 52, 56, 1986.
18. **Echeverria, P., Seriwatana, J., Sethabutr, O., and Taylor, D. N.,** DNA hybridization in the diagnosis of bacterial diarrhoea, *Clin. Lab. Med.*, 5, 447, 1985.
19. **Kaper, J. B., Moseley, S. L., and Falkow, S.,** Molecular characterization of environmental and nontoxigenic strains of *Vibrio cholerae, Infect. Immun.*, 32, 661, 1981.
20. **Kaper, J. B., Campen, R. K., Seidler, R. J., Baldini, M. M., and Falkow, S.,** Cloning of the thermostable direct or Kanagawa phenomenon-associated hemolysin of *Vibrio parahaemolyticus, Infect. Immun.*, 45, 290, 1984.
21. **Groman, N., Cianciotto, N., Bjorn, M., and Rabin, M.,** Detection and expression of DNA homologous to the *tox* gene in nontoxigenic isolates of *Corynebacterium diphtheriae, Infect. Immun.*, 42, 48, 1983.
22. **Nataro, J. P., Baldini, M. M., Kaper, J. B., Black, R. E., Bravo, N., and Levine, M. M.,** Detection of an adherance factor of enteropathogenic *Escherichia coli* with a DNA probe, *J. Infect. Dis.*, 152, 460, 1985.
23. **Rubin, F. A., Kopecko, D. J., Noon, K. F., and Baron, L. S.,** Development of a DNA probe to detect *Salmonella typhi, J. Clin. Microbiol.*, 22, 600, 1985.
24. **Boileau, C. R., D'Hauteville, H. M., and Sansonetti, P. J.,** DNA hybridization technique to detect *Shigella* species and enteroinvasive *Escherichia coli, J. Clin. Microbiol.*, 20, 959, 1984.
25. **Small, P. L. and Falkow, S.,** Development of a DNA probe for the virulence plasmid of *Shigella* spp. and enteroinvasive *Escherichia coli,* in *Microbiology-1986,* Leive, L., Bonventre, P. F., Morello, J. A., Silver, S. D., and Wu, W. C., Eds., American Society for Microbiology, Washington, D.C., 1986, 121.
26. **Wood, P. K., Morris, J. G., Jr., Small, P. L. C., Sethabutr, O., Toledo, M. R. F., Trabulsi, L., and Kaper, J. B.,** Comparison of DNA probes and the Sereny test for identification of invasive *Shigella* and *Escherichia coli* strains, *J. Clin. Microbiol.*, 24, 498, 1986.
27. **Pal, T., Echeverria, P., Taylor, D. N., Sethabutr, O., and Hanchalay, S.,** Identification of enteroinvasive *Escherichia coli* by indirect ELISA and DNA hybridization, *Lancet,* 2, 785, 1985.
28. **Sethabutr, O., Hanchalay, S., Echeverria, P., Taylor, D. N., and Leksomboon, U.,** A non-radioactive DNA probe to identify *Shigella* and enteroinvasive *Escherichia coli* in stools of children with diarrhoea, *Lancet,* 2, 1095, 1985.
29. **Hill, W. E., Payne, W. L., and Aulisio, C. C. G.,** Detection and enumeration of virulent *Yersinia enterocolitica* in food by DNA colony hybridization, *Appl. Environ. Microbiol.*, 46, 636, 1983.
30. **Totten, P. A., Holmes, K. K., Handsfield, H. H., Knapp, J. S., Perine, P. L., and Falkow, S.,** DNA hybridization technique for the detection of *Neisseria gonorrhoeae* in men with urethritis, *J. Infect. Dis.*, 148, 462, 1983.
31. **Perine, P. L., Totten, P. A., Holmes, K. K., Sng, E. H., Ratnam, A. V., Widy-Wesky, R., Nsanze, H., Habte-Gabr, E., and Westbrook, W. G.,** Evaluation of a DNA hybridization method for detection of African and Asian strains of *Neisseria gonorrhoea* in men with urethritis, *J. Infect. Dis.*, 152, 59, 1985.
32. **Kohler, G. and Milstein, C.,** Continuous cultures of fused cells secreting antibody of predefined specificity, *Nature (London),* 256, 495, 1975.
33. **Diamond, B. A., Yelton, D. E., and Scharff, M. D.,** Monoclonal antibodies: a new technology for producing serological reagents, *N. Engl. J. Med.*, 304, 1344, 1981.
34. **Scharff, M. D., Roberts, S., and Thammana, P.,** Monoclonal antibodies, *J. Infect. Dis.*, 143, 346, 1981.
35. **Reading, C. L.,** Theory and methods for immunization in culture and monoclonal antibody production, *J. Immunol. Methods,* 53, 261, 1982.
36. **Galfre, G. and Milstein, C.,** Preparation of monoclonal antibodies: strategies and procedures, *Methods Enzymol.*, 73, 3, 1981.
37. **Carson, D. A. and Freimark, B. D.,** Human lymphocyte hybridomas and monoclonal antibodies, *Adv. Immunol.*, 38, 275, 1986.
38. **Melchers, F., Potter, M., and Warner, N. L., Eds.,** Lymphocyte hybridomas, *Current Topics in Microbiology and Immunology,* Vol. 81, Springer-Verlag, Berlin, 1978.
39. **Moller, G., Ed.,** Hybrid myeloma monoclonal antibodies against MHC products, *Immunol. Rev.*, Vol. 47, Munksgaard, Copenhagen, 1979.
40. **Kennett, R. H., McKearn, T. J., and Bechtol, K. B., Eds.,** *Monoclonal Antibodies: Hybridomas: A New Dimension in Biological Analyses,* Plenum Press, New York, 1980.
41. **Hurrell, J. G. R., Ed.,** *Monoclonal Hybridoma Antibodies: Techniques and Applications,* CRC Press, Boca Raton, Fl., 1982.
42. **Goding, J. W.,** *Monoclonal Antibodies: Principles and Practice,* Academic Press, London, 1983.

43. **Campbell, A. M.,** *Monoclonal Antibody Technology,* Elsevier, Amsterdam, 1984.
44. **Maniatis, T., Fritsch, E. F., and Sambrook, J.,** *Molecular Cloning. A Laboratory Manual,* Cold Spring Harbor Laboratory, Cold Spring Harbor, NY, 1982, chap. 3.
45. **Cohen, S. N.,** The manipulation of genes, *Sci. Am.,* 233, 24, 1976.
46. **Lewin, B.,** *Genes,* 2nd ed., John Wiley & Sons, New York, 1985, chaps. 14 and 15.
47. **Grunstein, M. and Hogness, D. S.,** Colony hybridization: a method for the isolation of cloned DNAs that contain a specific gene, *Proc. Natl. Acad. Sci. U.S.A.,* 72, 3961, 1975.
48. **Southern, E. M.,** Detection of specific sequences among DNA fragments separated by gel electrophoresis, *J. Mol. Biol.,* 98, 503, 1975.
49. **Arrand, J. E.,** Preparation of nucleic acid probes, in *Nucleic Acid Hybridisation. A Practical Approach,* Hames, B. D. and Higgins, S. J., Eds., IRL Press, Oxford, 1985, chap. 2.
50. **Radford, A. J.,** Comparisons using ^{35}S- and ^{32}P-labelled DNA for hybridization on nitrocellulose filters, *Anal. Biochem.,* 134, 269, 1983.
51. **Engleberg, N. C. and Eisenstein, B. I.,** The impact of new cloning techniques on the diagnosis and treatment of infectious diseases, *New Engl. J. Med.,* 311, 892, 1984.
52. **Kuritza, A. P., Getty, C. E., Shaughnessy, P., Hesse, R., and Salyers, A. A.,** DNA probes for identification of clinically important *Bacteroides* species, *J. Clin. Microbiol.,* 23, 343, 1986.
53. **Salyers, A. A., Lynn, S. P., and Gardner, J. F.,** Use of randomly cloned DNA fragments for identification of *Bacteroides thetaiotamicron, J. Bacteriol.,* 154, 287, 1983.
54. **Kuritza, A. P. and Salyers, A. A.,** Use of a species-specific DNA hybridization probe for enumerating *Bacteroides vulgatus* in human faeces, *Appl. Environ. Microbiol.,* 50, 958, 1985.
55. **Kohne, D., Hogan, J., Jonas, V., Dean, E., and Adams, T. H.,** Novel approach for rapid and sensitive detection of microorganisms: DNA probes to rRNA, in *Microbiology-1986,* Leive, L., Bonventre, P. F., Morello, J. A., Silver, S. D., and Wu, W. C., Eds., American Society for Microbiology, Washington, D.C., 1986, 100.
56. **Stewart, D. J.,** An electron microscope study of *Fusiformis nodosus, Res. Vet. Sci.,* 14, 132, 1973.
57. **Walker, P. D., Short, J., Thompson, R. O., and Roberts, D. S.,** The fine structure of *Fusiformis nodosus* with special references to the antigens associated with immunogenicity, *J. Gen. Microbiol.,* 77, 351, 1973.
58. **Stewart, D. J., Clark, B. L., Peterson, J. E., Griffiths, D. A., and Smith, E. F.,** Importance of pilus-associated antigen in *Bacteroides nodosus* vaccines, *Res. Vet. Sci.,* 32, 140, 1982.
59. **Every, D. and Skerman, T. M.,** Protection of sheep against experimental footrot by vaccination with pili purified from *Bacteroides nodosus, N.Z. Vet. J.,* 30, 156, 1982.
60. **Egerton, J. R.,** Surface and somatic antigens of *Fusiformis nodosus, J. Comp. Pathol.,* 83, 51, 1973.
61. **Claxton, D., Ribeiro, L. A., and Egerton, J. R.,** Classification of *Bacteroides nodosus* by agglutination tests, *Aust. Vet. J.,* 60, 331, 1983.
62. **Every, D.,** Purification of pili from *Bacteroides nodosus* and an examination of their chemical, physical and serological properties, *J. Gen. Microbiol.,* 115, 309, 1979.
63. **Mattick, J. S., Anderson, B. J., and Egerton, J. R.,** Molecular biology and footrot of sheep, in *Biotechnology and Recombinant DNA Technology in the Animal Production Industries, Reviews in Rural Science 6,* Leng, R. A., Barker, J. S. F., Adams, D. B., and Hutchinson, K. J., Eds., University of New England, Armidale, N.S.W., 1985, 79.
64. **Yong, W. K., Moses, E. K., and Riffkin, G. G.,** Characterisation of virulence associated antigens for differentiating virulent and benign strains of *Bacteroides nodosus,* in *Footrot in Ruminants,* Proceedings of a Workshop, Melbourne 1985, Stewart, D. J., Peterson, J. E., McKern, N. M., and Emery, D. L., Eds., CSIRO Division Animal Health and Australian Wool Corp., Sydney, 1986, 109.
65. **Stewart, D. J., Clark, B. L., Peterson, J. E., Emery, D. L., Smith, E. F., Griffiths, D. A., and O'Donnell, I. J.,** The protection given by pilus and whole cell vaccines of *Bacteroides nodosus* strain 198 against ovine footrot induced by strains of different serogroups, *Aust. Vet. J.,* 62, 153, 1985.
66. **Skerman, T. M., Erasmuson, S. K., and Every, D.,** Differentiation of *Bacteroides nodosus* biotypes and colony variants in relation to their virulence and immunoprotective properties in sheep, *Infect. Immun.,* 32, 788, 1981.
67. **Mattick, J. S.,** The molecular biology of the fimbriae (pili) of *Bacteroides nodosus* and the development of a recombinant DNA-based vaccine, this volume, chap. 12.
68. **McKern, N. M., Elleman, T. C., and Hoyne, P. A.,** Molecular basis of antigenic variation in pili (fimbriae) of *Bacteroides nodosus* — application to footrot vaccines, this volume, chap. 13.
69. **McKern, N. M., O'Connell, I. J., Inglis, A. S., Stewart, D. J., and Clark, B. L.,** Amino acid sequence of pilin from *Bacteroides nodosus* (strain 198), the causative organism of ovine footrot, *FEBS Lett.,* 164, 149, 1983.
70. **McKern, N. M., O'Donnell, I. J., Stewart, D. J., and Clark, B. L.,** Primary structure of pilin protein from *Bacteroides nodosus* strain 216: comparison with the corresponding protein from strain 198, *J. Gen. Microbiol.,* 131, 1, 1985.

71. **Moses, E. K. and Yong, W. K.,** Antigens of *Bacteroides nodosus,* this volume, chap. 7.
72. **Yong, W. K. and Gordon, L. M.,** Comparative antigenic analysis of extracellular protein of *Bacteroides nodosus* isolated from virulent and benign ovine footrot, *Vet. Microbiol.,* 12, 135, 1986.
73. **Stewart, D. J., Clark, B. L., Emery, D. L., Peterson, J. E., and Kortt, A., A.,** The phenomenon of cross protection against footrot induced by vaccination of sheep with *Bacteroides nodosus* vaccines, in *Footrot in Ruminants,* Proceedings of a Workshop, Melbourne, 1985, Stewart, D. J., Peterson, J. E., McKern, N. M., and Emery, D. L., Eds., CSIRO Division Animal Health and Australian Wool Corp., Sydney, 1986, 185.
74. **Emery, D. J., Clark, B. L., Stewart, D. J., O'Donnell, I. J., and Hewish, D. R.,** Analysis of the outer membrane proteins of *Bacteroides nodosus,* the causal organism of ovine footrot, *Vet. Microbiol.,* 9, 155, 1984.
75. **Thorne, K. J. L.,** Regularly arranged protein on the surfaces of Gram-negative bacteria, *Biol. Rev.,* 52, 219, 1977.
76. **Egerton, J. R.,** Surface and somatic antigens of *Fusiformis nodosus, J. Comp. Pathol.,* 83, 151, 1973.
77. **Stewart, D. J.,** Biochemical and biological studies on the lipopolysaccharide of *Bacteroides nodosus, Res. Vet. Sci.,* 23, 319, 1977.
78. **Stewart, D. J.,** Studies on the antigenic structure of *Bacteroides nodosus, Res. Vet. Sci.,* 24, 293, 1978.
79. **Every, D. and Skerman, T. M.,** Ultrastructure of the *Bacteroides nodosus* cell envelope layers and surface, *J. Bacteriol.,* 141, 845, 1980.
80. **Every, D. and Skerman, T. M.,** Surface structure of *Bacteroides nodosus* in relation to virulence and immunoprotection in sheep, *J. Gen. Microbiol.,* 129, 225, 1983.
81. **Sadoff, J. C., Wright, D. C., Futrovsky, S., Sidberry, H., Collins, H., and Kaufmann, B.,** Characterization of mouse monoclonal antibodies directed against *Pseudomonas aeruginosa* lipopolysaccharides, *Antibiot. Chemother. (Washington D.C.),* 36, 134, 1985.
82. **Stoll, B. J., Pollark, M., Young, L. S., Koles, N., Gascon, R., and Pier, G. B.,** Functionally active monoclonal antibody that recognizes an epitope on the O side chain of *Pseudomonas aeruginosa* immunotype-1 lipopolysaccharide, *Infect. Immun.,* 53, 656, 1986.
83. **Stewart, D. J., Peterson, J. E., Vaughan, J. A., Clark, B. L., Emery, D. L., Caldwell, J. B., and Kortt, A. A.,** The pathogenicity and cultural characteristics of virulent, intermediate and benign strains of *Bacteroides nodosus* causing ovine footrot, *Aust. Vet. J.,* 63, 317, 1986.
84. **Mitchell, G. F., Cruise, K. M., Chapman, C. B., Anders, R. F., and Howard, M. C.,** Hybridoma antibody immunoassays for the detection of parasitic infection: development of a model system using a larval cestode infection in mice, *Aust. J. Exp. Biol. Med. Sci.,* 57, 287, 1979.
85. **Ivanyi, J., Krambovitis, E., and Keen, M.,** Evaluation of a monoclonal antibody (TB72) based serological test for tuberculosis, *Clin. Exp. Immunol.,* 54, 337, 1983.
86. **Egerton, J. R. and Merritt, G. C.,** Serology of footrot: antibodies against *Fusiformis nodosus* in normal, affected, vaccinated and passively immunized sheep, *Aust. Vet. J.,* 49, 139, 1973.
87. **Merritt, G. C. and Egerton, J. R.,** IgG_1 and IgG_2 immunoglobulins to *Bacteroides (Fusiformis) nodosus* protease in infected and immunized sheep, *Infect. Immun.,* 22, 1, 1978.
88. **O'Donnell, I. J., Stewart, D. J., and Clark, B. L.,** Serological identification of pilus antigen and other protein antigens of *Bacteroides nodosus* using electro-blot radioimmunoassay after electrophoretic fractionation of the proteins on sodium dodecyl sulfate polyacrylamide gels, *Aust. J. Biol. Sci.,* 36, 15, 1983.
89. **Hellstrom, J. S.,** The brucellosis eradication scheme, *N.Z. Vet. J.,* 30, 33, 1982.
90. **Yong, W. K. and Edwards, L. D.,** Evaluation of the microtitre spin agglutination assay in the diagnosis of bovine brucellosis, *Vet. Rec.,* 119, 454, 1986.
91. **Ferrier, G. R., Smith, S. C., and Spencer, T. L.,** Serological responses to antigens of *Bacteroides nodosus* in footrot infected and vaccinated sheep, in *Footrot in Ruminants,* Proceedings of a Workshop, Melbourne, 1985, Stewart, D. J., Peterson, J. E., McKern, N. M., and Emery, D. L., Eds., CSIRO Division Animal Health and Australian Wool Corp., Sydney, 1986, 161.
92. **Rylatt, D. B., Wyatt, D. M., and Bundesen, P. G.,** A competitive enzyme immunoassay for the detection of bovine antibodies to *Brucella abortus* using monoclonal antibodies, *Vet. Immunol. Immunopathol.,* 8, 261, 1985.
93. **Sutherland, S. S. and Den Hollander, L.,** Comparison of an enzyme-linked immunosorbent assay using monoclonal antibodies and a complement fixation test for cattle vaccinated and infected with *Brucella abortus, Vet. Microbiol.,* 12, 55, 1986.
94. **Van Knapen, F.,** Detection and significance of circulating antigens and immune complexes, *Lyon Med.,* 248, 51, 1982.
95. **Turunen, H. J.,** Detection of soluble antigens of *Toxoplasma gondii* by a four layer modification of an enzyme immunoassay, *J. Clin. Microbiol.,* 17, 768, 1983.
96. **Van Knapen, F. and Panggabean, S. O.,** Detection of circulating antigen during acute infections with *Toxoplasma gondii* by enzyme-linked immunosorbent assay, *J. Clin. Microbiol.,* 6, 545, 1977.

97. **Araujo, F. G., Handman, E., and Remington, J. S.,** Use of monoclonal antibodies to detect antigens of *Toxoplasma gondii* in serum and other body fluids, *Infect. Immun.*, 30, 12, 1980.
98. **Anon.,** A stubborn Scot's tenacious battle with footrot, *N.Z. J. Agric.*, 149, 39, 1984.
99. **Parker, C. F., Cross, R. F., Gradin, J. L., and Hamilton, K. L.,** Selection for footrot resistance in Targhee sheep, *J. Anim. Sci.*, 57 (Suppl. 1, Abstr. 75), 164, 1983.
100. **Skerman, T. M., Erasmuson, S. K., and Morrison, L. M.,** Duration of resistance to experimental footrot infection in Romney and Merino sheep vaccinated with *Bacteroides nodosus* oil adjuvant vaccine, *N.Z. Vet. J.*, 30, 27, 1982.
101. **Egerton, J. R., Ribeiro, L. A., Kieran, P. J., and Thorley, C. M.,** Onset and remission of ovine footrot, *Aust. Vet. J.*, 60, 334, 1983.
102. **Emery, D. L., Stewart, D. J., and Clark, B. L.,** The comparative susceptibility of five breeds of sheep to footrot, *Aust. Vet. J.*, 61, 85, 1984.
103. **Stewart, D. J., Emery, D. L., Clark, B. L., Peterson, J. E., Iyer, H., and Jarrett, R. G.,** Differences between breeds of sheep in their responses to *Bacteroides nodosus* vaccines, *Aust. Vet. J.*, 62, 116, 1985.
104. **Skerman, T. M.,** Genetic variation and inheritance of susceptibility of footrot in sheep, in *Footrot in Ruminants*, Proceedings of a Workshop, Melbourne, 1985, Stewart, D. J., Peterson, J. E., McKern, N. M., and Emery, D. L., Eds., CSIRO Division Animal Health and Australian Wool Corp., Sydney, 1986, 73.
105. **Skerman, T. M. and Moorhouse, S. R.,** Broomfield Corriedales: a strain of sheep selectively bred for resistance to footrot, *N.Z. Vet. J.*, 35, 101, 1987.
106. **Outteridge, P. M., Windon, R. G., Dineen, J. K., Stewart, D. J., and Skerman, T. M.,** Breeding for parasite and footrot resistance in sheep using ovine lymphocyte antigen (OLA) markers, in *Merino Improvement Programs in Australia, Proceedings of a National Symposium, Leura, N.S.W.*, McGuirk, B. J., Ed., Australian Wool Corp., Melbourne, 1987, 377.
107. **Groth, D. M., Taylor, L., Wetherall, J. D., Carrick, M. J., and Lee, I. R.,** Nature and applications of DNA probe recognizing biological individuality in *Merino Improvement Programs in Australia, Proceedings of a National Symposium, Leura, N.S.W.*, McGuirk, B. J., Ed., Australian Wool Corp., Melbourne, 1987, 519.
108. **Franklin, I.,** Gene transfer for the Australian sheep industry, in *Merino Improvement Programs in Australia, Proceedings of a National Symposium, Leura, N.S.W.*, McGuirk, B. J., Ed., Australian Wool Corp., Melbourne, 1987, 495.
109. **Echeverria, P., Seriwatana, J., Taylor, D. N., Tirapat, C., Chaicumpa, W., and Rowe, B.,** Identification by DNA hybridization of enterotoxigenic *Escherichia coli* in a longitudinal study of villages in Thailand, *J. Infect. Dis.*, 151, 124, 1985.
110. **Echeverria, P., Seriwatana, J., Chityothin, O., Chaicumpa, W., and Tirapat, C.,** Detection of enterotoxigenic *Escherichia coli* in water by filter hybridization with three enterotoxin gene probes, *J. Clin. Microbiol.*, 16, 1086, 1982.
111. **Echeverria, P., Seriwatana, J., Leksomboon, U., Tirapat, C., Chaicumpa, W., and Rowe, B.,** Identification by DNA hybridization of enterotoxigenic *Escherichia coli* in homes of children with diarrhoea, *Lancet*, 1, 63, 1984.
112. **Seriwatana, J., Echeverria, P., Escamilla, J., Glass, R., Hug, I., Rockhill, R., and Stoll, B. J.,** Identification of enterotoxigenic *Escherichia coli* in patients with diarrhoea in Asia with three enterotoxin gene probes, *Infect. Immun.*, 42, 152, 1983.
113. **Georges, M. C., Wachsmuth, K., Birkness, K. A., Moseley, S. L., and Georges, A. J.,** Genetic probes for enterotoxigenic *Escherichia coli* isolated from childhood diarrhoea in the Central African Republic, *J. Clin. Microbiol.*, 18, 199, 1983.

Index

INDEX

A

B

G

H

I

K

L

M

Q

R

S

T

U